Digital Ske

PRACTICAL REVOLUTIONS: DISRUPTIVE TECHNOLOGIES AND THEIR APPLICATIONS TO BUILDING DESIGN AND CONSTRUCTION

Digital Sketching

COMPUTER-AIDED CONCEPTUAL DESIGN

John Bacus

WILEY

This book is printed on acid-free paper.

Copyright © 2021 by John Wiley & Sons, Inc. All rights reserved

Published by John Wiley & Sons, Inc., Hoboken, New Jersey

Published simultaneously in Canada

No part of this publication may be reproduced, stored in a retrieval system, or transmitted in any form or by any means, electronic, mechanical, photocopying, recording, scanning, or otherwise, except as permitted under Section 107 or 108 of the 1976 United States Copyright Act, without either the prior written permission of the Publisher, or authorization through payment of the appropriate per-copy fee to the Copyright Clearance Center, 222 Rosewood Drive, Danvers, MA 01923, (978) 750-8400, fax (978) 646-8600, or on the web at www.copyright.com. Requests to the Publisher for permission should be addressed to the Permissions Department, John Wiley & Sons, Inc., 111 River Street, Hoboken, NJ 07030, (201) 748-6011, fax (201) 748-6008, or online at http://www.wiley.com/go/permissions.

Limit of Liability/Disclaimer of Warranty: While the publisher and author have used their best efforts in preparing this book, they make no representations or warranties with the respect to the accuracy or completeness of the contents of this book and specifically disclaim any implied warranties of merchantability or fitness for a particular purpose. No warranty may be created or extended by sales representatives or written sales materials. The advice and strategies contained herein may not be suitable for your situation. You should consult with a professional where appropriate. Neither the publisher nor the author shall be liable for damages arising herefrom.

For general information about our other products and services, please contact our Customer Care Department within the United States at (800) 762-2974, outside the United States at (317) 572-3993 or fax (317) 572-4002.

Wiley publishes in a variety of print and electronic formats and by print-on-demand. Some material included with standard print versions of this book may not be included in e-books or in print-on-demand. If this book refers to media such as a CD or DVD that is not included in the version you purchased, you may download this material at http://booksupport.wiley.com. For more information about Wiley products, visit www.wiley.com.

Library of Congress Cataloging-in-Publication Data:

Names: Bacus, John, 1969- author.
Title: Digital sketching : computer-aided conceptual design / John Bacus.
Description: Hoboken, New Jersey : Wiley, [2020] | Includes index.
Identifiers: LCCN 2020020368 (print) | LCCN 2020020369 (ebook) | ISBN 9781119640769 (paperback) | ISBN 9781119640806 (adobe pdf) | ISBN 9781119640790 (epub)
Subjects: LCSH: Architectural design—Data processing. | Architecture—Computer-aided design.
Classification: LCC NA2728 .B24 2020 (print) | LCC NA2728 (ebook) | DDC 720.28/40285—dc23
LC record available at https://lccn.loc.gov/2020020368
LC ebook record available at https://lccn.loc.gov/2020020369

Cover Design: Wiley
Cover Image: John Bacus

SKY10022777_112020

CONTENTS

Prologue   ix

Chapter 1

SKETCHING FOR CONCEPTUAL DESIGN   1
   Introduction   1
   Creative Rigor in Design   1
   How Do You "Design"?   3
   Design Starts in the Studio   5
   How to Sketch   6
   Sketching Digitally   10
   Be the Designer   12
   It (Really) Isn't about the Tools   14
   Stay Agile   19
   Version Control Systems   31

Chapter 2

THE ELEMENTS OF DESIGN   33
   Designing the Elephant   34
   The Measurement of Space   35
   The Qualities of Space   40
   Geometry and Form   44
   Freeform Curves   52
   Surface   56
   Building Objects and Other Higher-Order Primitives   64

Chapter 3

REPRESENTATIONS OF SPACE   71
   Descriptive Geometry   73
   Architectural Scale   76

Cutting Planes   79
Orthographic Projection   80
Plans   80
Elevations and Vertical Sections   82
Oblique Projection   83
Perspective Projection   88
Cinematic Perspective   92
Free Perspective   93
Stereoscopic Perspective   95
Unfolded Projections   98
Sketching in Diagrams   101

Chapter 4

SKETCHING IN 2D   105

Your Sketchbook   105
Sketching with Purpose   106
Let's Get Sketching...   110
Physical Sketching   110
Materials for Physical Sketching   114
Carry a Sketchbook, Really   115
Tools for Digital Sketching   116
Physical and Digital Together   120

Chapter 5

SKETCHING IN 3D   123

There Is No Hiding in Models   123
Benefits of Sketch Modeling   123
Physical Modeling   127
Digital Modeling   132
How 3D Modeling Actually Works   133
Sketching Lines in 3D   141
Sketching Surfaces in 3D   144
Sketching Forms in 3D   147
Assembling 3D Models from Parts   150

CONTENTS

Chapter 6

SKETCHING IN CODE   157
   Rules-Based Design   157
   Materials for Digital Sketching   159
   Getting Started with Programming   162

Chapter 7

FROM SKETCH TO PRODUCTION   185
   Sketching for Presentation   185
   Sketching for BIM   189
   BIM Levels of Development   191
   Sketching in LoD 100 BIM   193
   Sketching for Construction   198

Chapter 8

EPILOGUE   205
   Technology Is Incremental – Even Revolutionary Technology   205
   Computation Is Powerful, But It Isn't Limitless   206
   An Imaginary Scenario about Construction   207
   How Close Is This Future in 2020?   208

Appendix A

WHAT COMPUTER SHOULD I BUY?   211

Appendix B

INTEGRATED DEVELOPMENT ENVIRONMENTS (IDES) FOR ARCHITECTS   223

Appendix C

SKETCHING IN VIRTUAL REALITY   227

Bibliography   233

Index   237

PROLOGUE

It is a magical thing to start from nothing and bring something new into the world.

For some, the primal act of creation is a moment of self-doubt and fear. What if what you make isn't good? What if people don't like it? What if it causes you to lose face, or money, or to waste a bunch of time and energy on a useless folly? To people who think this way, the best strategy is to follow some pattern, some standard, or some best practice that worked out last time. Adopting a vetted standard reduces your personal responsibility if something goes wrong. And, anyhow, if there is a single right way to do it, why would anyone choose a different way?

If you picked up this book, I hope you are already thinking differently. Real acts of creation must acknowledge the norms and standards of practice as an important historical precedent. But they must also look beyond them to imagine something the world hasn't seen before. I am optimistic that the best designs for the future are those that question acknowledged norms, critique the ideology of their inception, and look past them to imagine something different. Design is a progressive act. You must believe that the future you propose is better than where the project started. Embrace that and treasure any opportunity to take part.

There are people in the world who know how to do design. As a society, we know how to train them, and we generally recognize that the work they do is uniquely valuable. They can put themselves into a unique mental state, fusing reflection and action into a laminar flow state, which takes input from the world and produces as output proposals for how things should change for the better. In architecture, the most effective designs are those that are buildable, that perform well against all objective design criteria, and yet at the same time that add a mysterious quality to the final product that cannot be rationalized easily but is evident to all. If you are one of the people who know how to do this, this book is for you.

Technologists like me tend to solve the problems we know how to settle before the harder challenges that we know might be more important. Since the 1960s, people like me have been building tools designed to remove the drudgery of professional design work. First, we made 2D CAD systems to help architects draft more efficiently. Then we built rendering engines to help architects present their work more beautifully. Most recently, we have been developing full building simulation tools (building information modeling, or BIM) that help to describe the proposed building to fully constructible levels of detail before the first shovelful of dirt is moved on the construction site. We have built tools to help estimate cost, to manage the flow of resources on a project, and to widen the bandwidth of communication between all members of the team.

There is visible, measurable, and repeatable value in all such tools. Likely, your firm is using some variations of all of them somewhere in your practice. You may well have trouble imagining doing your professional work without them. Can you imagine, for example, leaving your email behind for a return to the fax machine? This is not the full story, however.

It seems to me there is a harder problem that we have forgotten in our industry's heady rush of technological enthusiasm. Simply defined, it is the primary and essential work of architectural design. Your technological toolbox helps you to drive a design through production and construction with greater and greater efficiency. But all of that improvement is irrelevant if you haven't got a good design at the beginning.

Since 2002, I have been a member of the product management team responsible for the design and development of a digital sketching tool called "SketchUp," which we designed to speak to the problems of designers (particularly architects) doing conceptual design. At its invention, few other tools existed to satisfy this need. That is still mostly true today. To this day, we have users who can't understand why we haven't developed more tools for CAD drafting, for rendering, or BIM. The answer is simple, often repeated, but widely misunderstood. We haven't built those tools because that isn't really what SketchUp is. It is for doing design, which is an inherently messy, abstract, and iterative process. SketchUp is where you do the design work that you will later document, render, and otherwise simulate to obsessive detail – design work that may, if you're lucky, translate through to construction. But the long journey to that result always starts with a simple sketch.

A sketch is often little more than a collection of lines and ideas, often captured on paper, that together suggests the future only in the most abstract way possible. A sketch captures the essence of an idea and exposes it for reflection.

At the Beginning of Every Sketch, There Is a Line

In 1974, shortly after graduating from the Cooper Union School of Architecture, George Chaikin invented a quietly elegant technique for drafting curved lines of remarkable complexity using only a compass and straight-edge. *Chaikin's algorithm*, as it is now known, approximates smoothly curved lines by progressively cutting corners off a series of connected straight line segments. If more smoothness is needed, the draftsman simply knocks off another set of corners. And so on and so on until the curve looks smooth enough (Figure 1).

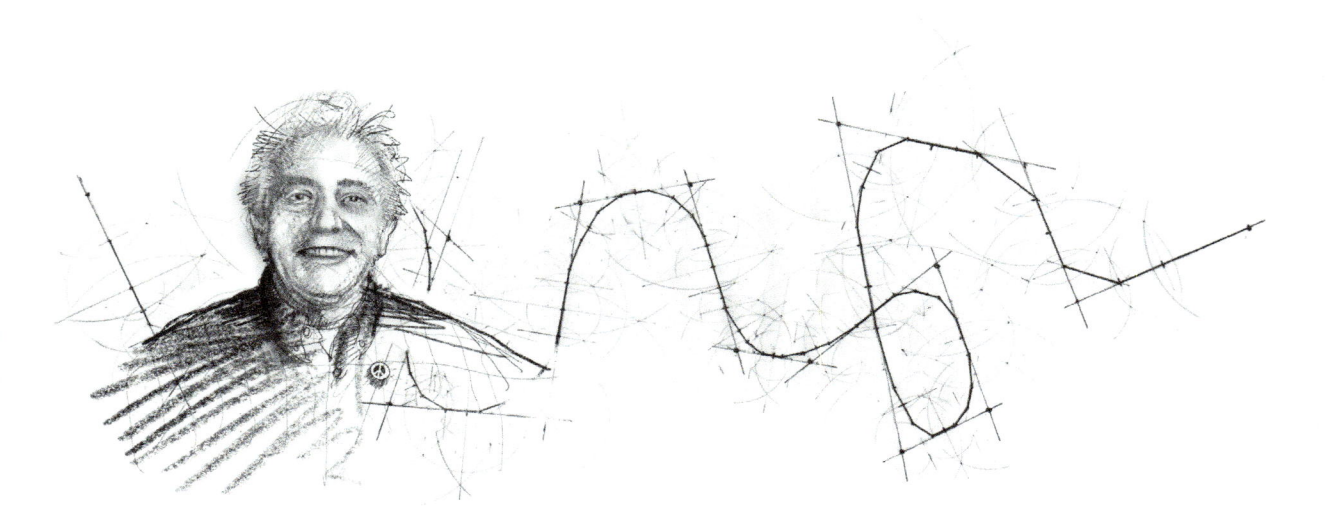

Figure 1:   George Chaikin and his algorithm, drafted by hand: "Cutting corners always works."

PROLOGUE

xi

To someone who has drafted by hand, this algorithm is intuitively apparent. The explanation probably sounds a bit silly. However, it happens that this algorithm also makes it possible for computers to represent smooth curves. Mathematically, Chaikin's algorithm produces quadratic Bézier curves. In other words, Chaikin's algorithm creates the same kind of curves that Adobe Illustrator does.

I had the good fortune to study with George at Cooper Union when I was a student there in the early 1990s. George's computer class (the only one taught at the time in the School of Architecture) was unusual – perhaps more like a design studio than a programming class. I learned, based on things like George's curve drawing algorithm, that a pencil is a pretty powerful piece of design technology, and there are not many forms that you cannot precisely describe using one.

Of course, computers can draw 3D form on-screen many orders of magnitude faster than I can with a pencil on paper. Graphics pipelines running on specialized hardware can transform and draw 3D models to the screen at 30+ frames per second. With a pencil, my (personal) frame rate is probably less than one frame per minute for a straightforward model. Computers are always faster, and faster is always better. right? To members of the first fully digital generation, this probably sounds pretty obvious. But if all I want is a single 2D view of a 3D object, paper and pencil are probably about as fast as modeling the entire object and letting the computer draw that 2D composition for me.

The transformative moment for me, when drawing on a computer changed the way I thought about design, came when I wanted to have a look at what I was designing from some point of view that I didn't know I was going to want when I started. I picked up the model and spun it around. As I moved my point of view, I got a smooth transition. I could watch the model move by in front of my eyes. I didn't get lost. I could have a cinematic sort of experience as I zipped from place to place in the model, experiencing multiple points of view as quickly as I could move my mouse. The computer transformed the experience of space by merely drawing fast enough that I could see past the individual frames and fall into the immersive spatial experience we call "3D."

In the late 1980s and early 1990s, right when I was completing my education as an architect, practices around the world were being sold computer-aided design systems that were affordable and (relatively) easy to learn. Architects were told that conversion from hand drafting to drawing on-screen would allow them to complete projects faster, and with considerable cost-efficiency. In the end, architects switched industry-wide to using CAD systems like AutoCAD. And the promise was fulfilled. Projects did become more comfortable to document, and changes did become easier to manage. The future, it seems, had brightened for the profession.

My first real professional experience with CAD was in Berlin, during the construction boom that followed Germany's reunification. Working together with young colleagues from around the world, I got my introduction to production drafting work. Working against a madcap weekly review schedule for Hochtief Construction, I helped document a million square meters of space for the Messe Berlin, working in the offices of O.M. Ungers, Walter A. Noebel, and (later, on other projects) Müller + Reimann, Architekten. We used a tool that was at the time the absolute apex of design technology, the "virtual building" simulation tool ArchiCAD,[1] from Hungarian software developers Graphisoft, running on state-of-the-art Power Macintosh computers (Figure 2).

When I returned from Berlin to the United States, I took a job with Boulder, Colorado-based

---

[1] ArchiCAD invented the idea of "virtual building" in 1984, over a decade before the first release of Revit popularized BIM.

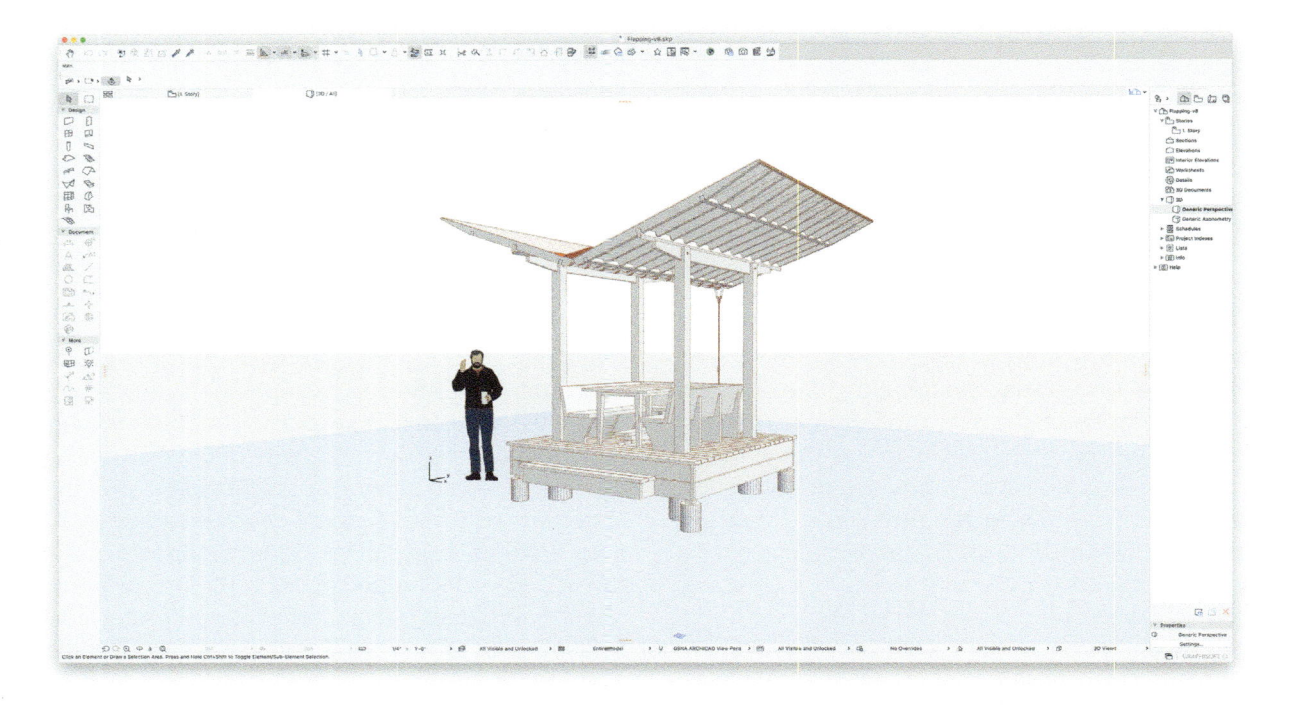

Figure 2: Modeling in 3D with ArchiCAD.

Communication Arts, Inc., where I worked for over six years as a designer on a wide range of large retail and other commercial projects. Still working on Macintosh computers, I traded ArchiCAD for FormZ[2]; traded the drudgery of construction documentation for the drudgery of rendering. But at Communication Arts, I had an opportunity to learn under some of our industry's most talented design communicators. I learned from guys like Mike Doyle (2006), whose book *Color Drawing* is now in its third printing – a bible for architectural rendering that is as relevant today as it ever has been.

From Communication Arts, I made a change. I knew that technology had an opportunity to change the way design work was done, but I wasn't satisfied that it was doing so in the right ways. Technologists usually automate the things they know how to automate, not always the things that are the most important to the users they are serving. The designers of those early AEC CAD systems rightly assumed that architects would know how to apply these new tools effectively and would protect the work they were doing that couldn't be automated. But I'm not sure this is exactly how it played out.

Technology, though automation of that which was judged to be manual drudgery, may have also reduced (inadvertently) the quality of design. Because if you can jump straight to the conclusion, to get a set of drawings produced that looks consistent and complete, then you can get paid quicker and move on to the next job. Admittedly, however, this isn't the fault of the tool. Magical buildings have been built that could only have been completed with computational help.

---

[2] I had many late nights and long days with FormZ, and I developed a genuine affection for it. In the mid-late 1990s, there was nothing more powerful on the market and it seemed like there wasn't anything I could model with it. It's still a great product today, especially for architects.

PROLOGUE

xiii

Leaving the traditional path to professional licensure and a full career in architecture at Communication Arts, I joined a snappy tech startup in Boulder with a dream that 3D modeling should be as easy as sketching by hand, and as powerful. Our company was called @Last Software, and in 2000, we released the first version of a product called SketchUp (Figure 3).

If the success of SketchUp has been any indication, the design tools of the future will not only work primarily in 3D, but they will be much simpler than what we're using today – simple like a pencil and paper is simple. As designers, we've got about all the "powerful" we need for a while.

If you're like me (and like most software users), you use less than 10 percent of the features available to you more than 90 percent of the time. The word processor I'm using to write this book has about a dozen buttons in the toolbar, of which I've used a total of three. I downloaded it for free from the web, and this book is the first time I have ever used it. This is an excellent tool for writing. Not particularly "powerful" from a feature perspective, but it does every single thing I need a word processor to do. Simple, complete, and powerful.

Simple tools are not just easier to use. They are also easier to share. When we released the first free version of SketchUp back in 2006, we found that both architects and their clients started building and editing 3D models together. Some architects were threatened by this, fearing that it undermined their design authority. But I think it led to much better design work all around. If your client has an idea about where the front window should go,

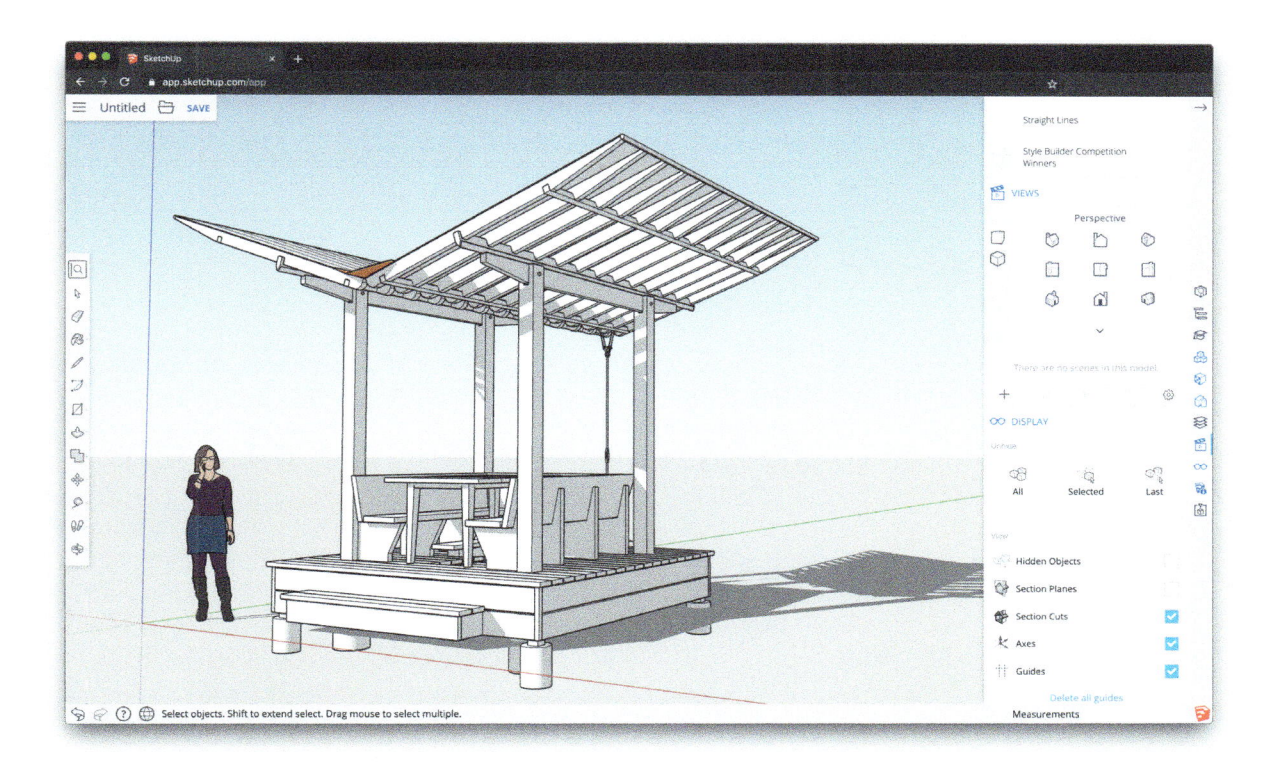

Figure 3:   Modeling in 3D with SketchUp.

allowing them to show you that idea in a 3D model is just so much more useful than listening to them try to describe it verbally. To be historically fair, clients also know how to use pencils to communicate their design ideas.

The great thing about sharing tools isn't really that projects move faster (although that tempts business-minded folks), but instead, that streamlined communication gives everyone more time to iterate. More time to think and to try things that don't work out. More time to improve the design and work the bugs out of it. More time to make the design a great design. The real democratization of design is happening right now, right under our noses.

The rising tide of information technology has brought some improvement to the construction industry. The now-ubiquitous tools for communication and collaboration afforded by the internet (email, the world wide web, search engines, and live videoconferencing) have forever changed the way we work. Designers are also learning that their design proposals can be fully simulated in the computer to constructible levels of detail before construction begins. BIM promises to encode into a designer's model not only the fully realized properties of a design but also a nongraphical encoding of every attribute, property, or even the chain of decisions that have gone into them throughout the lifecycle of the project.

But with this technology has come a breathless temptation to leap to constructibility too fast, glossing over the core design research and the traditional reflection-in-action that defines an architect's design process. As a profession, it has become all too easy to simply skip the careful thought and considered thesis/antithesis/synthesis that separates a competent design from a great one.

In this book, I will explore a category of design technologies that serve the needs of working designers in architecture. There are innumerable books available on the subject of BIM, and countless others focused on advanced architectural visualization (ArchViz). Neither of these popular categories attempts to cover quite what I'm interested in here, however. Simply put, I want to explore the myriad of ways that an architectural design process at the very earliest stages – from cocktail-napkin to a fully formed conceptual design proposal, can be supported by digital technology. I want to think through those first steps, from a blank page to a design concept.

This book is not a how-to guide that will teach you how to use any particular piece of software on the market today. Unavoidably, I will refer to critical components of technology that are contemporary to the time in which I'm writing. And I will inevitably talk about SketchUp – because that is my best answer as a technologist to the modern challenge of sketching in 3D for architecture. But, technology in general moves too fast for books, encoded in paper and ink, and subject to the lengthy timelines of a physical publication, to remain relevant for long. Where I need to go more in-depth on such topics, I've pulled that detail out to an appendix that can be updated more quickly in the future.

This book will explore the general principles of conceptual design in architecture and will propose a reframing of them in the context of modern digital technology. This book is for you if you have been confronted by monolithic software packages that cost more than your car payment every month but block your creative flow with every button click. Tools that are sold to you with fear tactics suggesting that if you don't buy them, learn them, and convert your work to their ways of thinking, you can't be an architect anymore. This book will help you learn how to use computers to expand your creative reach, to iterate faster, to learn more quickly, fail earlier, and be, in the end, a better designer.

chapter 1
Sketching for Conceptual Design

Introduction

Good design is easy to recognize but difficult to describe. When it is missing, you immediately notice even if you can't say precisely why. Design has consistently defied all attempts at logical rationalization since before the Enlightenment, though, not for lack of trying by generations of design theorists. We observe in the world examples of effective design every day, but the rules and standards that led to that design's realization are notoriously tricky to decode.

Armed with a rigorous knowledge of building science that has been informed by thousands of years of historical precedent, architects can engage in a uniquely iterative process of design and reflection that produces constructible, functional, and at the same time, inspiring, and delightful buildings. The good ones can do it on demand, skillfully dropping themselves into a flow state where sketches pour out in a flood of work. Once in that state, they can reflect-in-action with consideration for hundreds of individual requirements, including unstated but inferred ones and all their various opportunistic affordances. By accommodating both objective reality and ethereal poetics, architects can bring a rigor of thought to any set of inputs, framing and reframing the original problem as necessary and iterating collaboratively until a path forward has been reached.

Creative Rigor in Design

The design work of a professional architect is rigorous, in-depth, and (to outsiders, anyway) more than a little mysterious. Building construction projects are notoriously complex, with little opportunity to rehearse solutions before committing fully to them in construction. Once construction has begun, the cost of making changes to the design can be astronomical. Every detail must be considered before construction begins.

Architects are trained to start sketching from very little concrete information, perhaps only a brief list of requirements, a site, and some local building codes and ordinances. The project may have exciting opportunities that are evident or at least implied by the available information, but it may also include unrealistic expectations and hidden, self-contradictory dependencies. It might be challenging to make many formal propositions about the building at first, but architects are trained to dive right in regardless. In time, through a process of sketching, reflection, and reframing of the problem, they know how to figure it out. As professionals, architects are great at just "winging it," figuring out both the real issues (including unstated requirements that only later become evident) and their ideal solutions.

Some architects, particularly those who fancy themselves a bit closer to art than science, may actively resist any external attempt to rationalize their process. Design, to them, is a mystical art open only to those properly initiated. In truth, they may not really know how they do what they do, though few of them will readily admit this. But there's nothing to be afraid of if you are one of these kinds of designers. The magic of your work is not diminished because you can't rationalize it for others. Most designers, even the most rational of them, cannot describe how they do what they do or how they know what they know about the design. The intuitive leaps you make while working are not diminished because you can't entirely explain how you reached them. Your ability to make those leaps, correctly and effectively, is what defines you as a designer.

If they are honest with themselves, all designers have trouble describing how they reach the conclusions they reach while designing. They may talk about how individual decisions simply "feel" better than others, how one design direction is inherently better than another one because of where it leads the design next in its evolution. When pressed, they may be able to post-rationalize their past right decisions, even citing evidence and prior art to lend a sense of rigor to their overall design process. Skillful action almost seems defined by the fact that the professionals who engage in it know much more than they can say. And they can act based on more than they know, as well. This seems contradictory, irrational; perhaps even to the most rationally minded, it may seem irresponsible. If you can't say how you know something to be accurate, how can it be?

It happens as well outside of the design professions that the most skilled professionals, even among medical doctors, lawyers, and business managers, are actually at their best and most productive when they are working in ways they can least describe to others. A lawyer making an impassioned plea on behalf of a client in front of a jury, while certainly well prepared and rehearsed, may improvise new arguments on the spot as they watch the faces of their jurors. A trader on Wall Street may "feel the pull" of a barely visible market dynamic. Expert accountants can see through a wall of numbers in a spreadsheet and make accurate conclusions about the health of the organization it represents or the opportunities ahead of it in the market. Even baseball players "know" intuitively what they need to do to win a game . . . though they have no way to explain how they will pull it off. None of this makes any rational sense; we think of it as the "art" in their practice and (rightly) respect it deeply.

Since the Enlightenment, our culture has relentlessly applied the structures and norms of scientific, rational thought to all professional practice. "*Ars sine Scientia nihil est.*[1]" wrote Jean Mignot, fourteenth-century architect of the Milan Cathedral. Without a basis in some kind of rational theory, the practice of any art is without worth. That theory must be logical, consistent, and reducible to first principles that are universal enough to provide a common foundation beneath all similar work. In professional practice, trained architects should be able to apply the theory to real-world situations and come up with logically consistent design results.

Contemporary architectural theory isn't consistent in the same way that the underpinning theories in physics or philosophy are. Where the traditional work of architects could be rationalized, that rationalization has often led to the identification of a new profession that specialized in just that more rational act. For example, Structural Engineering. Or Construction Management. Perhaps even Interior Design, with its increased dependency on premanufactured lighting and furnishings, could be thought of in this way. In time, as more and more of the rationalizable components of the

---

[1] "Art without science is nothing."

traditional matrix of responsibilities of an architect have migrated to new professions, only the mysterious, intangible design work of the architect has remained behind.

Once everything that can be rationalized and specialized has been accounted for by others, it is tempting to assume everything of merit has been covered. What a heartless and cold world we would build for ourselves if this were, in fact, the final solution. The sum of all rational processes in architecture will not give you a complete design, not even close. It is intuitively obvious to people when the spaces in which they are living their lives do not delight them, even if they can't say why or how. They, of course, also know if some obvious box wasn't checked in the design if the roof leaks, the garage has collapsed, or the driveway washed out in the last storm. But it is easy to tell when the poetic concerns haven't been met, too. Few people will care that the roof doesn't leak, because they won't care about the building at all.

The problems faced by architects may be different from those faced by doctors or lawyers. They may be more ambiguous and less consistent. Every building inevitably presents a different set of problems, a different set of requirements, and a different set of opportunities. There are undoubtedly universal principles that apply. The building must provide light and air to its occupants, must protect them from weather, and must support them against the pull of gravity, for example. But there is little guidance for architects seeking to add truly delightful, life-changing space into the world. It is tough to rationalize this part of the work.

How Do You "Design"?

So how is it done? How does great design happen? It is certainly wrong to assume that design practice can in no way be defined. Even the purest of abstract expressionist painters acknowledge there is significant science underpinning their practice. The physics of light, the chemistry of pigment, and the biology of seeing all have roles to play; without them, there could be no poetry. They also know how to get themselves into the mental space where the work can happen. Maybe it is about waking up early or maybe staying up late. Or perhaps it is about just the right music on the stereo. Somehow, it is repeatable for them. And no matter what you think of architectural designers as professionals, anyone who has been trained through an architectural education intuitively "gets" the processes and practices that lead to being able to accomplish a great design. Design thinking is reproducible and persistently observable in human behavior. There must be some way to describe it.

Architects speak of their work as being part of their "professional practice," and I think we can find the answer we're looking for in that language. There are norms and standards that all architects must know – from building codes that you must meet to pass inspection before occupancy to the universal principles of gravitation that you must know so you design buildings that won't fall on their occupant's heads. Architects have to become the people who can answer questions about the building that nobody has thought to ask yet, and they must be able to do so consistently, decisively, and (even) defensibly. The best architects have fully internalized these and many more physical and philosophical truths. But to work exclusively within their mandates is to be blocked before you've even begun.

I think the *practice* we refer to when speaking about architectural work is more akin to the practice of a professional musician. Like an architect, a professional musician must know some physics, some historical precedent for their chosen music, and many other things. They must (probably) memorize the music they plan to perform – or at least internalize it enough that they don't need to waste time in performance reminding themselves what notes to play next. They must do all of these things in preparation for their performance, and they must reinforce them and drive them into instinctual behavior through relentless repetition. They "practice" endlessly, obsessively, repeating the same themes over and over again, testing variations, exploring newly discovered nuance and expression. They learn their subject so profoundly that they "become" it, physically embodying the music they will soon perform in public.

Malcolm Gladwell claims you need 10,000 hours of practice before you can be any good at something new (Gladwell 2008). I think most professional architects would say that they are just beginning to get warmed up by then. John Hejduk, who was dean of the Cooper Union School of Architecture when I studied in there in the early 1990s, taught that few architects were able to accomplish works of any real merit before their 50th birthday. There is just too much to learn, too many experiences to be earned in the field, and too few opportunities to try something genuinely new. The knowledge of a professional architect must be driven so deeply into their beings that it is instantly accessible, "…as easy as riding a bike." And it takes time and experience to do that. It takes lots of practice.

A well-practiced architect knows a collection of artistic themes that have worked well for them in the past, and they can synthesize new interpretations of them that respond to differences in programmatic conditions in a profoundly intuitive manner. They are practiced at recognizing new opportunities in the design as they are uncovered and know well how to "listen" to the design while they are working when it starts to "talk back" to them. They know the basic universal theory, where it exists, so well as to be able to take intuitive leaps beyond it into new ideas, new territory for exploration.

In his work *The Reflective Practitioner*, (Schön 1983) described the work of architectural design as closer to jazz improvisation than to engineering. The architect is like one of a group of musicians who are working together to improvise new music based on a shared theme, a jazz standard. The musicians all know the tune and their own instrument deeply. They have confidence in their necessary skills that permits them to explore the space around the melody, responding to new interpretations from their fellow musicians as they appear. We love listening to jazz music precisely because of the tension between the standard and the musician's interpretations. Everyone in both the band and the audience probably knows the melody. There is a basic musical structure (play the tune together, take some individual solos, then play the song together again) that is also understood by all. The music is exciting because of its interpretation, the musician's exploration, and the unique expression in the moment of this particular performance.

A skilled jazz musician is unlikely to be able to describe precisely why a particular sequence of notes was chosen in their last solo. And it is equally unlikely that they planned their solo out much in advance of the performance. Upon reflection, after the performance, they might be able to name a few of the events that lead to their decision in the moment, perhaps something they heard from the preceding soloist, or maybe just a memory of something from farther back than that. If forced to fully post-rationalize, a logical and consistent story could emerge based on history, physics, and prior practice. But in the moment of the performance, none of this would be in the musician's mind. They would be too busy playing.

In a more rational problem-solving profession, a practitioner might consult standard practices or industry norms to model the problem, then optimize a set of variables until a fixed set of performance criteria are met. The basic requirements (e.g., "the building must not fall down") can be modeled using real physical simulations and adjusted until the criteria have been objectively met. There is some room for interpretation, of course. You might choose steel over concrete for some complicated set of reasons. A unique soil condition might be discovered on-site during excavation that inspires a different solution for the foundation. Skilled engineers, well-practiced in their art, will know how to reflect-in-action just like a jazz musician.

Schön found in his research that all professions – even those most commonly assumed to be driven wholly by the most rational of scientific standards, have a component of intuitive, creative, design-like thinking in their practice. A doctor, for example, when confronted by a patient presenting symptoms for some illness, must intuit a diagnosis quickly and accurately. The majority of symptomatic conditions are unique, likely something the doctor has never herself quite seen before. In the moment, relying on years of practice to inform an intuitive response, she can respond with a plan of action. A life may depend on the accuracy and speed of response, so it has to be right.

Medical school prepares doctors by embedding in their minds an exhaustive collection of theoretical knowledge in biology, chemistry, anatomy, pharmacology, etc. Then, in grueling years of residency, they practice the application of that theory to real-world conditions under the guidance of a professional with many more years of experience than they have. Finally, they graduate into their professional practice, prepared for a career diagnosing and curing illness.

It is the "art" of medicine that is the most difficult to teach to medical students. Their primary knowledge may be profound, and their ability to recall scientific precedent may be without peer. But it is only through practice, in a professional setting on real patients with real problems and under the guidance of a doctor who has more experience than they do, that students learn how to do the work they do. The best doctors are those who know the science inside and out but can synthesize unique solutions for unique problems immediately, intuitively, and creatively. *"Scientium sine Ars nihil est"*[2] may be an appropriately enduring counterpoint to the reforms of the Enlightenment that favor rational thought over intuition.

Architecture doesn't have anything approaching the body of scientific or philosophical theory that underpins medicine. Still, it does have an intellectual rigor of its own that blends art and science in a practice of design, which is both reproducible and trainable. Like doctors, skilled architects can make decisions for reasons that they cannot describe in the moment of their making. Unlike doctors, this ability is taught right from the beginning of their training. The education of an architect is distinct from that of other professionals in that it actively teaches both science and art right from the outset. A skilled architect is comfortable from an early age with the process of reflection-in-action that allows them to work pragmatically as both scientist and artist.

Design Starts in the Studio

The key to this training is in the studio environment, where students work side by side for long hours on poorly defined problems in design. In the studio, young designers learn to think with their hands, to draw by seeing, and how to apply a unique intellectual rigor in framing and reframing problems as unanticipated opportunities present themselves. In a range of contexts from the most intimate of desk critiques to the most public of project reviews, young architects learn how to practice their profession. All architects have this experience in common, and for all of them, it is life-changing.

---

[2] "Science without art is nothing."

From academic design studios, young designers graduate to professional practice, where the studio process begins again. Desk critiques with a professor are replaced by desk critiques with a partner, only this time with the added dimensions of time and money bearing down as well. It is easy to forget the theory and the art in professional practice, to double down on project delivery and efficiency, and to select out all the time spent on design exploration in favor of a faster deliverable. That is, after all, the most rational thing to do. It may not, however, be the best thing to do.

Most software is built to automate rationalized processes, based on an offset understanding of what is the most important thing a future user will want to do. Product managers in technology firms are trained to interview their customers and ask them pointedly about what it is they most find onerous about their work so that valuable tools can be built to assist them with those tasks. This works well for accounting packages, but it is difficult to create software that supports a process that the user cannot describe. Design thinking, particularly during its earliest, least definable sketching phases, is difficult to explain by even its most practiced professionals.

How to Sketch

"We see with the eyes, but we see with the brain as well. And seeing with the brain is often called imagination."

—Oliver Sacks (2009)

What is the first thing you decided about your last project? And what was the thing you decided before that? And before that? What was the decision you made before all the other decisions that you made? (See Figure 1.1) Can you remember what it was?

The composer John Cage advised, "If you don't know where to begin, begin anywhere."[3] For architecture, everything starts with a sketch. It might be a sketch of an idea, a sketch of a proposal, or just a sketch to fill the emptiness and uncertainty. Nothing is more terrifying, more exciting, more filled with risk, and opportunity than a blank sheet of paper. In the beginning, anything is possible. Maybe, even everything is possible. If that moment isn't exhilarating, you might want to take up a different profession. For most creatives, this moment is pure, solid gold (Figure 1.2).

To be sure, no design ever starts from a completely blank piece of paper. Particularly in architecture, design always begins with a site and a set of other requirements and opportunities. Requirements are functional. They help to put some structure into that terrifying void; that is what your project looks like at the beginning. If you feel like your project has no constraints, you haven't looked closely enough. And if you absolutely can't find any, you're going to have to make some up on your own.

The moment you make the first mark on the project, whether a literal line on a piece of scrap paper or something else, it will be wrong. If it doesn't seem that way to you, you're probably deluding yourself about something. Maybe about a lot of things. Usually, the first idea you capture is the one that's been standing in the way of all the better ideas still to come. The best thing to do is to purge it out of your mind quickly. Clear the road and get ready to move on.

Your first sketch on a project might be a physical sketch (to be sure, I recommend that), or it might be a formula in a spreadsheet. Maybe it is a great conversation over coffee with your client. Or it might be a haiku. Or an interpretive dance.

---

[3] This is a broad theme for Cage, published in many places. For a canonical reference, begin with his "Diary: How to Improve the World (You Will Only Make Matters Worse)" (Cage 2019).

HOW TO SKETCH

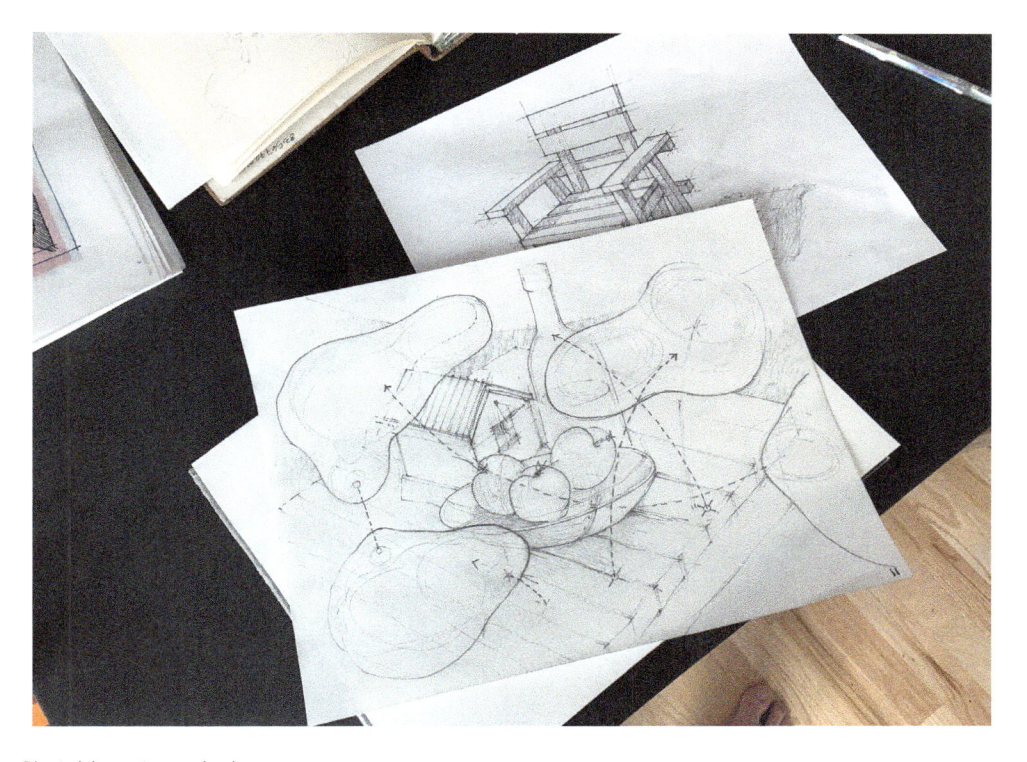

Figure 1.1: Sketching at my desk.

Figure 1.2: "If you don't know where to begin, begin anywhere."

Figure 1.3: Sketching an ensō to check my status.

Whatever gets your blood moving conceptually, whatever purges the bullshit that was occupying your attention before you got started on this new thing. Keep it simple. Move fast (Figure 1.3). There's better stuff ahead.

If you are relatively inexperienced with your creativity, you are almost sure to hang on too tightly to your first sketch. Because it is hard to make things, it is hard to be creative. You're exposing yourself to critique the minute you pick up your pencil. What if the idea is terrible? What if the concept is OK, but the drawing doesn't communicate well? Or maybe the sketch looks like it is saying something completely different than what you meant it to say. What if your client looks at the sketch and sees that you didn't know what you were doing this whole time and you're nothing but an imposter[4]?

Experienced creatives know this moment well, and they recognize it for what it is. It is just something to be acknowledged and moved beyond quickly. Take that first sketch and throw it away. Crumple it up, toss it on the floor. Set it on fire, maybe. Poof, it is gone. Mourn its passing quickly and move on (Figure 1.4).

Now, make another sketch. And another. And another. Throw most of your work away – that helps you move past initial ideas and get faster to the good stuff. Keep your inner critic behind you as much as you can. Every sketch you make is going to be better than the last one was, but you're going to want to learn from the mistakes you've made. Every sketch has a purpose, even if it is only to prove to you that it represented a dumb idea (Figure 1.5).

Be prepared for unexpected consequences and happy accidents. They can come from anywhere, at

Figure 1.4: Letting it go.

Figure 1.5: Iterate, iterate, iterate.

any time. It is your job to see them when they appear and to understand what they might mean to the project. The simple truth that ideas sometimes come from unexpected places is singularly frustrating to less-creative people. If the breakthrough approach for a project came from somewhere other than the

---

[4] "Imposter syndrome," the cognitive bias that afflicts creatives with the fear that they are about to be exposed as a fraud, is incredibly common among all high-performing professionals. While it is possible that you are an actual imposter, you might also consider giving yourself a break.

designer's mind, why do we need the designer? Let's just save that money for something else, right?

Wrong, of course. It is a delightful moment in every design when the project takes on a life of its own. Experienced designers observe that the project begins to "talk back" to them. That decisions made begin to naturally lead to one another as the project as a whole begins to fall into sharper focus. The window you placed above the foot of the master bed just happens to frame the client's favorite view of the mountains perfectly. You didn't plan that, but *bam*, there it is anyway. Embrace it and carry it forward.

You should also be prepared to see something great but not to be able to explain what makes it so. Great designs are found at the intersection of many smaller good ideas. Often, transcendently good designs can only be experienced holistically. Like art, the best architecture usually defies a simple description. Construction teams are frustrated by the fact that you cannot always explain your decisions rationally. If you describe your building to them without being able to explain everything you have designed, they will think you're nuts. If you have a great client (and the best projects come, as we say, from great clients), they might be willing to take your word for it, but you might be the only one on the team who sees it.

Are you feeling stuck? It happens to everyone at some point. Try changing media – if you were drawing with a pencil, pick up a marker. If you were nose-deep in a spreadsheet, go for a walk with your camera. If you were doing an interpretive dance, well, you're on your own with that one, Martha Graham. The important thing is to establish some movement in the flow of ideas, to get past that initial excitation and into the flow state of real, productive, creative design. You trained for this work. Your client hired you to do it. You are special, different, and unique on the job because you can do this work. Relish opportunities to do this work when you find them.

Once you find yourself in this state, you'll never want to leave it. Hours will go by like minutes. Before you know it, the streetlights are on outside the office, and everyone else will have gone home. And you'll feel like you're just getting rolling. Time to stoke up the espresso machine and get serious about this thing. Time to start asking the tough questions, time to get outside your comfort zone.

But then, there's a deadline. You need to come to some conclusions, prepare some concrete recommendations for your client. You're only getting paid for a tenth of the time you just put into this thing. It is time to wrap it up and move on to the next phase. Time is money, champ. Let's make some of that this time, shall we?

But it is so hard to put down the pencil, to come down from that magical flow state. You know that you can still improve the design, always make it better for your client, more comfortable to build, more magical for its future inhabitants. The end of conceptual design is going to make you feel some mourning, some melancholy, some doubt. Beware the inevitable bargaining phase, a pivotal moment in every grieving process. Maybe if you skip lunch today, you can squeeze in one more hour of work. Perhaps if you only explain how important this is, you can move the deadline. Just a little bit. And for the last time, you promise (Figure 1.6).

No creative ever admits to being done with a design. You're just out of time to work on it further. Design doesn't have an obvious single right answer, just a series of decisions, optimizations, and opportunities to be explored further down the line. This notion frustrates people who don't understand it. Your client has in mind (probably) that they would like to have a building to occupy. They would like it to be a magnificent building (they hired you, right?), but they would also like it to be done someday and for a price that won't put them in bankruptcy. And your contractor, guaranteed, just wants to, ". . . get this project done."

Figure 1.6: Put. The pencil. Down. You're done. The charette is here for your work.

And so, you have to move your unfinished, incomplete, sketchy preliminary work on to the next phase, with all of its half-explored opportunities and practical concessions. If you're lucky, you'll get to keep working on it through the next phase; refining, detailing, and documenting the design for construction. Maybe if you're fortunate, you'll have a chance to track your design through the construction process to make sure it stays true to your intention and (maybe) to take advantage of unexpected opportunities that only appear after construction has begun. Beware, however, that construction teams universally hate change, even if the change you want to propose improves the project significantly.

All the rationalists on your project hate change, though, and will try to beat you down every time you propose one. Be strong, be professional. Stay agile; stay safe. Your business is to change things for the better, but most people hate all change passionately. Embrace that, hear it, and help move people past their fear. Every architect who has proposed a change in the field thinks they are doing so because the outcome will be better. Every contractor believes this is just moving backward in the process, and if it was so damn important, why didn't you ask for it that way at the beginning.

And they are right. The most significant design decisions you are going to make on the project are made during the conceptual design phase. Before anyone else has woken up their teams to get started, you will have explored a hundred possible variations and navigated through a densely interconnected web of uncertainties, dependencies, and opportunities. You will know not just what the project will become, but also all the other things that it could have been instead. It is your responsibility to defend that until the job is done.

But the design is never done, and the building is a living organic thing that will continuously evolve throughout its decades-long lifecycle. It is a delusion to believe the building is ever complete. You know better. Maybe you can convince the team to give you one more day to work up a couple more sketches to show them how good the building could become. And so, the process begins again. To a designer, the work is never done, the design could always be improved, and there is still a next stage to consider. With time, energy, and the right practices, you can move the world in an ethical and positive direction. And at the heart of your work, no matter how you think about its purpose, you'll always find a sketch – sketch of a new plan, a further detail, a new way to reframe the original design problem. A sketch of the next thing to try.

Sketching Digitally

This is a book about sketching, particularly with digital assistance, and about drawing as a way of working. It is about sketching as a mindset and an attitude during the early phases of work. Sketches, particularly those done in the beginning when ideas are fresh, and opportunity abounds, are where all the most impactful design decisions are made. It is critically important to figure out how to sketch effectively in the context of ever faster-paced project delivery schedules and tighter budgets. And I don't think we have really figured that out yet. I certainly don't believe we have figured out how to do it digitally. I wrote this book because I think that sketching is getting lost in the digital age, and I believe it needs to be recaptured.

Digital tools, in the end, are as good or bad as their users make them. In the remainder of this book, I will walk with you through a range of digital technologies common (and less common) in architectural practice today to explore their relative merits and usefulness to the problems a designer must face. Pencil and paper remain the most useful, durable, and scalable design technologies for most designers. No question about that – I love pencils. But there are more benefits to be had from digital tools than you might think, more value in particular in learning how to use them in a sketchy way. And the most significant benefits may come in surprising forms-not the ones you expect because they are what "everybody says." Let's take some time for introspection, some time for really thinking about what the right way to work might be.

Technologists like me are great at solving the problems they know how to answer. Like the man searching diligently for his wallet under a streetlight far from where he probably lost it because "the light's better here,[5]" technology develops where it can be built more often than where it should be. Computers are great at solving problems that are rational and clearly defined. They can, with bloody-minded efficiency, solve those problems over and over again, faster and more efficiently than you ever could on your own. And that is a great asset to your work. In what other contexts in your life do you have an assistant who never tires, which works fast and never questions your demands. Adequately considered and with the right expectations, digital sketching can take you and your work farther than you could ever go alone. But it would be best if you never abdicated responsibility for the original design to a machine. They just aren't going to be as good at it as you are. Your intuition, your irrationality, and your ability to deal with abstraction cannot so easily be replaced. They cannot be replaced with any yet known technology, and in the end, they may be computationally irreducible.[6] Not without some fundamental breakthrough in our understanding of how the mind works.

Because computers as "thinking machines" are designed from the particular knowledge and theories of cognition that crystallized during the Enlightenment. Computer "intelligence" is modeled on our own imperfect understanding of human intelligence. The core optimism in the design of computers is that human thinking is somehow rational, understandable, and therefore can be modeled, simulated, and made more efficient. But on an ontological level, we don't really know how the mind works. There is good evidence, and theoretical interpretations abound, but at the root, there isn't yet much consensus. Particularly when we consider the fluidly interpretive states of intuition and abstraction. The functioning of the human mind is still mostly irrational. Despite that, with training and experience, it becomes irrational

---

[5] There have been many variations on this joke, but my reference comes from Abraham Maslow's "Motivation and Personality" (Maslow 1954).

[6] "Computational irreducibility" is a concept I learned from Stephen Wolfram in his book "A New Kind of Science" that refers to the fact that many computational systems, even quite simple ones, create output that cannot be predicted in any way other than simply running the system iteratively over time. (Wolfram 2002).

in the most practical, repeatable, and rigorously effective ways imaginable.

In May 1888, in a letter to his brother Theo,[7] Vincent Van Gogh taught what has become for me the most critical lesson I know about sketching. Confounding the dominant pictorial paradigms of painting in his time with its ever-increasing photographic realism, he wrote, "I'm trying now to exaggerate the essence of things, and to deliberately leave vague what's commonplace."

This is a powerful statement about abstraction, about simplicity and about sketching. If I'm able to convince you of nothing else, I hope I teach you how to take the time to draw again – only this time, with computers. To sketch is to forgive yourself for having some bad, unformed, and ill-considered ideas. Ideas that are OK to explore, then OK to throw away to make way for the better ones to come. Sketchy ideas. You don't have to nail the whole design right away, and you don't have to do it as fast as you can. Take the time, do the work, and iterate, iterate, iterate. Learn how to recognize the right level of detail – the essential message you want to express. Your clients and the world they operate within will thank you for it in the long run.

Be the Designer

Among building construction professionals, there is an understandable desire to compress construction schedules and make them more predictable. Faster, cheaper, better is ever the mantra of industrialization. Craft often suffers in the name of progress, and not always inappropriately so. To adequately house the world's ever-urbanizing population, we have to build more efficiently. To build faster, we need to optimize project delivery schedules. Construction schedules compress everything that can be shortened, leaving behind only the truly irreducible elements of the overall work. "Design," identified as a distinct phase, given a name, and defined with fixed deliverables, appears to rational schedule-optimizers like a reducible component of the overall project. The ephemeral, never-ending work of the designer must be put in a box and quantified. You can't change how long it takes for concrete to cure on a job site, but you can tell the architect to spend less time picking out the wallpaper.

If you understand design as a process, a continuous one that is never completed, shortening the amount of time spent on it does not lead to better buildings. You can surely spend less time navel-gazing about wallpaper, but there is so much more to the process than that. By reducing design to trivial things, construction teams are missing the big stuff, too. The absence of design is bad design, not no design or less design.[8] Time spent on design is fundamentally about making the project better than anyone might have expected it could be at the outset. What are we saying about the future if we reduce the time we're willing to spend making it better?

In the construction industry, the word *design* is overloaded, misused, and poorly understood. At its best, design satisfies both functional, known requirements ("keep the roof from leaking, please") and the more intangible, human, emotional needs that make you feel like a better person when you're wrapped up in them. Somewhere, however, *design* picked up pejorative connotations that are wholly unflattering. Design became associated with the "pretty" things in a project; the colors, the furniture,

---

[7] In a personal letter from Vincent to his brother Theo Van Gogh, Arles, Saturday, 26 May 1888. Reprinted in (Jansen 2014).

[8] Restated from Douglas Martin, "The alternative to good design is bad design, not no design at all." (Martin 1991).

the finishes. Practically oriented teams began offering design as a "value-add" above and beyond the cost of just making a building. Something you might add to the project if you have a little extra money. And of course, in this context, it is an easy thing cut from the construction budget.

But if instead, you think of design as the activity of figuring out what we will build before we build it, you can't avoid doing it anymore. There is no such thing as a building that has not been designed . . . but there are plenty of buildings that were built by teams who didn't spend enough time during a design phase to really work things out well. In many cases, real design decisions are made passively by professionals with no practical training in the art. Design decisions may be made by people who don't even really know that they are doing design work at all, leaving critical decisions up to ad-hoc memories of best practice or just going with "what we did last time." There is nothing common about common sense – it always carries with it with unanticipated baggage. Not acknowledging that "design" happens on every project isn't a way to avoid doing it altogether. It is, instead, a way to ensure that it is done poorly and not being given enough time and attention.

I don't much care how you go about the essential practice of design. There are as many design methodologies are there are designers in the world. I'll, of course, have some suggestions in this book, but in the end, I think you want to come up with your own way of doing it. Architecture schools are good at training people to do "design thinking." But if you think there is a single best practice for doing it, a unique "right way" (and by association then also a much longer list of "wrong ways"), you are doomed forever to a life of doing derivative works. Your work might end up looking like the work of whoever convinced you to follow their process. I want to persuade you to just keep doing it. Whatever it is, by any means necessary.

Design can be thought of a something that happens in a phase at the beginning of the project, but that isn't really the complete answer. If you take the broad definition I'm promoting, the design isn't done until the project is complete. Not, that is, so long as someone is still making decisions about what the building is. Maybe, even, the design is never done – not as long as the building is still standing. Someone somewhere will always be thinking about what to change next.

Architects are the people who know what the building is. Their particular knowledge comes from an understanding of the project's unique design requirements, the wants and needs of stakeholders, the affordances and limitations of the site, and the thousands of other things that all go together into the designer's process. And that process is what really matters. No design is hatched fully complete from the brow of the architect. Great designs take time, mainly time spent making bad decisions, culling through loose ideas, and iterating to a final decision about what should be done next.

You are responsible to your client to make the building as right as it can possibly be. You may be alone at times, maybe most of the time, in truly understanding what the building wants to become. Be quiet, be confident, and, when necessary, be persistently immovable on the decisions that really matter. Because you may be the only one who understands just how much of a difference it will make to do it the right way.

But as soon as you can, as often as you can, be the expert in a room full of people who care about the same great outcome for the project that you do, but who are experts at other things that you don't know anything about and who represent points of view that complete the project in ways you can't. No project is built by a singular vision, nor should it ever be. Even the smallest of plans have a team of people working on it. Learn how to take feedback as easily as you give it. Even harsh critique is a gift. Someone thought enough of you and your work to tell you it should be better. Shut up and listen to that every chance you get.

You get better for your next job by learning from the experts around you on this job. Learn how to present your ideas, even if they are only poorly formed sketches so far and remain open to feedback. Critique is a way of finding the problems you are blind to yourself. Challenges lead to reframing the question and to unblocking the flow of improvement and advancing the design.

In the end, the design isn't yours. It belongs to the project, and if you have done your early work well, the rest of your team will pick it up and run with it. All those hours you spent getting the conceptual foundations laid will pay off when the next level of detail comes into focus, right between the lines you laid down in your first sketch. It is a beautiful moment when it happens. The design has, in a sense, a life of its own. It knows by itself what it will become, and your job is just to let it go.

It (Really) Isn't about the Tools

Most designers at some early point in their career are confronted with the feeling that they aren't going to get hired unless they know how to use some digital design tool. Maybe it is AutoCAD or Revit, or maybe SketchUp; perhaps it is the latest Hollywood rendering engine, or maybe it is something else. Before computers, young designers had to prove their worth with pencils, gauche, and portfolios. But now, internet forums are crowded with questions about which tool is the "best" one, which tool is the "industry standard" for their particular chosen field. Students are clearly worried that they won't be able to break into the profession without knowing some tool that they have heard ". . .everyone uses." And, if you look, you'll quickly find people who will feed those worries.

To be sure, employers aren't happy about paying to train newbies on the job. This is true in really any profession. Architects are no different, though perhaps they run on thinner profit margins than other professions. Every professional has some quantity of drudge work to get done, and most are only too happy to foist that off on someone they hire just out of school. Maybe it was manual drafting once. Now it may be CAD work or architectural visualization. Or organizing the sample library and just keeping the coffee pot warm.

True virtuosity with any skilled task comes only after many hours of practice doing that skill. We become experts at what we practice, and so choosing the right thing to practice does make a real difference. The tools you choose to help you accomplish the skill do matter, but maybe not as much as most people think. A virtuoso violinist will be able to make any instrument sound great, even a broken-down fiddle from a garage sale. Surely, they can make even more beautiful music with a Stradivarius violin if they have one, but their skill is in them, not the instrument. Conversely, an amateur playing that same Stradivarius will not automatically become a master through playing it.

Many architects suffer a kind of tool-oriented identity crisis at some point. They might lose an important job to someone who uses a different tool than they do. Clearly, that other guy is no smarter, so it must be the tool that made the difference. Or they might find that some celebrity whose work they admire in the profession has adopted something new. It is easier to believe that it was the tool that made the difference, not the designer who used it. Classically, all that was really needed to express an architectural idea to someone who will help to build it is a piece of paper and a pencil. You don't need software to make architecture. You certainly don't need someone else telling you what software you need.

Technologists build software tools that address an identifiable need among customers in a market willing to spend money to improve their work. For architects in the 1980s, it was the manual drudgery of drafting that seemed most in need of improvement. It takes a lot of concentration to put together an effective, complete, and well-coordinated set of construction documents by hand. This is especially true when the design that is being documented is still being refined every day. If what you really care about is the design of the thing you are building, you naturally come into conflict with the static documentation of that design. You really need a lot of your decisions made before documentation starts… and that lends a stiff formality to the ambiguous certainty of your conceptual work. Wouldn't it be great if there were a tool that magically solved all those problems for you? Heck yeah, it would.

The software industry responded by designing and launching computer-aided drafting tools like AutoCAD. In time, these more primitive tools were followed by whole-building simulations that even automated the basic drawing layout to some degree. Computers are really great at sweating the details of your design for you. It seems like a perfect match, even. Let the human do the squishy organic conceptual thinking and leave the computer to handle the details. Unsurprisingly, many software companies have recognized this opportunity and launched products that aim to support it. And architects have correctly embraced them for the quantifiable performance improvement they bring.

But none of these tools, even those that cost thousands of dollars per year and promise a single all-inclusive solution, can by itself turn a marginally worked out design into a masterpiece that changes the lives of its inhabitants. Nor should you want them too. If you are trying to learn to sing opera, these tools are like the autotune for your quivering beginner's voice. They'll make you look better at a glance, but they can't really make you a better designer all by themselves. And those who claim expertise in these tools are often masking their underlying inexperience with design. I'm sure you've seen drawing sets that look great but represent designs that can't actually be built.

Ditto for rendering tools, though, as a class of applications, they are even more commonly associated with the design process. Buying the latest rendering engine, the most expensive Hollywood animation program, or the hottest new hardware to run it all will not make you automatically into a great artist. Again, a genuinely great architectural illustrator can make a compelling rendering with only a stick of vine charcoal and a piece of paper. You'll find no shortage of people who claim that the success they have had in their practice has come because they bought some new rendering tools. Always look closer when people tell you that. What are they not telling you?

Most software is sold on the promise that it is simpler to learn, easier to use, faster, and more powerful than its competitors on the market. This is true for free software as well, which has the advantage of costing you nothing but the time it takes you to learn how to use it. These messages are honed to resonate with an architect's deepest set professional fears – the fear of obsolescence, the fear of missing out on future success, and the fear of being exposed as an intellectual imposter. Every professional, particularly the creative ones, has these fears and needs to come to terms with managing them. But new software tools won't help you overcome these fears automatically.

The way to overcome your professional fear is found in work. Real, hard, constant practice to become better at what you do. The tools you choose should support your practice, not define it. Examine everything in your toolbox regularly and honestly appraise its position in your process. Be quicker to reject new things than you are to add them. Always refine your process to fit your needs. And commit to the tools you can't think of a reason to reject in the end fully, thoroughly, and enthusiastically.

Take input from others you respect and let them mentor you where they can. This includes the experts within software companies. Put aside your natural paranoia and fear of exploitation by people who are making a living by selling you something. Inside every software development team are people who genuinely care about making your practice a better one. They have your back, always. Regardless of how it may seem to you from the outside at times. But don't ever listen to, or even worse, take direction from the mob on the internet.

Virtuosity with pencil and paper, with charcoal and newsprint, with pen and trace, comes with practice to most people. The cost of tooling up is modest, well within reach of anyone interested in learning to use them. Digital tools are different in that they usually cost real money. Even free-to-buy tools will end up costing you money in the end. You still need to buy a computer to run them. You still need to spend time learning them. You still need to put in the hours of practice required to reach that reflection-in-action state where you get the real design work done.

Additionally, digital design tools come with fixed costs of obsolescence. Eventually, you are going to have to buy a new computer to upgrade your software to the latest version. You might get frustrated at how quickly you run through paper and pencils, but digital tools don't last forever, either. In the software industry, we refer to this as "bit rot."[9] Basically,

this is an acknowledgment that digital technology is still in its infancy in many respects – that it still makes radical leaps in performance and capability from year to year. And those leaps into the future often come at the expense of last year's technology. Like a shark who needs to move forward to keep water flowing through its gills, technology must continue to advance to stay alive.

Logically, you might think you can just hop off this constant upgrade train at any time you want. Just freeze your computer, your operating system and all the applications that are installed on it right where they are, right where you like them today. And to be fair, some people do try to do this. In the end, though, I wouldn't recommend this strategy to anyone serious about their tools. Digital tools are far from maturity. Unlike a pencil, whose underlying technology has remained unchanged for centuries, digital tools still have a long way to go before they stabilize. This is especially true with software from companies that stay in business by selling upgrades. Every year, they must add some new feature that captures enough attention to drive customers to buy again. Can you imagine how your pencil might look if you kept adding new features to it every year?

True virtuosity with any tool comes only after lots of practice. Good habits are usually earned by kicking bad ones. It is easy to conclude that a piece of software that changes out from under you should never be upgraded again. There's a terrible loss of agency that comes from a change to a tool that is inflicted on you for reasons you don't understand by a development team that doesn't know you, doesn't appreciate you, and doesn't listen to your needs. How can you really hope to gain virtuoso-level skills with a tool that changes all the time? Stradivarius-playing violinists never had to deal with that kind of change, right?

It is both a problem and an advantage that digital tools are changing all the time. New hardware

---

[9] "Bit rot" refers to the fact that software, if it isn't updated for compatibility as the hardware and operating systems required to run it are improved over time, will eventually fall into obsolescence and become impossible to run any more. Documented in *The Hacker's Dictionary* (Raymond 1996).

advancement affords new opportunities (features, performance, stability) to the developers of software tools that run only on the new hardware. Good software developers are always looking both forward (the latest, hottest, newest opportunities) and backward (the features our users already know how to use) at the same time. But there's no responsible way to stem the flow of technological change. While you may well rage against it, it is always best to embrace the constant cycle of improvement (even when it doesn't feel like an improvement to you) and keep moving forward.

So how can you be a virtuoso on an instrument that is always changing? By keeping your attention focused on the simple bits that still matter the most. Details in the implementation of a particular capability may change from release to release, but the core paradigms change only very slowly. Consider Photoshop, for example, which I have been using on and off since its initial launch in 1990. It has always had a conceptual ability to modify the brightness/contrast levels in a photograph. The build of Photoshop I'm using today is running on hardware almost unimaginably more capable than the Macintosh SE I first ran it on. It has hundreds of additional features but is still rooted in its core can change the brightness and contrast levels in a photograph. My monitor is better, my computer faster, the camera that took the photo is profoundly different, too. But if I knew how to judge the right brightness and contrast for a photo then, I still remember that now. But I can do it with much more control, much more precision and much more interactively than I could back in the 1990s.

If I kept a frozen system, with the original hardware, software, and operating system to run that original version of Photoshop, I could still use it. But think of all the other things I would be missing out on. No internet, no digital cameras. No color, even. If I move forward with the technology, I have more time to compare versions, more time to try different shots from the same photography session. More time to iterate and interactively find just the right solution. More time to collaborate with others and come to a consensus on the final product. Quite a lot has improved about Photoshop since that first version.

We are 50 years into the digital revolution, give or take. Fifty years ago, the first TCP/IP packet traveled across the internet, the first CAD system was demoed, and the first human being walked on the moon. AutoCAD, the first CAD system of merit to run on personal computers, grew to popularity in architecture in only the mid-1980s – just 30 years or so ago. Desktop publishing, popularized with Apple Macintosh computers, the Postscript language, and cheap laser printing systems, really came into its own only 20–25 years ago. And 3D modeling, available widely to ordinary people on their personal computers (notwithstanding the meteoric rise . . . and subsequent fall of Silicon Graphics), really only came to life in the late 1990s with the advent of cheap GPU-carrying graphics cards from companies like NVIDIA. SketchUp, born in 2000, rode that wave side-by-side with first-person 3D shooter games like Quake. Even now, as I'm tapping away at a state-of-the-art 2018 Mac mini, this is all still very new. Architects in their fifties all still remember drafting by hand. And they are the ones who will be doing most of the great works of the next two decades.

You might be thinking that you can just build your own software, just perfect for you and your needs without having to compromise in any way. Software development projects are expensive to maintain. You need a talented team of developers, product managers, UX experts, and quality assurance testers – not to mention the sales, support, and marketing folks who all work together to keep a company's lights on every day. As an individual designer, you can't afford their work by yourself. For purely economic reasons, your needs are going to be bundled together with the similar needs of a bunch of other folks like you. Together, if you all like what you see and are willing to pay for it, vital and capable digital tools can be brought to life.

Outside the insulation of a small number of super-titanic companies like Adobe, Microsoft, and Autodesk (and of course others), a software company's lifespan may be shorter than you imagine. I was reminded of this during the 2006 opening ceremonies for Denver's newest art museum, a crystalline structure designed by Daniel Liebeskind's office. In the building's opening exhibition, curators had added a display of the technology used in the building's construction.

They showcased an (excellent) 3D modeling application called FormZ. Like many architectural designers in the late 1990s, I had been using FormZ in my professional work, and at one time I knew it quite well. When the Denver Art Museum opened its new galleries to the public in 2006, SketchUp usage had become quite common inside design studios around the world. One of the curators asked me, upon discovering my association with SketchUp, if it had also been used on the Denver Art Museum. I confessed that it had not even really existed when that project was in its conceptual design phase . . . six years previous to the opening ceremony. Now, 10 years or so later, it is more likely that an architectural designer would have used SketchUp than FormZ. This is no slight to FormZ, which remains an excellent tool for architecture, substantially better than it was when I used it, even. But companies rise and fall in the technology world. Faster, in many cases, than the large construction projects on which it might be employed.

I maintain a personal archive of old 3D modeling software, especially their manuals, that I find enormously useful. The first software on any new market sets norms and practices that all others must either follow or reestablish on their own. The basics that we all rely on today are all in there, right at the beginning. In fact, if you look farther back, to Ivan Sutherland and his "Sketchpad" demo,[10] you'll see even more precedent-establishing basics outlined for others to follow.

Learning to be a digital sketching virtuoso relies on a few basic principles. Simplify your practice such that your basic tool needs are simple and universal. But don't ever let yourself become dependent on any single feature on a particular tool or any individual company's services. Never let your practice become so brittle that something as simple as moving a button or changing the name of a menu item will break your flow. Change is inevitable in software development, and you don't really have much practical agency over it. But you can manage the way you work, your practice, and your expertise. With an appropriate focus on the result you really want – which is presumably (as an architect) a way to design and build better buildings – you can build mastery, even virtuosity, over any collection of tools you may be confronted with.

Some aspects to digital practice stand a better chance of lasting than others. The internet, for example, is likely to outlast us all. It was built for resilience and designed to last. The internet has grown so large since its birth 50 years ago that it almost defies imagination. There is something new on the web every second of every hour of every day. And yet, the core principles behind its conception remain available to all as well. I am a strong proponent in the reliability of the internet. Hardware failures and local disasters have become, for me, only annoyances for my digital archives. My digital work, I believe, is going to outlast everything else I have done.

---

[10] While the idea for similar systems existed in science fiction prior to his work, Ivan Sutherland, of MIT's Lincoln Lab, is generally acknowledged to have demonstrated the first functional Computer-Aided Design (CAD) system. His demonstration application "Sketchpad ('the Robot Draftsman')" set the standard for all subsequent CAD systems (Peddie 2013).

Stay Agile

At the beginning of any design project, you will know almost nothing about what the design will ultimately become. Even the best and most carefully researched projects will have a minuscule fraction of even the most critical requirements defined. You may think you know a lot about the project, but you will have only begun to discover what the project will eventually become.

This is not unique to architecture. In fact, all design projects share this problem. When you are making something truly and profoundly new in the world, you don't have much precedent to fall back onto for guidance. For example, writers working on an original work of fiction commonly admit that they don't know how their story will end until they get there. Certainly, they may have a rough outline to guide their work, but the best writers remain open to make changes in response to new things they discover about the story during the process of writing.

In mathematics, the phenomenon of emergence in complex systems is well known and has been widely studied. Stephen Wolfram describes entire classes of mathematical problems that are computationally irreducible in his book *A New Kind of Science* (Wolfram 2002). They can only reveal their future states through iterative development over time. Even the simplest of computational systems, for example, the one-dimensional cellular automata most often associated with Wolfram's work on complex systems, are capable of showing results over time, which are wholly unpredictable from their initial conditions (Figure 1.7).

Designers of a more engineering mindset often question this point, believing that if the requirements of a project can only be defined completely enough at the outset, then there is only one possible path through the design process. Leading,

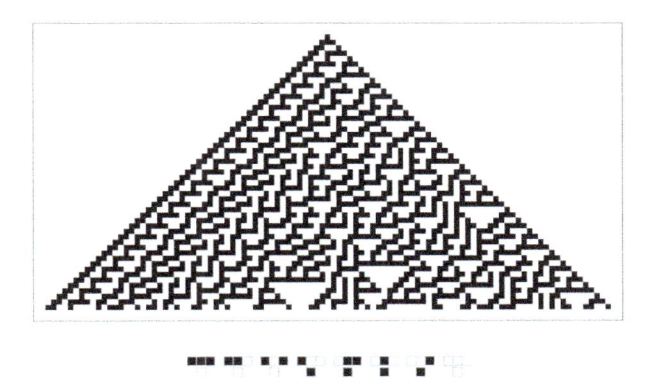

Figure 1.7: Rule 30, an elementary 1D cellular automaton, evolving from a single seed.

eventually, to a single best solution as a highly predictable outcome. To be sure, a complete understanding of all known requirements at the start of a project will help everyone on the team considerably.

But you shouldn't despair at your inability to meet your engineering colleagues' high expectations. Design, like it or not, is a wicked problem. As defined by H.W. Rittel (1973), wicked problems are those without a clear best answer. Their solution seldom follows a predetermined path, and there is no deterministic conclusion to the process of their solving. Every wicked problem is unique, and the formal definition of the problem may only be apparent after a solution has been proposed.

Rittel and his colleagues found wicked problems in domains as diverse as politics, economics, and environmental science. Natural disasters and disease epidemics pose wicked problems to be solved. On closer inspection, there are wicked problems everywhere you look. But design is undoubtedly among the most wicked of them all.

While wicked problems resist most attempts as rational systematization, some practices can help you and your team to cope. In software engineering (which, despite the name and many outward appearances, has much in common with other design professions), a tactical methodology for managing the wickedness of software development has become widely used. Called *agile development*, it is a system by which development teams can deal iteratively with changes that become apparent only over time. Design decisions made early in the development of a piece of software can be reworked or replaced in time without driving the entire team insane.

I think there is much to learn from Agile development in software, and there are enough shared problems between software and construction that I feel a little "agile" thinking during design is quite valuable. There have been many books written about agile as a software development methodology, but there's little application of that thinking in architecture and design. Think of this as a proposal for something that might work for you, not a silver bullet that guarantees your success.

Like all great social movements, Agile started with a manifesto. The core principles were initially published in the Manifesto for Agile Software Development: "Our highest priority is to satisfy the customer through early and continuous delivery of valuable software" (Beck, Beedle, and van Bennekum 2001). Agile favors individual interactions over the adoption of tools, working software over documentation, customer collaboration over contract negotiation, and the ability to respond to change over blind adherence to a predefined plan. If that sounds familiar to you, I think it should. You may already be practicing according to at least some of these principles. Agile software development adds some ideas that will seem a little more radical, though, as well.

What might an "agile" sketching practice look like for architecture? Let's imagine what an agile practice, based on the original Manifesto, might look like when you're sketching on the design of a construction project. The following sections are adapted from the Manifesto (Beck, Beedle, and van Bennekum 2001).

Client Satisfaction by Early and Continuous Delivery of Valuable Design

Every project begins with a single line in a sketch and, if all goes well, ends with the construction of a building that meets and exceeds the owner's expectations. But no design is hatched fully formed and perfect from the mind of a designer. It takes time and iteration to reach the best solution for construction. Traditional construction practices follow "waterfall" project management schedules, where each discrete task is assigned a duration, with dependencies, and then is tracked through to completion. This takes too much time, and it is way too rigid for effective design practice.

Instead, you should plan on delivering a little work all the time. Don't collect all your decisions into one giant presentation that hits both your client and the rest of your team with hundreds of new pieces of information all at once. Instead, think in a continuous delivery process. Every day, every minute, you should have the best current representation of the project ready for review. There will always be unsolved problems, and there will always be unfinished parts. But there should always be a design "build" ready to share, at any time.

Welcome Changing Requirements, Even When They Come Late in the Design

"Agile processes harness change for the customer's competitive advantage" (Beck, Beedle, and van Bennekum 2001). Design work requires iterative reframing of the problem as new requirements emerge.

You know the value of this without even having to acknowledge it, though it is a natural part of every project to be angry when things change. Especially if the changes cause you to rethink decisions you already thought were complete. Or, even worse, if they cause you to throw away something big that you worked hard to complete. An agile practice recognizes that every design begins with imperfect requirements and that, inevitably, conditions change from day to day. Embrace that and become a champion of changes as they appear in the project.

Because, sometimes, you will be the one who wants to change something. If you have been open and flexible about things that are changing out from under your work, you set an example for others on the team to follow. Changing requirements indicate the team is learning and understanding the problem they have set out to solve more clearly and in greater detail.

Be responsible, however, and always be professional when things don't change in the direction you want them to. You are always working to make the best of situations as they manifest; don't complain about things that can't be helped by your complaining. Sometimes, especially in the latter stages of a project, it is appropriately hard to make the changes that you might want to make.

Deliver Complete Design Proposals Frequently

From Beck, Beedle, and van Bennekum (2001): "Deliver working software frequently, from a couple of weeks to a couple of months, with a preference to the shorter timescale." Agile software teams usually set up automated "build" servers that continuously rebuild the software every time someone checks in a change. Once a day, all the day's changes are collected up, and a "daily" build is made that might be ready for closer testing. And maybe once a month, a real ready-to-release to the user community build might be done that is carefully tested and documented, then sent out into the world for people to use.

In your design practice, you can think in similar terms. If your clients happens to drop by your office, you should always have something to show them; today's best and most complete version of their design. If you are genuinely integrated with other design consultants, you might have a daily meeting where the latest design can be shared, and issues coordinated. You'll need a "daily" version of your work for that. And maybe you have a weekly meeting with your team as well, where you have some real-time to dig into deeper issues. You need a "build" for that as well.

You should be thinking at all times about the complete project. Don't get hung up in excessive detail on any one part of the project before bringing everything else to a similar level of development. You should always know what the best current information is on any particular question about your design. Forgive yourself for not having detailed answers to every question, though. There will always be some aspects of the design that are not complete, even that might be self-conflicting or clearly out-of-date. By definition, some part of the design is not complete until your design work as a whole is complete. And given the "wicked" nature of your work, there will always be some new thing you could work on a little bit more.

Maintain Close, Daily Cooperation between Stakeholders and Other Designers

"Business people and developers must work together daily throughout the project" (Beck, Beedle, and van Bennekum 2001). The more often you communicate with stakeholders on your design, the less likely you are to uncover a huge catastrophic problem that you drop onto the team like a bomb. Frequent communication about incremental issues and the details of their development reduces the stress of everyone on the team. Ideally, you should be talking to your client, your contractor, and the other consultants and trades on your job daily. You want them to think of you as the best source of information at all times. Don't be annoyed when they contact you for help – that is exactly what you want them to do. And it is evidence that they respect you and your design work.

Projects Are Built around Motivated Designers Who Should Be Trusted

"Build projects around motivated individuals," Beck, Beedle, and van Bennekum (2001) write. "Give them the environment and support they need, and trust them to get the job done." Great design projects are built around great designers. With no fixedly defined ideal design solution in mind at the outset, your client is really depending on the professional skill of the team of people they assembled to get the design work done. Fundamentally, they are paying for a design team to work on the design for their project until the clock runs out.

It is commonly believed that design is a fixed deliverable, an asset to be delivered to the construction team fully detailed, rehearsed, and perfected before construction begins. I prefer to think of design as an activity that is performed for as long as the client's budget will allow. Among the many assets created during design will be "final" documentation of the design for construction.

In an agile mindset, the final set of construction documents is only the last in a continuous series of design documents created as the design evolves through iteration over time. This is the service you are providing, and if you are trusted to do it well, your team can, together, work with much more velocity.

The Best Design Teams Are Integrated and Co-Located

The concept of a co-located design team is not foreign in the construction industry. Integrated project delivery (IPD) contracts often stipulate that teams must be co-located for maximum efficiency. Wherever possible, you should seek out opportunities to share physical studio space with everyone on the construction team. "The most efficient and effective method of conveying information to and within a development team is face-to-face conversation" (Beck, Beedle, and van Bennekum 2001).

If you accept the Agile notion that the best project communication is continuous, then it makes sense to work in a situation where constant contact is possible. Rather than spending 10 minutes wordsmithing an email, what if you could just pop over to your neighbor's desk and have a quick chat? The benefits to overall design quality and the velocity of its delivery are apparent.

Think back to your design education and to the friendships that you formed in the studio environment. Studio environments are at the center of every functional design team, though they may be mysterious and (occasionally) frustrating in how they work. If you can create a studio environment across the team responsible for your project's design, you will have caused an enormously positive change to traditionally isolated and suspicious design team dynamics.

Credible Design Proposals Are the Best Measure of Progress

There is nothing quite like an automated software build for a design project in the construction industry. However, you can still have a continuously updated current best design available to the team at all times.

The essence of this point in the Agile Manifesto is that delivered outcomes are a more robust measure of success than anything else you might spend your time doing. Nobody is impressed if you met a delivery deadline with a design that stinks. What matters, universally, is delivering excellent design work over and over again.

I rewrote this point to include the term *widely credible*. If you are the only person on the team who thinks your design is a great one, then you really are in trouble. To be truly useful on your team, you need people to think of your design contributions (in particular) and of the current best state of the design (broadly) to be credibly great. Credibility means different things to different people. To an engineer on your team, it might mean that you have correctly anticipated how long you will really be able to span structurally between columns. Or, maybe, that the story you are telling about environmental sustainability is groundable in fact.

If the design proposals you champion aren't regarded as constructible, efficient, or in any way diverging from norms and standards of practice, you will need to do a better job selling them to the team. Or, maybe, you just need to do a better job with your design.

Agile Processes Promote Sustainable Practice

Find a pace you can sustain. I use the term *sustainable* here to refer not to green building practices but to give a name to the *sustainability* of your design practice. At no point should you be expected to cause yourself physical harm to meet a deadline or a delivery target. Beck, Beedle, and van Bennekum (2001) write, "Agile processes promote sustainable development. The sponsors, developers, and users should be able to maintain a constant pace indefinitely."

Agile development is designed to be something sustainable as a practice for as long as you want to keep doing it. By switching from a rigid waterfall management style with big deliverables (and big catastrophes if the deliverable falls short) to one of continuous gradual improvement, agile designers can learn how to manage their time more humanely.

Presumably, you got into professional design because it is something you want to do for a long time. And you get better at doing it the more you practice, so there's a distinct value in sticking with it for a long time. If you can't learn how to sustain yourself physically and emotionally, not to mention financially, you won't be able to do that. You're either going to burn out, exhaust yourself, or just plain run out of money to stay in business. You have to make your practice a sustainable one, one that you do every day for weeks, months, and years at a time.

Continuous Attention to Technical Excellence and Good Design Enhances Agility

The construction industry is distressingly full of distrust and self-preservation behavior. And yet, with very few exceptions, every single person on every part of the extended team is likely motivated more by a desire to do great work that they can be proud to say they did. Nobody willfully derails the process, and nobody intentionally sabotages the project. If they did, they wouldn't be in business for very long.

Your animal brain, the lower order primate thinking that kicks in when you are afraid, might lead you to think someone on the team is out to get you. It is highly unlikely that this is true. Instead, they are just like you, trying to do the best job they can and make the most meaningful contribution possible in the time available. Some people may be more naturally motivated to avoid effort than others, but laziness is easily overcome by the motivation to work together to deliver great work.

Look for the best in your team, and in return, give your best as well. A team continuously committed to excellence works faster, more efficiently, and is a joy to collaborate with. You can contribute to the positive team dynamic by committing to bring your best effort, holding nothing back, at all times. Every single project you work on should be the best one you have ever done.

Good design makes space for more good design. Excellence in your practice inspires others to give their best as well.

Simplicity, the Art of Maximizing the Amount of Work Not Done, Is Essential

In principle, you should seek to maximize your working efficiency at all times. Don't get distracted by shiny things that aren't related to the project at hand. You know what they are like – an idea that you have always wanted to try, even though it really doesn't have anything to do with the project your client wants you to make for them. You have to show discipline, and restraint, in all these situations. The best design work is concise, clear, and refined. Throw away more from the design than you commit to keep. The simplest ideas are the easiest to explain, the easiest to share and (ultimately) the most powerful when they finally make their way all the way through the construction process.

In design, there is another meaning to this rule. Simplicity in design is more than just a way to maximize the efficiency of your work. You are also responsible to simplify the work of all those who will follow your design through execution. And usually, it takes a little extra work during design to make a project simpler to execute. Practice the art of simplification every chance you get. You will find that simplicity usually comes only after exploring more complex ideas and throwing out everything about them that isn't directly aligned with your main idea. As Antoine de Saint-Exupére famously advised, *"...perfection is finally attained not when there is no longer anything to add, but when there is no longer anything to take away."*[11]

The Best Requirements and Design Emerge from Self-Organizing Teams

The most motivated people are the ones who feel a sense of agency over the work they are doing. When you ask someone, or worse when you tell them what they should work on next, you reduce their agency and diminish their motivation. Just a little bit, maybe, but it is there, nonetheless. The most motivated and productive designers are those who love what they are working on and who are so committed to the overall idea that they can make their own decisions about the things they are working on without much oversight or review. They make the right decisions because they understand why, profoundly and personally, those are the right decisions to make.

You won't always have a say about who you are working with, but you can always make an effort to allow teams to self-assemble. And when a team on your project self-assembles for the right reasons, you can recognize it, celebrate it, and then get the hell out of their way.

---

[11] *"Il semble que la perfection soit atteinte non quand il n'y a plus rien à ajouter, mais quand il n'y a plus rien à retrancher."* (de Saint-Exupéry 1967).

At Regular Intervals, the Team Should Engage in a Constructive Design Critique

Every design process includes a process of reflection and critique. In your own design process, you are likely reflecting-in-action continuously every time you pick up a pencil. Your team needs an opportunity to review progress as well. "At regular intervals, the team reflects on how to become more effective, then tunes and adjusts its behavior accordingly" (Beck, Beedle, and van Bennekum 2001).

The importance of critique and process review is likely the one point in the Agile Manifesto that needs no explanation to an architect. It is an unusual way of working for most software developers, but it is built into the experience of every designer during their design education. Peer critiques over a drafting board, desk critiques with a partner, or even large-scale pin-up critiques with a group. You know that it is vital to get outside of your own head and get some feedback from others as often as you can.

Critique is a skill that needs to be learned for many people who have not had a design education. It is far too easy to take a critique of work you have done as personal criticism; that you are a terrible person because of the design work you presented. You can set the tone for critique by always keeping it positive and upbeat, even when you have devastating comments about the design proposal at hand. I always look for a little laughter in a fierce critique. I let the designer who is presenting know that I think they are just fantastic as a person, even though their design really stinks.

There are some straightforward guidelines for a successful, productive, constructive critique that I like to follow. When I am working with folks who aren't accustomed to either giving useful critique or to receiving it themselves from others, I will often lay out some ground rules before we begin.

Critique Is Not the Same as Criticism

Critique is delivered after careful consideration of the design proposal as a whole. It is intended to suggest ways that it could be made better or more appropriate to respond to the stated project requirements. A practical critique is one that is recognized by the designer as being correct, even insightful, based on a unique perspective that the designer did not have themselves.

Criticism, on the other hand, is only aimed at finding fault with the proposal. Criticism seeks to degrade the validity of a proposal rather than finding a way to build it up or improve it. Criticism has little value in a critique and should be avoided or ignored.

Critique the Design, Not the Designer

While the design work under critique was certainly created, championed, or at least pitched by a designer, any comments should be targeted at the design, not at the designer. Save comments about the skills of the designer for another context, or just keep them to yourself.

Be Specific, Be Relevant

There's not much value in general statements like, "This design is really nice. I like it." Your general statement of support may be welcome to the designer, but it won't help them advance the design to its next level. Specific comments about what works, and why that is relevant to the project are useful and may be expanded on by others. Equally, possibly even more valuable are comments pointing out something that doesn't work as well, and why that matters.

Always Ask the Hard, Obvious Questions

Just because it is obvious to you, it may not be obvious to everyone else in the room. Never keep something important to yourself. Everyone else may have just missed it.

Critique Takes Practice

It is hard to give useful, actionable, and appreciated critique. You shouldn't be frustrated if you are bad at it at first. Try to remember that you are critiquing the efforts of another human being if you can remember nothing else. The person who is sharing their work is exposing themselves in ways that are usually pretty hard. Acknowledge their bravery and give them the help they need.

The Design Team Owns the Design, Not an Individual Designer

While a single person is, of course, responsible for each individual design asset that you are reviewing, you should always claim collective ownership of the design as a whole. By entering into a real critique of the work, you are exposing yourself and your design ideas almost as much as the person who pinned their work up on the wall in the first place. You are all in this together, so you should act like that.

Practice Your Agility

Many aspects of an agile methodology may seem familiar, and maybe some are a little uncomfortable. Agile practices demand a level of transparency that makes some people quite uncomfortable. You are going to have many more of your flaws exposed, along with many more of your successes. The most important thing to remember, though, is that both the risks and rewards are more frequent, but they are also much smaller. You will find yourself subject to major course-correction much less frequently. Instead, you'll receive more frequent little taps and minor corrections. In the long run, you'll find you and the rest of your team are much more aligned and resilient to external changes.

Agile design practices are an excellent match for wicked problems. With Agile, you always have the ability to adapt to something new that is discovered only after you've completed some prior work. Every day, in fact, you can tear up something you did the day before in favor of something better that emerged overnight. Since you haven't had time to do much work since the last time you shared, you aren't going to have to reverse much of your work most of the time. And that is a much more sustainable way to work.

Stay Safe

Even with the most efficient of agile practices, the conceptual design phase is a chaotic time for most designers. Sketches are flying around all over the place, with the ideas they embody rising and falling in favor with dramatic speed. As the hours melt away in your reflection-in-action flow state, it is easy to lose track of progress. It is also easy to forget the work that you have left behind.

Finding that one great sketch that you think you remember doing last Thursday, but you're not sure can be quite demoralizing. In physical sketching and modeling, all of this stuff is usually left lying around the studio in a happy (but eminently discoverable) pile of clutter. But digital processes can quickly feel like they leave you at arm's length from your stuff making it incredibly easy to lose things. Sometimes irretrievably and forever. You can't just shuffle through that stack of discarded sketches until you find the one you remembered.

There are excellent desktop search technologies you can bring to task when you feel like you've lost something, and I encourage you to take the time to really learn how to use them. For design work, however, they have limited capability. Much of the digital design work you're going to do will be stored in data formats that are hard to index and, as a consequence, hard to rediscover once lost. This is especially true for 3D model assets. You might only be able to find them if you remember the filename and the approximate date on which you made them. Search technologies, including the ones from Google that claim to be able to "find and make useful all the world's information" are really only good at searching for strings of text, not images or 3D models.

There are specialized search engines available to architecture firms today that do promise to improve on the search-ability of nontextual digital assets. Probably by the time this book is in your hands for reading, there will be a dozen architecture, engineering, and construction-specific collaboration platforms on the market that you might consider using to store your project data. In their own ways, they are all excellent and up to the task of managing your project data. Most importantly, they are beginning to learn how to index building model information. And indexing is the first step toward making assets re-discoverable. There isn't a single best solution for this problem yet.

Back Up Everything

If you have been using computers for anything at all (even just sending and receiving email), you have probably also encountered crashes with data loss. And if you haven't lost data to hard drive failure, you eventually will. Digital data can feel much more ephemeral than "real" drawings on pieces of paper. If you store it only in one place, on a single fragile spinning hard drive platter, it is. Properly managed, however, digital data is far more reliable, far more secure, far more discoverable, and (ultimately) far more durable than a piece of paper in a flat file in your office. You just have to get it off that one old hard drive under your desk.

Now that the internet (and cloud computing) has matured, there's really no excuse for losing data ever again. Once you give your data over to the cloud, you'll never want to go back to the old way you stored things. It is really that much better. Cloud backups are easy to set up, cheap to manage, and absolutely bombproof in their reliability. The internet was literally designed to withstand nuclear attacks.

Let me emphasize this point. You should be backing up everything you do to the cloud. This is safe, reliable, and performs like a champ. It doesn't matter which service you choose, and I'll leave it to others to argue the relative merits of Google Drive over Amazon or Apple's iCloud. If you don't have a backup right now using one (or, better, several, redundantly) of these cloud data storage providers, you should put this book down right now and go set it up. It's OK, I'll wait. Come back when your backup has started, because you'll have a bit of time on your hands while you're waiting to read more of this book.

. . .

There are two kinds of digital creatives in the world today – those who have lost data to hard drive failure and those who will lose data to a hard drive failure in the future. Which one are you going to be?

. . .

OK, now that you have that out of the way, let's talk about why storing your data in the cloud is the better choice than keeping it on your local computer's hard drive. The benefits are considerable, though you may have some lingering doubts. Lots of people do, though fewer and fewer outside of the AEC industry. It is OK. I'm here to help you get over your cloud aversion.

To begin with, you never want to keep all of your data in one place. Especially not on a single device. What happens if that device fails? Crisis. At one time, setting up a backup system that included some offsite storage was an expensive and complicated process. I still remember rotating magnetic tape backups for my office that had to be shuttled offsite to a bank safe deposit box once a week. That was enough of a chore that it probably only really got done once a month. A device failure in this context was still a catastrophe, just a modestly smaller one. Today, you can set and forget an offsite backup with a simple system extension and a monthly subscription. I pay about five dollars per month to keep unlimited backups of each of my personal computers. It is the most comfortable bill I pay every month.

If anything happened to my computer, like if my house burned to a cinder in a wildfire, I would go down to the computer store, buy a new computer, connect it to the internet, turn it on, restore from my backup, and I would be up and running in an hour or so. Quick, simple, and very safe. I have used this capability dozens of times and thanked the gods of the internet every single time.

The simplest possible way to store the physical sketches you produce in your design work is probably just to dump them all in a big box. I'd wager that all creatives have some version of this hiding in a closet, garage, or studio. I call mine "The Boneyard." Minimally, it is waterproof, and I've done some basic things to keep mice from nesting in it. Loosely organized by project, or at least according to some broadly thematic project-like structure, my boneyard has everything in it I ever thought was valuable in even the most minor way. I am, admittedly, a bit of a hoarder in this respect.

There are several advantages to my physical boneyard. I know I probably really only have one place to look for an old drawing. If I remember something, but can't find it in the boneyard, then I have to get on with the grief of knowing that it is probably lost forever to me. It wouldn't have been stored anywhere else. Sorting through a bunch of old drawings can actually be a pretty rewarding experience, even if you don't find what you were looking for initially. I often find not what I was looking for, but instead something serendipitously close to it that I wasn't looking for . . . but actually gets the job done much better. Every old project is full of ideas that didn't make it into the final project. Maybe they can find a new home in my next project?

You have to adjust your thinking some when you add digital sketches into your practice. Of course, if my garage burned down, as in the same wildfire that melted my computer in the previous example, this work would be lost forever. My digital backups, dispersed redundantly across the internet, would likely survive a nuclear holocaust. The internet was designed for redundancy, for reliability, and for resilience to disaster. It is much more reliable than a box full of paper in my garage.

Separate from my cloud backup solution, which just keeps a snapshot of my entire system tucked away safely in the cloud, I also keep a digital equivalent of my physical boneyard. Like the box full of paper in my garage, my digital boneyard contains digital assets from projects I have done going back to the very first computer I ever owned. Every time I get a new computer, I just roll it forward. Every new project gets its own folder, and I just keep piling them in. Luckily, persistent storage capacity is still advancing faster than my ability to make new digital data. Every new hard drive I buy has at least double the capacity of the one it replaces.

This folder gets a little bit extra backup. I simultaneously sync it to several different internet cloud storage providers, which gives me additional redundancy and protection from the rising (and falling) fortunes of Silicon Valley. Today, I use Google Drive, Apple's iCloud, Dropbox, and Adobe Creative Cloud. I also use Trimble Connect, which provides me unlimited cloud storage on top of a collection of other more exciting things specific to the AEC industry.

The contents of my digital boneyard may include files in formats that are unique to editing products long since disappeared from the market. There's no perfect way to prevent that. In general, I try to choose digital tools that I think will be around for a long time, or that have formats I know I'll be able to open somewhere else in the future. I've made good choices about this and bad ones.

And there's an important lesson learned. Beware the specter of data lock-in. Be suspicious of any software ecosystem that has an easy entrance, but no apparent exit. Every digital content creation tool has its own proprietary file formats, for good and responsible reasons. If you want to be able to do something unique with data, some special modeling techniques, or whatever, you will have to adapt to a specialized representation of that data. This is particularly true with 3D modeling tools. No two modeling applications are precisely alike, unfortunately. As a consequence, you shouldn't expect to be able to open 3D models from one in another without going through a lossy translation process.

Luckily, there are open standards that make it easier to go through that lossy translation process. With SketchUp, we have always been careful to support both import and export for the most critical exchange formats. Standards vary by industry, but for us, it is Autodesk's DWG[12] (for 2D CAD data, primarily), IFC[13] (for BIM data), COLLADA[14] (for 3D rendering models), and STL[15] (for 3D use with 3D printers). Both imports and exports are inherently lossy, so you need to adjust your expectations appropriately. 3D model data is tough to standardize.

But most other data types you will find yourself working with are much more settled into commonly accepted standards by now. With some exceptions, any image editor can open any raster image file. Any text editor can open any text file. And so on. And every application worth your attention can open files from older versions of itself without any hassle. Photoshop v.20 (2019) can open files created in the very first release of Photoshop almost 30 years ago. SketchUp 2019 can open files created right back at the first release as well (Figure 1.8).

I have seldom lost data irrecoverably to obsolete software, but it has happened. For a year or two in the early 1990s, I did a lot of writing on a NeXT computer. I loved that machine – just a beautiful piece of tech. I used a simple little word processor called WriteNow, and I stored my files on a shared network drive. I still have all of those files today, rolled forward dozens of times to new storage systems and

---

[12] Autodesk's DWG ("DraWinG") format is a proprietary binary data format invented by Mike Riddle and licensed for use as the native file format for AutoCAD since 1982 (Walker 1989).
[13] IFC (Industry Foundation Classes) is a platform neutral, open file format developed by building SMART to aid in the exchange of canonical building design data in the construction industry (buildingSMART International 2020).
[14] COLLADA ("COLLAborative Design Activity") is a 3D data interchange format managed by the Khronos Group designed to facilitate asset exchange in the media and entertainment industry. (Khronos Group 2020).
[15] STL ("STereo-Lithography") is a low-level 3D CAD data interchange format created by 3D Systems to facilitate exchange of CAD data for rapid prototyping and 3D printing. (3D Systems, Inc. 1989).

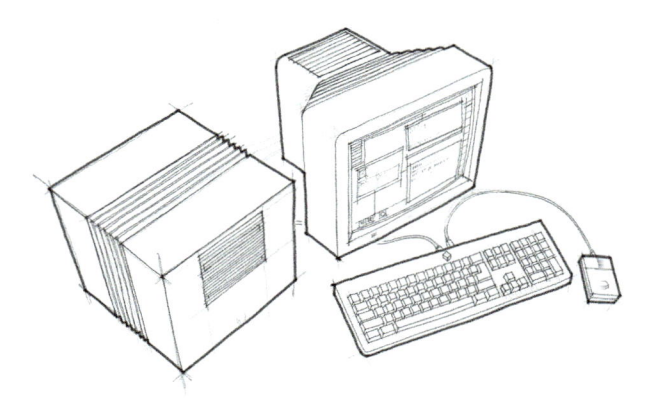

Figure 1.8: My once-beloved NeXTCUBE computer.

design redirection. "A place for everything and everything in its place" won't help your design mojo during the conceptual design phase of your project – whatever it may be. You don't know what the "everything" you're making is yet. How on earth could you be expected to know where it should be put?

Don't feel ashamed that your previous attempts at a priori classification of the design stuff you haven't yet started to make have failed. The problem is that it is almost impossible to classify drawings, models, and digital data before you have made them for the first time. Eventually, your design will settle down into a set of assets that you can tuck away neatly into a taxonomy of your choosing. Still, in the beginning, I recommend just giving in to the chaos. Just dump it all in one folder and figure it out later if you have to. And if you have a good enough search tool at your disposal, you may never really have to.

backups. WriteNow, on the other hand, shipped its last release in 1992. I have the documents I created, but I'll never be able to open them again. Unfortunately, there is little you can do to recover from this. Mourn the loss and move on, as I have done.

Over time, as your key design patterns for the project start to emerge, you'll be much better equipped to assert some organizational structure. Ultimately, you're going to need some system that helps you reference things you've made later. By the time your work is ready for construction, it will be classified richly, logged into formal version control, and changes will be formally handled in team meetings. But when you're just getting started, there's no need for all of that. It will only get in your way, and there is plenty of natural attribute data on most modern computer storage systems to get you started. You don't need an a priori taxonomy in addition to that. Just the usual folders in your computer's filesystem.

Don't Overorganize

Of course, in time, your personal boneyard of digital data will become large enough that you will start to have trouble finding what you are looking for inside it. Time to get organized, right? Every office I have ever worked for or with has pushed a highly formalized project directory where every single piece of data, real or imagined in the future, has a place to be stored. Usually, the work of a committee (at least in the larger offices where there are enough people to organize committees) is expected to be able to solve the problem of finding data again in the future. If everything on a project has a place where it should be stored, only irresponsible people will store it anywhere else, right?

I have to be honest. I hate this kind of thinking. Templates pre-suppose that every project is the same. Hierarchical systems (like a rigid folder structure) almost always fail to accommodate unexpected

If you were active on the internet during the early years of the World Wide Web, you probably remember finding information through extensive human-curated indexes like Yahoo! Sites had to be carefully added to these indexes by a human being, much like a traditional reference librarian, who would have to be able to classify everything according to some rigid preordained taxonomy. Once organized, users of the system could browse the taxonomy, progressively clicking headings and subheadings to drill down to the place where, eventually, they could find the website they were looking for.

Now, it is not easy to find browsable indices of the internet. Can you imagine a browsable index that contained over a billion individual sites? That just isn't practical anymore. Search engines have replaced top-down taxonomies, and it isn't hard to see why. Search is much more direct than it was to surf an index, hoping to find what you needed.

Version Control Systems

The simplest of systems are unquestionably the best ones when it comes to organizing your data. You can and should (right now if you still haven't; how many times do I have to tell you?) have regular "set and forget" backups running for all your digital data, and you don't need to have a right folder structure in place before you get your backups running. But sometimes a backup alone isn't enough. Design projects aren't particularly linear in their development, despite what project managers might hope for them. Sometimes, a change you made must be rolled back to a previous state and taken in a different direction. Sometimes you need to consider two different variations of a design in parallel. For anything like this, you need a version control system.

Luckily, architects aren't alone in needing to be able to manage versions during design. Software development projects have very similar needs. And for once, as an architect, you can take advantage of another industry's hard work more or less for free. Thanks to the community development of large software projects like the Linux kernel, reliable and straightforward version control is readily available for any computer. My favorite system is Git,[16] created by Linus Torvalds to help him manage the Linux kernel project.

Version Control Systems like Git operate on a simple principle. Every time you save a file (any file) in a traditional filesystem, you overwrite the previously saved version of the file. That previous state of the file is then lost forever, replaced by your latest version. If you replace your traditional file system with a version control system, you can keep both the previous version and the new version, two separate snapshots of the file from two different times. If you decide later on that an older version was better than the one you're working on now, you can revert back to that earlier version. Or, if you want to keep both the older version and your newer one, you can keep them both, forking the design into two separate version trees. Once a version has been committed to the system, it can be preserved forever, always available if needed.

Every designer in any industry should learn how to use version control systems to support their work. They are easy to set up, easy to operate, and very secure. You can easily integrate them into your existing backup system without any special setup. They also make it easy to share your design responsibility with others. Version control systems allow you to invite multiple contributors to make their own versions of a project, offering up improvements that can either be reviewed and discarded or integrated into the rest of the project's design work.

---

[16] "Git is a free and open source distributed version control system designed to handle everything from small to very large projects with speed and efficiency" (Software Freedom Conservancy 2020).

Version control systems are very accepting of changes in the way you store data in them over time, making them ideal systems in which to allow a project organization to emerge gradually. You can commit changes to your project's folder structure as quickly as you can check in any other change. Versions of the project organization are snapshotted in time like everything else. And a simple system like Git is likely going to last a really long time. If it is used to manage a project as vast and meaningful as the Linux kernel, it is probably going to outlast your needs.

You owe it to yourself to learn more about using version control to keep your sketches organized. It isn't hard, it doesn't cost a lot of money, and it is enormously powerful in the long run. You can be confident that the basic premise of version control is here to stay. Give it a try, and you'll never return to your previous way of doing things.

chapter 2 The Elements of Design

There is an old Buddhist parable – a story about an encounter between a group of blind men and an elephant in a marketplace square, illustrated in Figure 2.1. None have ever encountered a creature like this before, and they must each attempt to understand it using only their limited senses and their understanding of the context around them. One of them touches the elephant's trunk and proclaims he has encountered something that can only be a new kind of snake. Another, reaching first the elephant's ear, decides the first is wrong – this creature must instead be some sort of living fan. A third, grasping the elephant's leg, declares both are wrong – this creature is clearly some kind of tree.

All of them are right, but none of them entirely so. Only through a sharing of information, combining

Figure 2.1: The parable of the blind men and the elephant (Hanabusa 1888).

their unique insights together, will they understand what the elephant truly is.

Designing the Elephant

In a way, this is a lesson about design, especially in architecture. Like the blind men in the market, none of us have the totality of the design in mind. It is only through exploration, discovery, and iteratively reframing the question that we can truly understand the building. On initial inspection, it appears to be made of discrete elements, each of which seems familiar to us due to our prior world experience and the mental context in which we operate. By sketching the building from different points of view, we may discover its true nature and come to understand it as a synthesis of parts into a more satisfying whole.

The best way to eat an elephant is one bite at a time. When sketching out the initial concepts of a building, you must be careful not to get weighed down prematurely by too much detail. Detail is important, but too much too early will add constraints to your design that you don't need. If you understand only the detail and have no sense of the overall structure, you have no context for the detail yet. All projects have unsolvable interdependencies between competing interests and other technical realities. You are going to have to optimize to make compromises to reach a constructible level of detail eventually. But in the beginning, you need to know very little about that.

Supporting even the most complex of design systems are a small number of universal geometric elements. These are the "trunk," "ears," and "legs" of the metaphorical elephant. Their division into separate elements has been argued endlessly over the ages, subject to the challenges of history, culture, and context. The elements we may think of as simple universal truths have embedded in them such rich histories and deep traditions. Deep enough as to be mostly invisible to us ordinary practitioners.

When pressed for an origin point to the elements of architecture, I typically turn to the geometry of Euclid. The 13 books of his *Elements*[1] contain definitions of geometry and the logic of thinking about forms that have been widely studied. Euclidean geometry offers a set of axioms that can be manifested as forms (made of some material), spaces (formed between and around forms), and the logical systems that bind them together. As a system of elements on which to base your architectural sketching practice, you can't go wrong by starting with Euclid.

Architectural beauty has much to do with the harmonious arrangement of basic Euclidean form, but alone that's not enough to carry a building design through to completion. Kant, in his *Critique of Aesthetic Judgement*,[2] reminds us that aesthetic beauty is found only when geometric formality (symmetry, proportion, relation, etc.) is combined with a practical fitness of purpose beyond the geometry. A beautiful work of architecture is, in other words, only so if it is both harmoniously formed and efficiently useful. For those of you hoping to focus your practice solely on the generation of beautiful forms, perhaps you should consider a career in the fine arts. Alternately, if all that concerns you are the functional requirements of the building, maybe you should dig for a little more meaning in your work, too. Truly exceptional works of design thrive on the tension between these two polarities.

Conceptual design, the sketching phase of a project, needs to concern itself first with fundamental

---

[1] The first six books of Euclid's *Elements* have been widely reproduced in the 2000 years or so since their original publication. Here, I refer to Oliver Byrne's 1847 edition (Byrne 1847).
[2] This subject is discussed extensively in Kant's third critique, the "Critique of Judgement; Part I. The Critique of the Aesthetical Judgement" (Kant 1892).

principles of form, space, and order. These principles must be explored in the context of a client's design requirements and with knowledge of the natural affordances provided by the project's site. As a project pushes further into detailed design phases, where increasingly specific questions are asked and answered by an increasingly tactical construction team, it is more and more critical that the conceptual design has been well considered. As the level of detail increases, it becomes harder and harder to make fundamental conceptual changes. It is also easier to make detailed decisions on the foundation of good conceptual design. With a strong enough conceptual foundation, details can almost seem to resolve themselves.

It is not uncommon for your construction team to appear uninterested in the conceptual design for a project, though, frankly, that is really when their input is the most valuable. Much of the conceptual work you do will be ignored completely. Your contractors are focused only on the part of the elephant that matters to them. It is your job, as the designer of the building, to represent the whole elephant as much as you can. Even if you can't really describe what makes your design a good one, everyone appreciates a well-considered space.

The Measurement of Space

In a building, it is the space between the formal elements of the building that matters most to people. Construction teams, by contrast, are mostly concerned with the material form of the building. The interplay between form and space is mysterious, complex, and elusive to define. Architectural space has ineffable qualities that are the product of the materials and structures that enclose it and provide it with context and scale. Architectural design creates space but only through indirect means. The qualities of space have, for centuries, defied direct rationalization despite observable truths about it that we surround ourselves with in everyday experience. We can study space, and we can give it name and measure. This is where an understanding of all geometric forms must begin.

In the 1600s, Rene Descartes[3] developed a coordinate system wherein locations are indexed on an infinitesimally subdividable grid of numeric addresses. In three-dimensional "Cartesian" space, we typically use the letters X, Y, and Z to address coordinate locations. This system is incredibly convenient and highly practical to use. Project teams need only come to a consensus on a common coordinate origin (0, 0, 0) and a calibrated unit of measure, and all work from any contributor can be quickly coordinated spatially. Locations in the project can be either relative to that origin point or may have their own subservient coordinate systems. But all of them exist in a shared conceptualization of space.

The simplest of Cartesian spaces is one-dimensional, the essential space of linear measurement that is encoded on every ruler or tape measure. Our ability to measure distances between things in a way that is agreed upon by everyone on a construction team cannot be undervalued. Can you imagine how it would be if everyone on your team had a different way of measuring things? Of course, it really doesn't much matter what standard you choose. But you'll have absolute chaos if you don't come to a consensus.

Linear measurement can exist in any single dimension. Length is the most obvious, of course. Time, also a single axis, can also be considered a linear measure. Weight, as a vector measurable in an axis to the center of the Earth, could be regarded

---

[3](Descartes 1637)

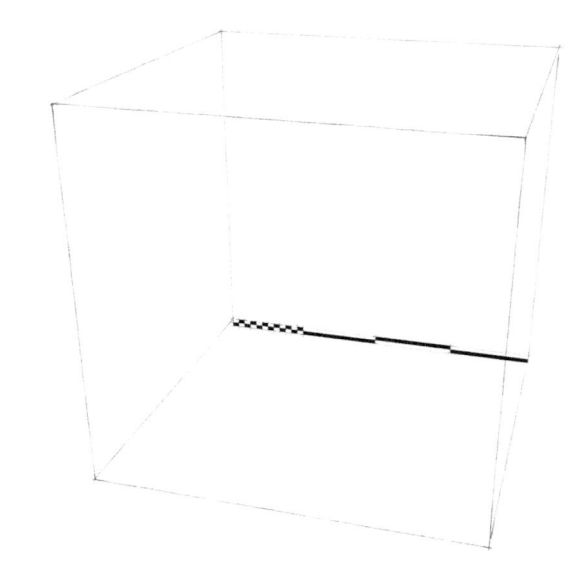

Figure 2.2: 1D space.

as a linear measurement. Even temperature might be thought of in these terms. But for sketching purposes, the linear measurement of space is undoubtedly the most practical to consider. (Figure 2.2)

When sketching, you should work as often as you can in a measurable scale of some kind. Objects in perspective projection cannot be scaled except in proportion with some reference of intuitively known size. It is for this reason that SketchUp, for example, includes a human scale reference figure standing next to the coordinate origin in every new model.[4] In every case, these represent a real person on the SketchUp development team, measured and drawn to their actual height.

Orthographic projections can and should always be sketched in scale. You should get good at estimating dimensions in scale when you're sketching. Keep a standard architectural scale at your elbow if you need to recalibrate your estimates while you're working. But learn to move as quickly as you can from diagrammatic representations (proportional, maybe, but not yet dimensional) to something with a credible scale.

However, don't go overboard. Approximate dimensions are the most useful when you're sketching. It is possible to have too much precision at the beginning. While you are drawing, you should know that the stairs you've drawn are credible, but if you work out their rise and run to the tenth decimal place, you're really wasting your time. Measurements, even the most unbelievably precise of them, have a threshold of precision. You should learn to use that to your advantage. Think of the increasing certainty in your design as adding decimal places to the geometric precision of your design. Maybe at the beginning, you are only worried about integer feet of precision in your design. By the time you reach your constructible level of detail, you may be precise to 1/16th of an inch. To what standard of precision will you hold your contractor?

Looking beyond linear space, we should next consider planar space. The two-dimensional Cartesian space of a plane is easily understood by any secondary school student with a piece of graph paper, subdivided into evenly spaced divisions in length and width. As you likely remember from your math classes, by convention, we place the "origin" somewhere on the page (traditions vary[5]), then measure distances horizontally (increasing values from left to right) and vertically (increasing values from bottom to top) from that origin, as shown in Figure 2.3.

---

[4] By tradition, the scale figure in SketchUp changes with every major version release. Each represents a real person, a member of the SketchUp development team. Which SketchUp team member do you have in your copy of SketchUp?

[5] In trigonometry class, it is usually at the bottom left of the page, or possibly in the middle of the page to accommodate negative and positive values. In 2D computer graphics, it is often in the top left corner of the screen. Like many 3D modeling applications, SketchUp follows the "right-hand rule" for axis orientation in three-dimensional space.

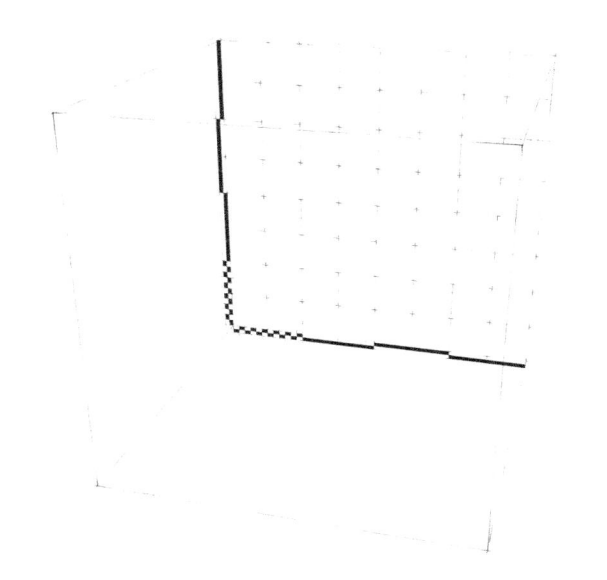

Figure 2.3: 2D space.

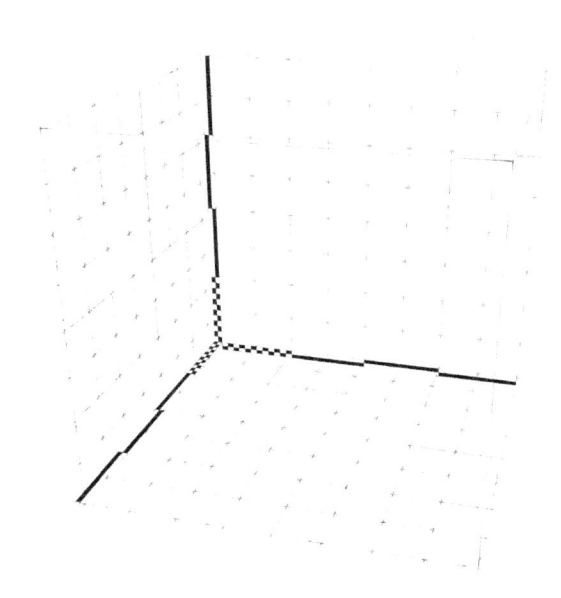

Figure 2.4: 3D space.

2D space can also be represented on a flat computer display. Where graph paper includes a ruled grid of squares, your computer's screen has a regular matrix of pixels. Each pixel has a unique coordinate, individually addressable in the space of the screen. 2D Cartesian space is sufficient to describe any flat object, and buildings have many flat objects in them. It is also enough for the most straightforward and most diagrammatic architectural projections, plans, sections, and elevations.

To find the true space of architecture, we need to add another dimension. We can no longer adequately represent the space on a flat piece of paper without employing some kind of spatial projection system. The representation of 3D space in orthographic, perspective, or some other projection system is a deep enough topic that I've given it a chapter of its own.

For most design work, a three-dimensional space is sufficient, as shown in Figure 2.4. That space needs a scale, and for architecture, it is essential to understand which of the axes points "up" (perpendicular to gravity) to more effectively represent the world.

Every computational space has a window of precision, from a lower bound to an upper bound. The designers of CAD systems usually try to set the upper and lower limits of precision as unobtrusively as they can for the scale of an object they expect you'll make with their tool. But when you model architectural-scaled projects that span from the size of a single anchor bolt to the scale of a city, you can begin to see compounded precision-related errors in floating-point math. This is especially true if you are modeling far from your model's coordinate origin.

In 3D Cartesian space, parallel lines never meet. The grid is infinitely expandable, infinitely subdividable, and can index the location of anything you need it to describe. It matches well with the discrete nature of computational math, but at the same time is easy to understand and rational to operate within. It is a simple, logical system of the mind, and as such, it is among the most universally convenient spaces imaginable for sketching a building.

However, when confronted with the organic forms of the natural world, there are problems to consider. Earth is not, of course, flat. Cartesian coordinate systems begin to have apparent errors at city-sized scales. Usually, you can get reasonable precision for only about 50 mi, as any cartographer can attest. The oblate spheroid form of Earth, when projected into the two-dimensional space of any map, has glaring distortion at the edges.

This is not just a theoretical problem. Faced with the challenge of apportioning land newly acquired by treaty after the American Revolution, a system of measurement had to be established before land ownership could be established and taxes levied. Following in line with the principles of the Enlightenment that so shaped our country's founding, a Cartesian coordinate system (named the "Public Land Survey System," or PLSS) was devised in 1785 that sprung from a "point of beginning" near the town of East Liverpool, Ohio (de Ruijter 2019). From that origin, the grid spread west into the great planes, subdividing the land into "quadrangles" of about 24 mi square (Figure 2.5).

Today, this grid is easily visible in the alignment of public roads across the Midwest. Roads often follow the edges of quadrangles or the townships

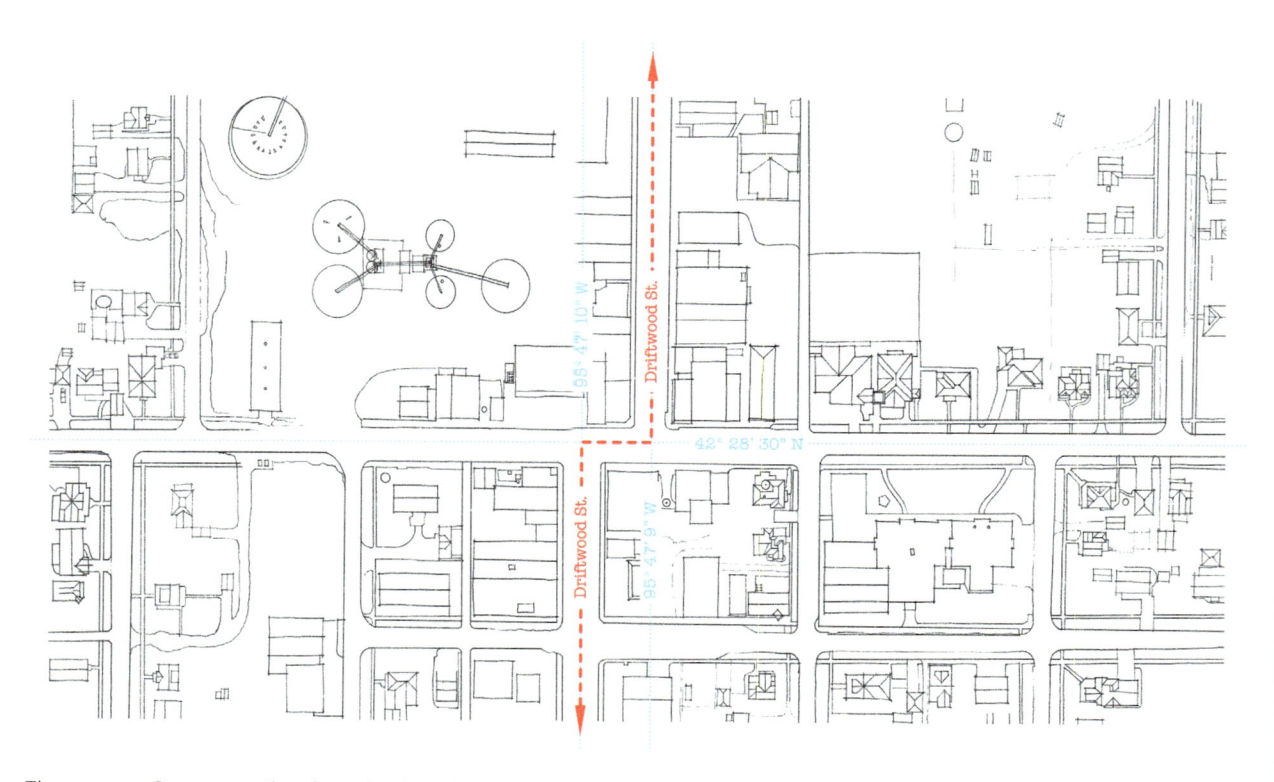

Figure 2.5: Compensating for spherical distortion in road alignments; Driftwood Street in Correctionville, Iowa (de Ruijter 2019).

THE MEASUREMENT OF SPACE

Figure 2.6: Steps in an animated rotation of a hypercube.

and sections that subdivide them. Every now and again, these roads take 90° jogs for a 100 ft or so, seemingly at random. Actually, this is evidence of a necessary correction to accumulated positional error inherent in the conflict between a planar Cartesian coordinate system and the spherical curvature of the Earth. Surveyors and civil engineers working on road and rail projects cannot use Cartesian space to describe their work. Instead, they use a system of spherical coordinates measured in latitude and longitude. A system in which parallel lines-at least lines of longitude, do, in fact, converge. Sorry, Euclid.[6]

The natural world, not constructed by human industrial processes, tends not to be quite so formal as the heady abstractions of Euclid. Natural forms (landscape, plants, people, and other animals) are, by definition, "organic" in form. Organic forms are made through natural processes of growth or decline, cellular growth, crystallization, or the operations of geological erosion and folding. Organic forms are not well defined by Euclid, nor are they necessarily well measured by Cartesian means. The math best used to describe them is different, more fractal, more continuous, and as a consequence, much harder to model with a computer. Organic form is easy to sketch by hand, though.

There is also a fourth dimension of space to consider – the dimension of time. Architecture exists in time, but few spatial projection systems can help us encode it in static drawings. We can easily project 3D space onto flat surfaces like paper or computer screens, but to add a time axis, you need something more – traditional media can't help in any but the most poetic ways. Film and digital modeling offer some alternative projection systems that point in the right direction. Performance simulations, as well, usually provide a time-series projection of performance mapped over days, weeks, months, and years. Assembly instructions are another type of time-encoded spatial representation. And, I would argue, the freely manipulable point of view found in 3D modeling applications on a computer screen offers another kind of time axis for consideration (Figure 2.6).

While it becomes increasingly difficult to visualize spatial dimensions higher than three, mathematically, it is possible to deal with spaces of infinitely many dimensions. Each additional dimension just adds another axis of measurement. Considering other axes of data in a model can be quite useful – and the dimensions need not always be understandable in strict geometric terms.

For example, construction convention sometimes refers to "cost" as a fifth dimension in building information models. As measurable and as variable as any other measured dimension, you may find you are already managing projects in 5D space. A sixth dimension is occasionally discussed as well – one representing the ongoing lifecycle of the building

---

[6] ...and thank you Lobachevsky.

post-construction. For sketching purposes, though, the projection of these nongraphical parameters into the same visual space as your model might not make much practical sense.

The Qualities of Space

In the Museum of Modern Art in New York, I once studied a set of three paintings by Italian Futurist painter Umberto Boccioni. The paintings represent the departure of a train from a station, each from a different observer's state of mind. The first is a depiction of the departure, the movement of the train, the throngs of people boarding and exiting the passenger cars, and the steam, light, and motion of the locomotive itself. In the second, Boccioni depicts the same scene from the perspective of an observer who is leaving on the train. With cloudier, more muddled color the state of mind of the travelers' enters into the space of the painting, with a hint of color suggesting the future, the destination to be reached at the end of the journey to come. In the final painting, a third spatial dimension is explored. This time, it depicts the same scene as experienced by those left behind. The moody, sorrowful palette of color is overlaid on an even more abstract field of wavy, vertical lines. You can feel the weight of emotion, the sorrow of being left behind, in what might almost represent a foggy, cold rain.

Using only the relatively simple materials of oil paint on canvas, Boccioni opens new spatial dimensions for consideration in his train station. The traditional dimensions of length, width, and height are almost entirely obscured by his depiction of motion (a dominant theme for the futurists). And also, by his sense for the ephemeral, emotional qualities of that same space. By opening for consideration the non-physical qualities of a train station, he allows the viewer to more fully inhabit the experience of that time and place.

If you think back on the places you most remember in your life, what are the qualities about them that left the most lasting impression? Do you remember the analytical dimensions of their floor plan, perhaps, or the thickness of a wall section? A unique handrail detail, perhaps, caught your eye. Or, maybe, what you remember most is the way the particular building you were in was situated on its site?

I'm betting that what you most remember was something difficult to describe in words. The most memorable places for me are those which I have associated with something beyond the physical characteristics of their construction. My design training leads me to be ever aware of the physicality of the constructed spaces around me, of course. Still, there is another, more intangible, quality of space that slips past the rationality of construction straight into my subconscious mind. The qualities of space are rich, complex, and memorable. What I always seem to remember is how a space "felt" to me (Figure 2.7).

Perhaps it was also tricky to depict in a sketch, but as the designer on a construction project, there is nobody more qualified to take care of the ultimate feel of the spaces you will help to design. The ephemeral qualities of space are derived from the multitude of design decisions made by you and others on the project, beginning right from the very first line on the very first piece of paper.

A line drawn with sumi ink and a brush feels differently than a line drawn with charcoal (Figure 2.8). The ink, laid down in liquid form, leaves a sharp edge on the paper. If your hands shake, as mine do, that line will betray every quiver of your emotional state you were feeling when you drew it. A charcoal line, on the other hand, feels dry and rough when laid down. Charcoal leaves an imprecise line, and it can easily be brushed away with a hand or an eraser. Two lines drawn, perhaps both representing

THE QUALITIES OF SPACE

Figure 2.7: My own "states of mind" remembrance of the Forum Romanum.

the object to be constructed (a wall, or a handrail, or a fence line), do so in very different ways.

The means of representation you choose when sketching can bring to light new opportunities and new considerations about your project that you might otherwise have trouble thinking through. Your construction team may understand that a wall is built either from metal studs or cast-in-place of concrete. The line you use to represent that in a set of construction documents may be more or less the same, but their ultimate spatial qualities will be vastly different.

Walking into a space enclosed by concrete walls feels different. The smell of raw concrete on a job site is unmistakable, of course, but there are other qualities to consider as well. A concrete wall reflects sound differently, making the space sound more reverberant, more "live." A space enclosed by studs and sheetrock might transmit sound through the wall, making the space feel sonically transparent.

A space surrounded by concrete holds heat differently, as well. In the summertime, you might appreciate the thermal mass of concrete enclosure, as it slowly heats through the sunny hours of the day and releases heat back into the space in the

Figure 2.8: Wet line (Sumi ink and brush), dry line (vine charcoal).

coolness of the evening. Or, in a humid climate, the same construction might become dank and covered with mildew.

When you are sketching, you can keep these ephemeral qualities of space in mind, of course, but you can also encode them in your actual sketches. Just as Boccioni encoded deep states of mind in his paintings of a departing train, you can capture the feel of the spaces you design in graphical form.

The most straightforward place to start is color. When representing a red wall, you can use a red pen. A red line drawn next to a black line confirms that one, or maybe both, of the things you are representing, will have color as a significant attribute. If all of your lines are drawn in the same color, this is less apparent. If your entire sketch is drawn in red, perhaps color isn't what you have on your mind – maybe that red pen happens to be the one you grabbed. Through contrast, the one red line drawn in a field of black lines clearly has some special significance (Figure 2.9)

The materiality of your sketching medium matters. A freshly uncapped marker, for example, has a distinct smell that might take you and your client to an unexpected new space in the project.

What will the space you are designing smell like? Likely, it will not smell like markers, but maybe it will smell like freshly baked bread? Or wet grass in springtime? Or perhaps it will smell like motor oil and gasoline. Many people only really recognize the smell of the spaces around them when they become objectionable – perhaps the trap in a drain line is dried out, and sewer gas is making its way in. Or maybe there is something else musty under the sink. But almost everyone recognizes the smell of their home and has an emotional response when they smell it.

I love the smell of art supplies, of oil paint, charcoal, paper, and fresh canvas. It makes me feel ready for creativity and work. But the materiality matters in other ways as well. The feeling of wet media, like watercolor opposed to dry media like pastels, has a distinct impact on my understanding of the space I'm sketching. Nowhere is this more apparent to me than when I am sketching in a desert landscape. On a hot, dry day, working with pastel feels like an extension of the desert onto my page. A splash of watercolor feels like an unexpected waterhole or trickle of spring water. I can, through careful leverage of these qualities, tell my viewer a story about the overall dryness and unexpected pleasure of wetness in the desert that is quite different from the visual depiction of space I might otherwise be trying to capture (Figure 2.10).

Architectural space is not just experienced in a single room at a time, of course, and this leads to another intangible dimension that you can explore through your sketching practice. The space of your project might be best understood as a sequence of experiences, strung together like beads on a string. The space of a kitchen is experientially different if you have entered it from the garage with arms full of groceries rather than in pajamas early in the morning on your way to the coffee machine. The sequence of uses in the space is much more closely related to the way the space feels than the

Figure 2.9:   Red line, black lines.

THE QUALITIES OF SPACE

Figure 2.10: Sketch of the desert, from Ghost Ranch in Abiquiú, New Mexico.

physical construction of its walls might be. A space surrounded by the darkness of midnight is a different one from the same space filled with the light of a late autumn afternoon when you have just kicked off your muddy boots and begun to warm the kettle.

Spaces that have the most impact on people's lives are the ones they can fill with their own stories. You are designing a canvas on which they will paint their lives. It is difficult to explain this in the earliest conceptual sketches, you might think, but I would argue that there is, in fact, no better time. Even the first quick moves on a piece of paper can begin to suggest the ephemeral qualities of space as quickly as the physical qualities. A line can depict the location of a wall. But in doing so, it also suggests the adjacencies that space will have to other nearby functions. A line drawn with a wash of color next to it can begin to explore the play of light across a surface made of some material. Pastel, perhaps, with its dry chalkiness, might represent concrete, where an ink line could represent hard surfaces like steel or glass. Every material choice you make when sketching can be a trigger for another thought about the design.

You may have been making choices like these without really knowing you are doing it. Maybe, even, it just feels to you like this is the natural and most common way for you to work. It bears reconsideration, however, in the context of digital sketching. Much of the functional materiality of sketching by hand can feel as though it is lost when you are sketching on a computer.

I would argue, however, that digital media have just as much materiality as any other drawing medium. Perhaps instead of sketching with the dragging smudginess of soft graphite on toothy paper, you are instead sketching with the cold glow of pixels on glass, but there is still a material quality to everything you draw. And with every new medium come some unique representational affordances that you can bring to your design with purpose.

For example, glowing computer screens are, in my experience, really effective when depicting glowing things like neon signs or the inviting evening glow of light through a window (Figure 2.11). Aided by photorealistic rendering engines, the play of

Figure 2.11: Glowing neon line, sketched digitally. By printing this image on paper, the innate "glow" is lost; you have to imagine this glowing on a computer screen . . .

light as it bounces from surface to surface, picking up color and changing character is easy to explore in-depth. With a single 3D model, you can test lighting conditions at different times of the day and in different seasons. Doing this by hand would be too cumbersome by far.

The detailed interplay between light and physical materials like glass, stone, or plastic can be easily simulated in a digital rendering as well. If your design is one that depends on the materials from which it is built, a digital rendering can be very informative as a sketching tool. And ultimately, what design doesn't depend on the materiality of its construction?

A digital sketching expert knows where to apply the right tool for the job, and knows that their choices, even right at the beginning, will have a measurable, tangible, and emotional result in the final design. The spaces that truly delight people, the spaces that embed forever in their memories and in which they play out the stories of their lives, are those that were designed with the fullest spectrum of consideration possible. A space that feels great is one that is not only well constructed but also fits into the fullest possible human experience of space.

Geometry and Form

Every language has a grammar and syntax that governs the ideas that it can encode. We're starting from Euclid, caveats aside for now. Cartesian space is sufficient for the majority of building-scale

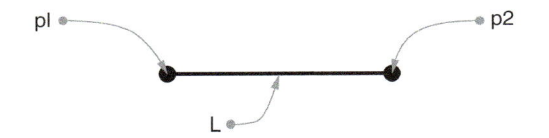

Figure 2.12: A straight line (L) can be drawn between any two points (p1, p2).

geometries. Consider his postulates for plane geometry (Figures 2.12–2.15):

With this set of geometric postulates, architects have been able to rationalize the forms in their imaginations and make them constructible from antiquity through our modern age. Every Greek temple, every Roman amphitheater, every soaring Gothic cathedral, or the modernist city of Brasilia can all be described with the rationality required for construction using only a palette of geometric forms derived from these postulates.

Euclid had a fifth postulate, named by tradition the *parallel postulate*, shown in Figure 2.16.

If two lines are drawn that intersect a third in such a way that the sum of the inner angles on one side is less than two right angles, then the two lines inevitably must intersect each other on that side if extended far enough.

Euclid's first four postulates are self-evident and provable as theorems. They are practical and useful on any construction site. You can demonstrate them with a compass and straight edge. Or a chalk line and a tape measure. Collective everyday

Figure 2.13: A straight line segment (L) can be extended infinitely in a straight line.

GEOMETRY AND FORM

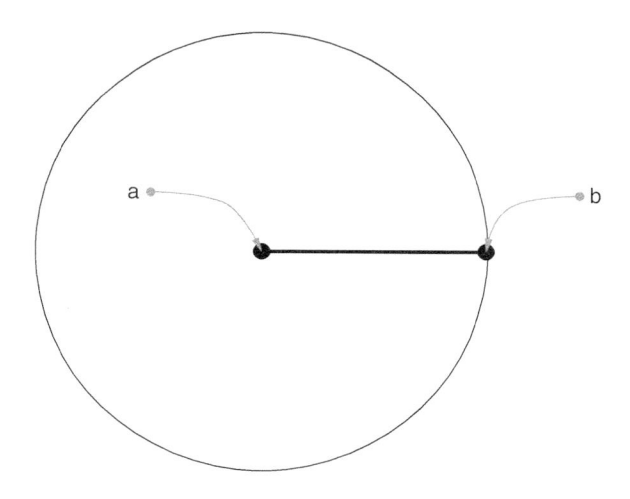

Figure 2.14: A circle (C) can be drawn from any line segment (L) given one endpoint (p1) as its center and the other (p2) defining its radius.

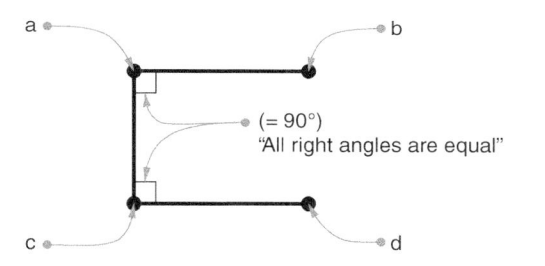

Figure 2.15: All right angles (90°) are the same.

experience suggests you are working with these geometric axioms all the time, perhaps not even recognizing how you know them anymore. They just feel like common (geometric) sense.

The parallel postulate is less self-evident than Euclid's first four propositions and more difficult to consistently demonstrate in everyday life. Evidence suggests even Euclid was not entirely comfortable including it, and he was able to make 28 of his 29 propositions in Elements without requiring it. Euclid stated that parallel lines extending into infinity will never cross. But what if they did? In the 1860s, the parallel postulate was restated by Lobachevsky, Bolyai, and Gauss[7] as follows (Figure 2.17):

For any infinite straight line (l) and any point not on it (p), there are many other infinitely straight extending straight lines that pass through (p) and do not intersect (l).

This is significant to architecture as projects expand in scale, and the impact is known to any land surveyor. Parallelism, much like Cartesian coordinate space that derives from it, is only valid at local scales. When you have a project that is larger than a few thousand feet, you may find the error becomes significant. To land surveyors for whom most projects are larger than this limit, plane geometry can only be considered an approximation of the hyperbolic space of their true measurements.

---

[7] Lobachevsky's seminal work on this subject is his "Geometrical Researches on the Theory of Parallels", originally published (posthumously, in Russian) in 1909. My reference comes from Lobachevsky (2019).

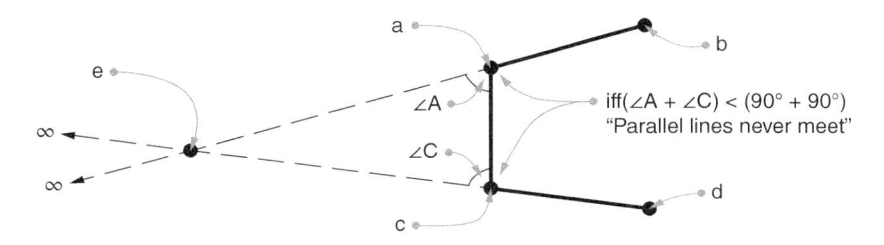

Figure 2.16: Euclid's fifth postulate, the "parallel postulate."

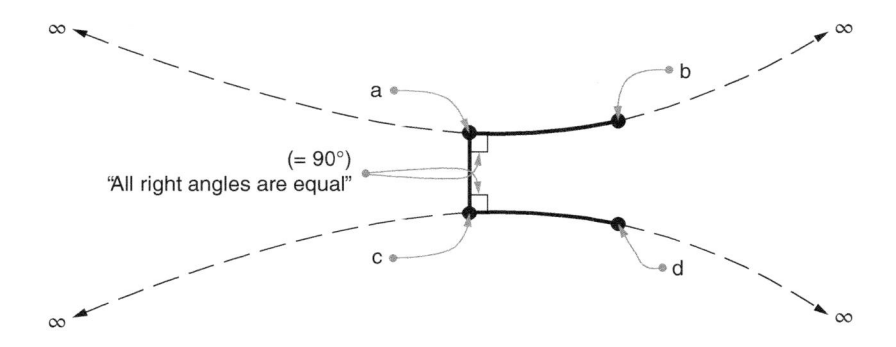

Figure 2.17: Parallelism, as restated by Lobachevsy, Bolyai, and Gauss.

But at local scales, even at the scale of a building, basic Euclidean geometry applies well. And within it, we can find a basic formal geometry for architecture: point, line, surface, and form.

Point

As the joke goes, *"Without geometry, life would be pointless."* Points, drawn as a single dot on paper, located with a single coordinate, are the simplest of geometric constructions for the conceptual designer. Euclid defined a point as being *". . . that which has no parts"* (Figure 2.18).

A point can either represent a singular object in space or form one of a set of coordinates for some larger construct.

Mathematically, points have no dimension on their own – no length, width, nor height. A point drawn on a piece of paper is the simplest of possible sketches. When representing a point on a digital screen, a single pixel is the smallest and most compact representation possible.

As a coordinate location, points can index any object's position in space. A point can be the endpoint of a line, the vertex of a plane, or a form. A point can operate as a control vertex for a curve or a freeform surface. A point can index the abstract constructs of a point of view or a perspective vanishing point. Consider the practicalities of a coordinate origin point for a project. Or maybe just the location of a piece of furniture in plan. In architecture, everything you sketch has a point.

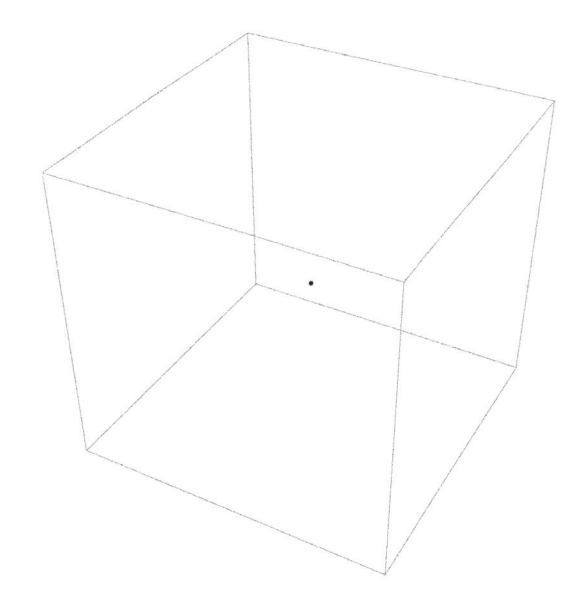

Figure 2.18: A single point.

GEOMETRY AND FORM

Line

Conceptually, by sweeping a point through space, a line can be described. Euclid defined a line as "... length without breadth." In planar geometry, a line is a straight path formed by continuous variation in coordinates from one endpoint to another. Euclid adds that "... the extremities of a line are points" (Figure 2.19).

Lines have length, but no width nor height. In the real world, linear elements can be found everywhere – from plumbing systems to highways. The great axial vistas of classical architecture are linear in design. Any collection of objects arrayed linearly in space like a colonnade can be seen as a line.

It is easy to draw a line on a piece of paper – only slightly harder than drawing a point. In fact, most 2D drawing centers around the artful arrangement of lines in 2D space on a page. When sketching, you will make lines of all kinds. Freehand lines or hard-lined lines sketched against a straight edge. You might draw squiggly lines or dashed lines. Your lines might be scaled or only proportional to their contextual neighbors. Any kind of line is possible when sketching by hand (Figure 2.20).

The finite mathematical definition of a line, which is needed for a computer to accurately represent it, is a bit more complicated. Line segments, which require only two endpoints defined, are considerably more straightforward than curved segments. In algebra, a straight line segment can be drawn between any two Cartesian coordinates, in any number of dimensions. Or you might use a slope-intercept function definition ($y = mx+b$) to define all the possible points along a line. Or maybe you have a different algorithm to define a line in code.

When drawing a line with a pencil on paper, friction between the pencil and the paper's abrasive tooth causes minuscule motes of graphite to adhere to the paper. If enough of them can be seen, the brain interprets them as a continuous line. Digital representations of lines must be rendered to your computer screen in some way as well. Typically, this is done by rendering them into a raster frame buffer – a two-dimensional array of pixels with individually defined color values. The pure function definition of a line, perfectly smooth in its mathematical expression, must be rasterized to pixels for your computer to display it. On low-resolution displays such as Figure 2.21, it is quickly apparent how approximate this representation can be.

Digital lines drawn on a low-resolution screen will render as though composed of stair steps. This rendering artifact is known as "aliasing." Computer displays gained the ability to display millions of different colors for each individual pixel well before they reached the high ("retina") resolution we see on the market today. Methods of anti-aliasing have been invented that are the best solution to fool your eyes into thinking the screen resolution is

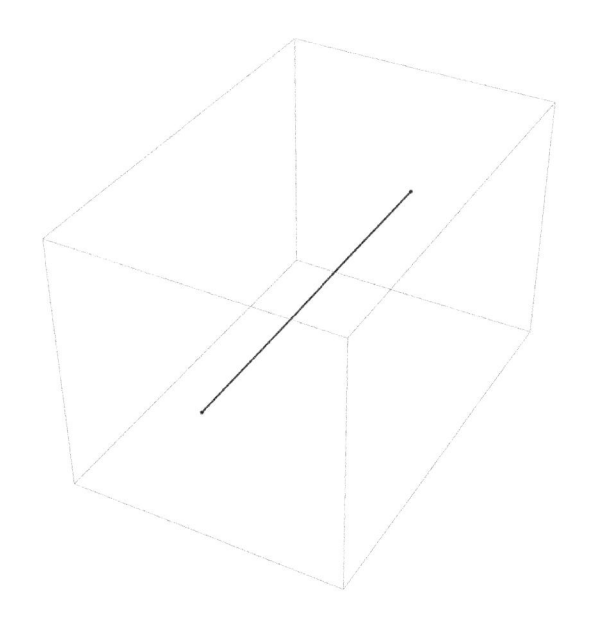

Figure 2.19:   A line.

48 THE ELEMENTS OF DESIGN

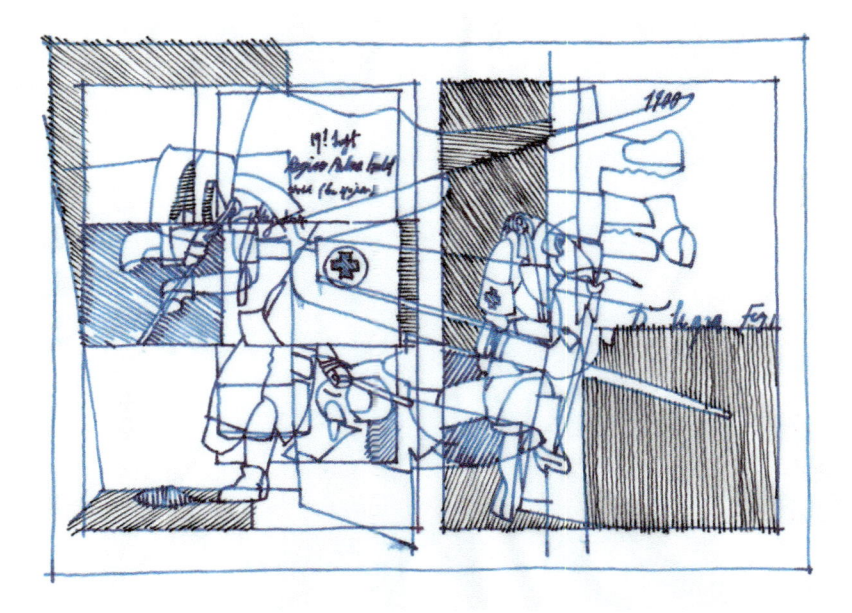

Figure 2.20: Sketching lines by hand; after an image by Valerio Adami (Adami 1973).

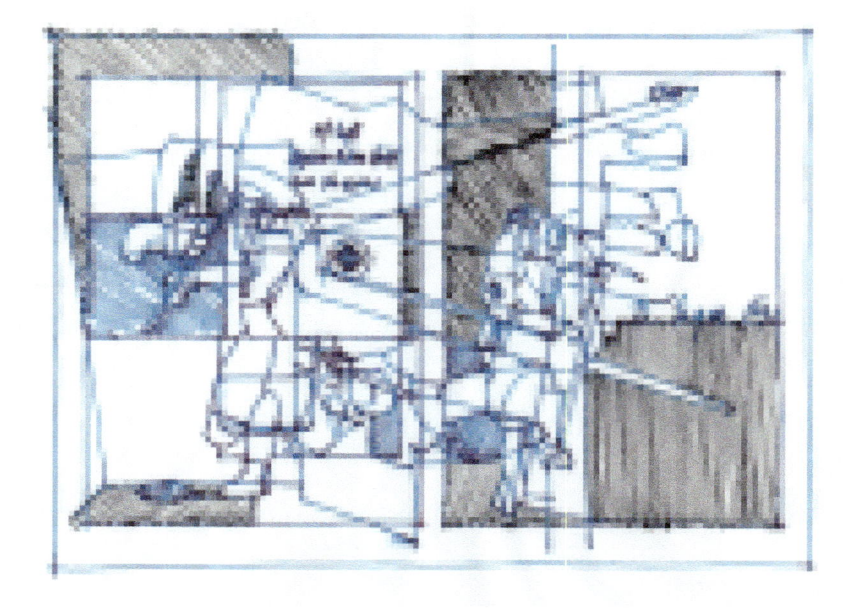

Figure 2.21: Rasterization on a low-resolution digital display.

GEOMETRY AND FORM

higher than it actually is. They work by selectively blurring pixels around the stair-step aliasing of straight lines, tricking your eyes into thinking they aren't able to focus fully and engaging your brain's ability to interpolate from the bad data.

No matter how you choose to represent it, the mathematically pure perfection of a line can only ever be approximated when it is drawn. Even when drafting by hand with a straight edge and the most precise of technical ink pens, lines always have a thickness, implying a level of precision that is relative to the scale of your drawing. Sketched lines only make the approximate nature of your design's dimensionality more apparent.

Arc

Not all lines must be straight ones, and perhaps it is appropriate to recognize that there are almost no perfectly straight lines anywhere in nature. Arcs, derived from a point swept evenly around a fixed center, are a special kind of line drawn with a compass instead of a straight edge (Figure 2.22).

It is surprisingly hard to draw a perfect circle by hand without some mechanical assistance. Georgio Vasari, in his "Lives of the Artists," tells a story about the painter Giotto, who, when asked for evidence of his qualification for a new commission, said nothing and simply drew a perfect freehand circle. According to Vasari, he got the job. In Japanese Zen Buddhist practice, a circle drawn with a brush and ink called "ensō" is used in meditation. In its creation is a full encoding of the moment of its making, capturing in one brush stroke the fullness of absolute enlightenment in that fleeting moment. Here is one from me (Figure 2.23).

Mathematically, Euclid said, a circle is *". . . a plane figure, bounded by one continued line, called its circumference, and having a certain point within it from*

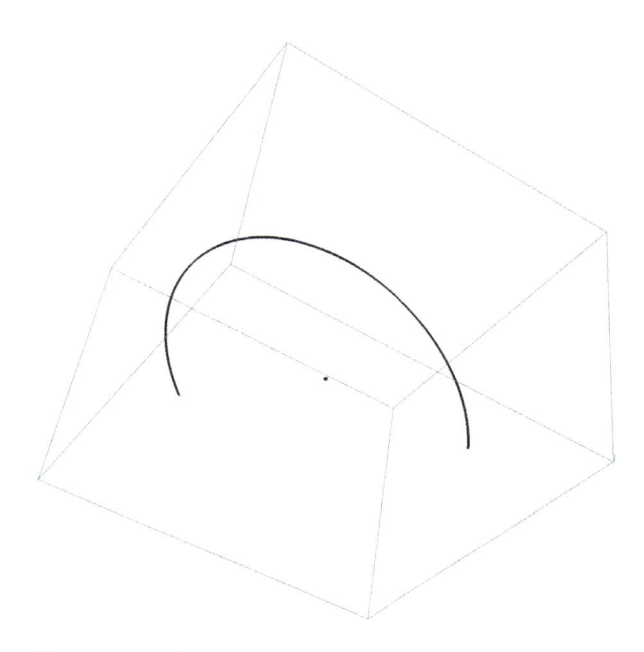

Figure 2.22: An arc.

Figure 2.23: Drawing a circle by hand, the ensō; too much coffee today.

which all straight lines drawn to its circumference are equal" (Figure 2.24).

A circle represents the set of all possible points equidistant from a center point. Circles can be

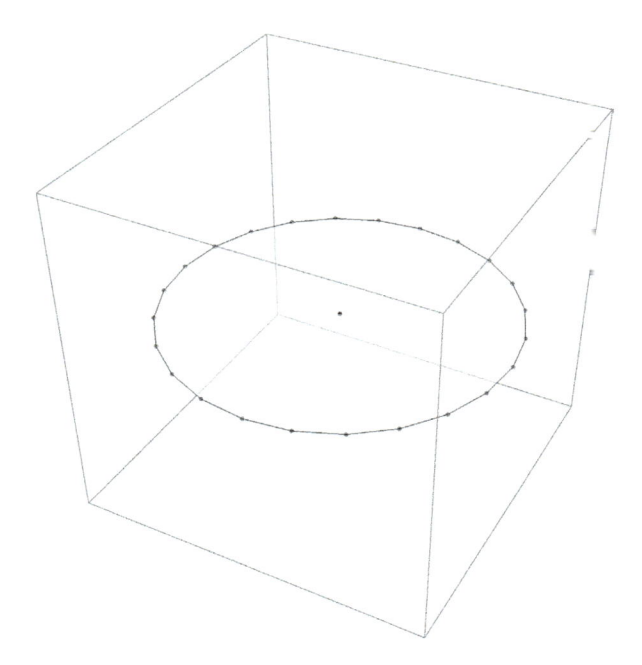

Figure 2.24: A circle drawn in 3D space.

found everywhere in nature, from the shape of a single soap bubble to the cross-section of a tree trunk.

The combination of arcs and straight line segments, which can be created by hand with only a compass and straight edge, can be used to construct geometry of incredible complexity. Compass and straight-edge constructions are at the heart of descriptive geometry, which is itself the basis for all orthographic projection systems. With only arcs and line segments, you can perform basic arithmetic – dividing and adding, even computing square roots. Albrecht Dürer's 1525 *The Painter's Manual* is full of such constructions (Figure 2.25).

There are many ways to constrain a circle geometrically, as evident in the diversity of methods for their construction in most CAD systems. Circles may be drawn through any three points, giving

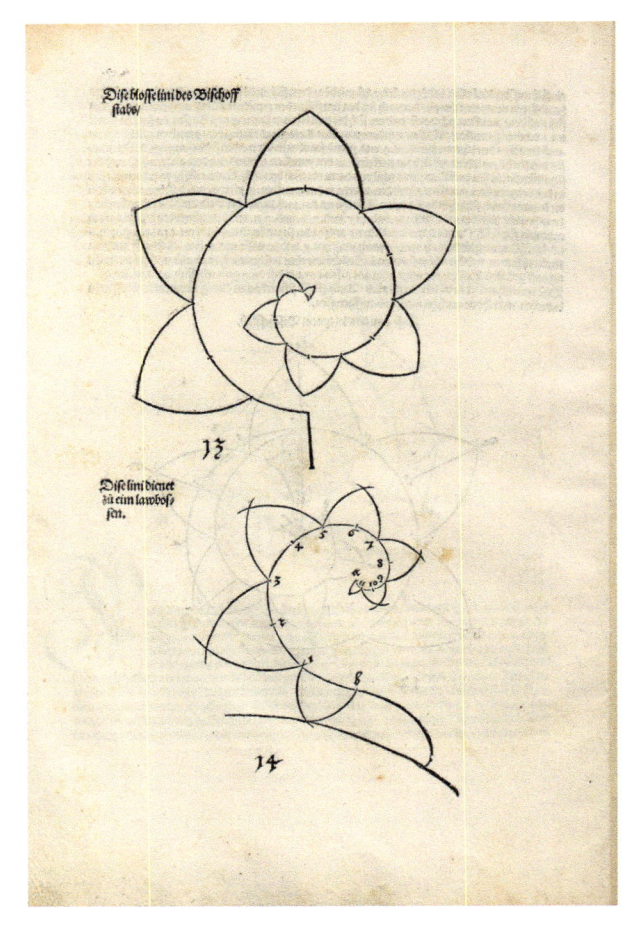

Figure 2.25: A compass and straight-edge construction of the complex geometry of a Bishop's Crozier; from Albrecht Dürer's "Underweysung der Messung" ("The Painter's Manual") (Dürer 1525).

them a special relationship with triangles. Most CAD systems include circle creation tools that allow you to define a counterpoint and radius, or any combination of three points on the circumference of the circle.

Every regular polygon can be either inscribed within or circumscribed by a circle. To most digital systems, polygons with high edge counts are an appropriate visual representation of a circle, especially when the resolution of the display is such that the user sees pixel aliasing before they can perceive the polygonal faceting (Figure 2.26). This approximation can be confusing when trying to use the polygonal visualization to compute mathematically precise points of intersection between arcs and other geometry.

Conic

Euclid understood circles as but one precise form of a more general set of curves known as the conic sections. Conics are two-dimensional curves found at the intersection of cutting planes passed through three-dimensional cones, which are themselves formed by sweeping a line around an intersecting axis of rotation.

There are four named classes of conic section, each dependent on the angle at which the cone is cut. The circle is found by cutting perpendicular to the axis of rotation. The ellipse is found on any cutting plane where the intersection creates any noncircular closed curve. The parabola exists where the cutting plane is parallel to the swept line. And the hyperbola exists where the cutting plane intersects both halves of the cone but does not converge.

Conic sections can be constructed by hand, but the process requires some setup. An ellipse, for example, can be drawn with a pencil, a length of string, and two thumbtacks (Figure 2.27).

Parabolas are easily demonstrated in nature by tracking the motion of a projectile over time. The water jetting out of a fountain nozzle describes a perfect parabola. A ball thrown in a game of catch follows a parabolic path from pitcher to catcher. Renaissance mathematicians who knew how to calculate this natural phenomenon were prized for their ability to precisely calculate the trajectory of cannonballs.

Figure 2.26: Segmented arcs, at high-enough resolution, are indistinguishable from perfect mathematical arcs.

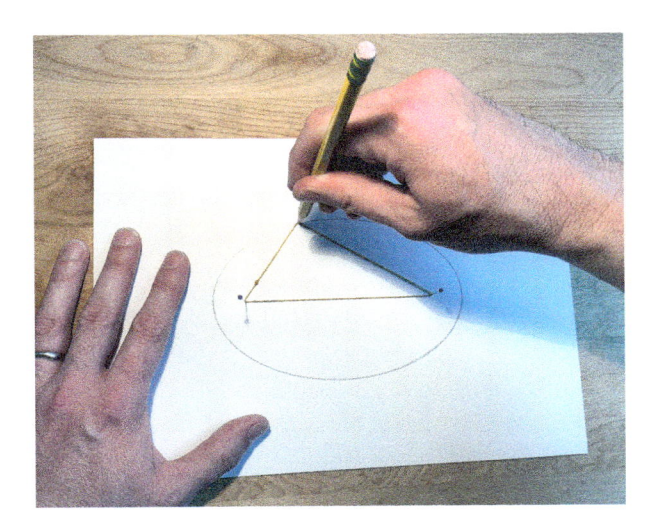

Figure 2.27: Hand drafting a perfect ellipse.

Figure 2.28: A sketch of a parabola in architecture; L'Oceanogràfic, in Valencia, Spain; designed by Felix Candela.

Parabolic arches are found throughout the history of architecture. For example, consider the unique structure from Felix Candela shown in Figure 2.28; its impossibly thin concrete roof functions structurally without any additional support because of its parabolic form.

Alternatively, consider the work of Antoni Gaudi, shown in Figure 2.29. His Basílica de la Sagrada Família is absolutely filled wsith them. Similar, though mathematically distinct, to parabolic curves are the curves formed by the natural drape of ropes and chains. They are called *catenary* curves, and you can see examples on any suspension bridge. While mathematically different, they are practically very close to parabolic curves. Gaudi recognized this and was able to physically model the complex structural system of his cathedral upside down. Once the system was balanced, he simply dipped it in plaster to harden it and flipped it right side up again. The catenary curves became parabolas, and the parabolas were structurally perfect for holding up the soaring roof (Huerta 2006).

Freeform Curves

It was not until the 1960s that designers at the French automotive companies Citröen and Renault[8] needed a better way to rationalize the streamlined sculptural forms than could be provided by arcs and conics. They could sculpt the forms they imagined for cars like the Citröen DS by hand. Still, to scale production of the cars, a mathematical model was needed to reproduce the sculptor's work on industrialized machine tools.

Of course, artists had for centuries been able to sculpt gracefully curved surfaces by hand, just as skilled graphic artists can easily sketch any imaginable curve. And those same artists had skilled

---

[8]The story of the invention of spline curves and NURBS is an interesting one. Though Pierre Bézier (working at Renault) and Paul de Casteljau (working at Citröen) may have invented NURBS at nearly the same time, Bézier was the first to publish his results. De Casteljau's technique remained a trade secret at Citröen until 1974, though he was likely working on it as early as 1954 (Peddie 2013).

FREEFORM CURVES

Figure 2.29: Hanging structural simulation of La Sagrada Familia. Gaudi's use of this technique is described in detail in Huerta (2006).

apprentices who could manually copy the work of their masters all day long, as needed. But they could do this only after thousands of hours of practice under the Master's careful guidance. This degree of craft is notable, but cannot scale easily. Mass production demanded automation and machine control.

In this case, it all starts with a curved line. The first curve drawing algorithm I learned, as described in the Prologue, was Chaikin's algorithm. I think it makes sense to go through an implementation of it in some detail here. I have found that learning how to do an algorithm by hand is immeasurably helpful in understanding how it really works. There are, of course, software implementations of this and other curve drawing algorithms that make them much faster and more practical to use. But a little time spent with the basics seems appropriate if you want to have full control over a tool you might learn later.

Chaikin's algorithm is easy to remember in its shortest form, "Cutting corners always works." Broadly speaking, you make a curve in this algorithm by beginning first with a connected set of straight line segments, as illustrated in Figures 2.30–2.34.

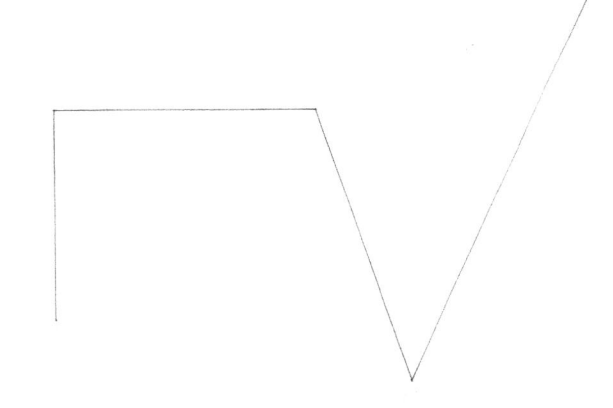

Figure 2.30: Chaikin's algorithm, step one.

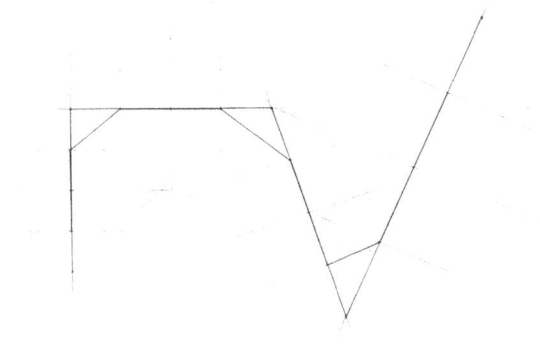

Figure 2.31: Chaikin's algorithm, step two.

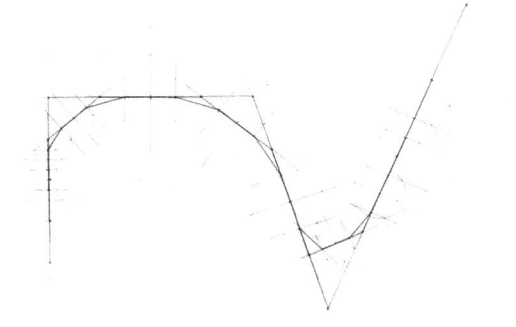

Figure 2.32: Chaikin's algorithm, step three.

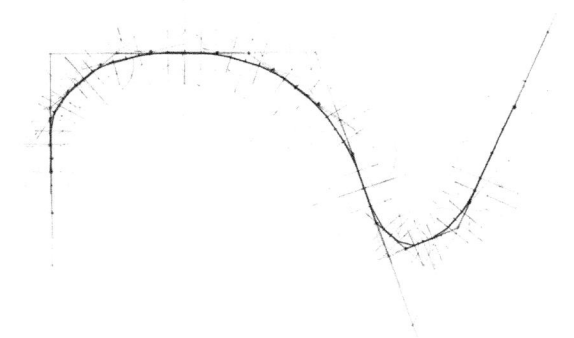

Figure 2.33: Chaikin's algorithm, step four.

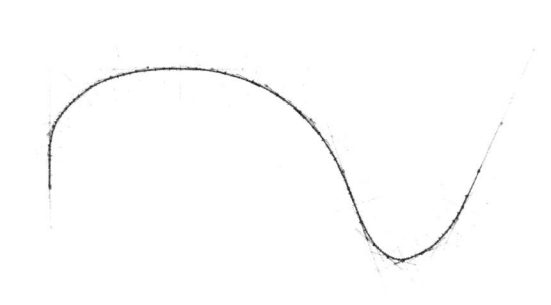

Figure 2.34: Chaikin's algorithm, step five.

Divide each line segment equally in quarters. Actually, any regular ratio will work, providing you some degree of control over how smooth the final curve will be in a general way. Draw new lines, knocking off each of the corners with new line segments. Repeat recursively until the curve looks smooth to you. Eventually, the subdivisions will be too small to practically subdivide further.

Congratulations! You have just drafted, by hand, a precise cubic b-spline curve. Chaikin's curve is a great way to get started, and it is wonderfully easy to implement by drawing. This is also the basis of all freeform surface modeling on a computer. And now you know how to do it by hand. There are a couple problems with Chaikin's algorithm, however, and for that reason, it may be worth taking this a little bit farther. There other algorithms available as well, some of which are only slightly harder than Chaikin's but are also capable of generating much more general geometric solutions. In other words, you can make more different kinds of curves with them, more expressive, or maybe more precise for your particular needs.

Consider another set of connected straight line segments. We're going to turn them into a smooth Bézier curve as well, through a different corner-cutting algorithm. This time, we will use Paul de Casteljau's algorithm, illustrated in Figures 2.35–2.39.

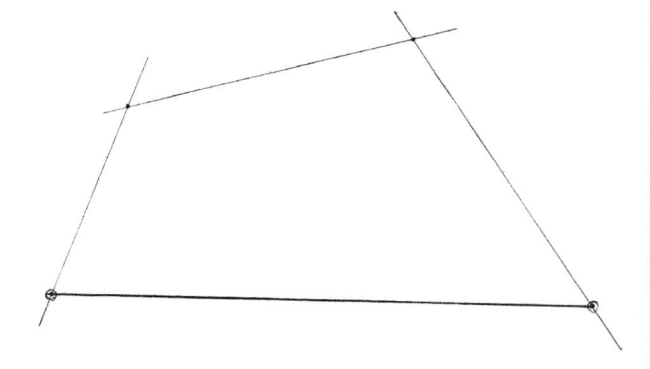

Figure 2.35: Bézier curve (de Casteljau's algorithm), step one.

FREEFORM CURVES

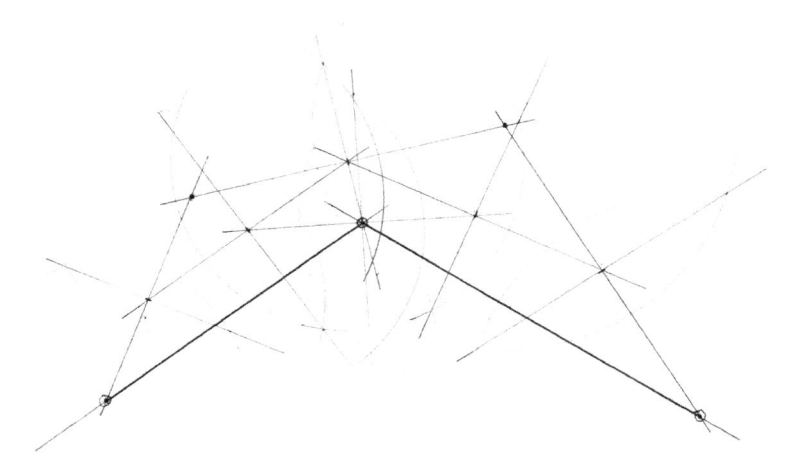

Figure 2.36: Bézier curve (de Casteljau's algorithm), step two.

This time, rather than thirds, we're going to divide each line in half. Now, connect the midpoints and divide each new line in half as well. Connect each new midpoint with a third line, and find its midpoint one more time. This new point is a unique point on the smoothed Bézier curve, along with each of the endpoints of the original figure. As with Chaikin's algorithm, keep repeating this, subdividing every time you iterate through the curve until you have enough points identified to draw a smooth-looking curve.

This drawing method, applied over and over again by your computer, is how it draws curves. There are many different algorithms, but all of them are similar in the end result. And now you know how to sketch them by hand.

Nature is full of curved lines – they are far more common than Euclid's abstractly perfect straight line segments. To rationalize an arbitrary curve in space with a computer, you need more advanced

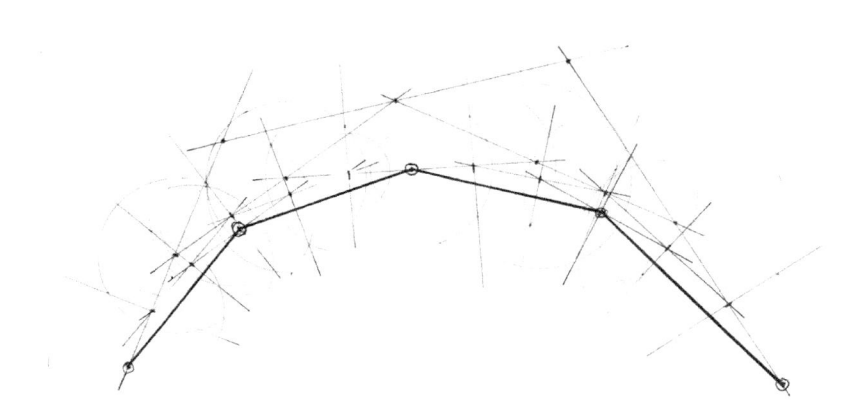

Figure 2.37: Bézier curve (de Casteljau's algorithm), step three.

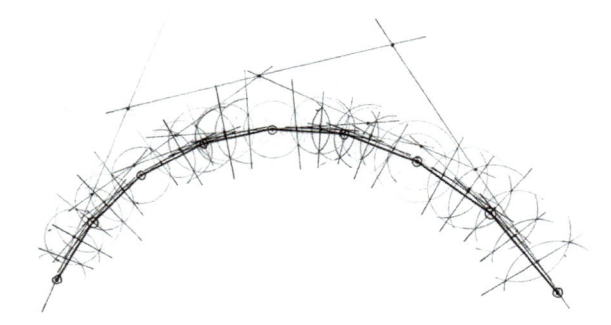

Figure 2.38: Bézier curve (de Casteljau's algorithm), step four.

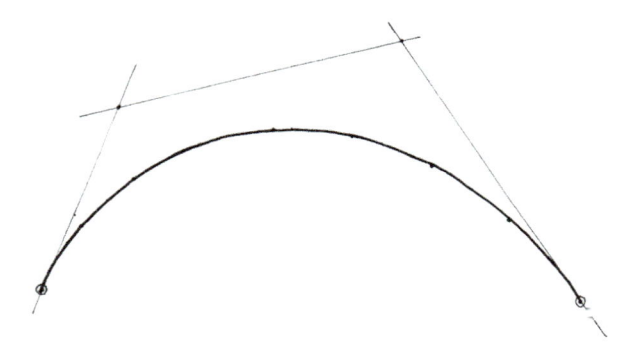

Figure 2.39: Bézier curve (de Casteljau's algorithm), step five.

math than Euclid had at his disposal. The simplicity of a freehand curve drawn by hand is deceptive when you try to define it mathematically. The Bézier curve is a function that can define visibly smooth curves at any resolution in a compact calculation that computers can easily accomplish.

Bézier curves can describe lines with arbitrary degrees of curvature, including straight lines, conics, quadratics, and beyond. They can describe lines of arbitrary length while at the same time ensuring continuously curved transitions from segment to segment. Applications like Adobe Illustrator depend heavily on Bézier curves to define the smoothly curved vector artwork and typography that they are known for.

Surface

A line segment swept through space describes a surface, defined by Euclid as ". . . that which has length and breadth only." Additionally, he stated, "The extremities of a surface are lines." Planar surfaces can be defined in Cartesian space between any three points, bounded by any three coplanar lines. These "triangles" are the most minimal definition of a surface in all digital modeling systems. They are easy and fast to draw, and with them, you can approximate any kind of surface (Figure 2.40).

Triangles have several special properties to Euclid, particularly related to their interior angles and the lengths of their component edges. As the simplest representation possible of a plane, they have special significance in computer graphics as well. Digital modeling systems, no matter how they represent form internally, display the forms you build in them with triangles. Everything the computer displays is rendered first to a triangular mesh, then to pixels to display on the screen. Solid models, freeform surface models, subdivisional surface models – all of them are drawn with triangles.

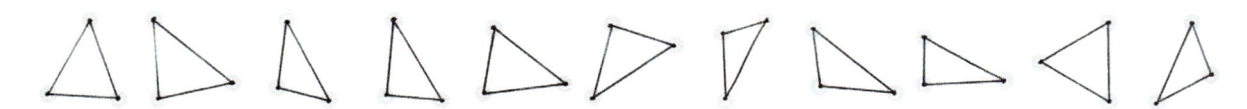

Figure 2.40: A diversity of triangles.

SURFACE

Freeform Surfaces

As Bézier curves are to straight line segments, so freeform surfaces are to planar polygons. Just as we drew a Bézier curve by hand to better understand it, so we can draw a Bézier surface. Let's begin with a field of points this time. I'll arrange them in a kind of grid to keep things easier to understand, as illustrated in Figures 2.41–2.45.

To begin, I'll connect the grid with vertical line segments. Each of these can be smoothed like a regular Bézier, but only the ones on the outside will end up with points on the surface, so I'll focus on them first. Now, do the crossing lines the same way. Again, the outside lines are the ones that will matter the most. And then smooth each of the outside lines.

You have just drawn a freeform curved surface, by hand. There are many different mathematical techniques for defining freeform surfaces. Still, if you are spending any time sculpting freeform surfaces on a computer, chances are, you're doing so with Nonuniform Rational B-Spline Surfaces, or "NURBS." They are a generalization of the algorithm we just drew by hand, more comfortable to implement in code, and with many additional surface parameters that can help them to shape a wide range of conventional forms.

NURBS are the most common method for defining mathematically precise freeform surfaces in CAD systems. They are easy to manipulate in a modeling environment because their control points are often directly on the surface that they help to define.

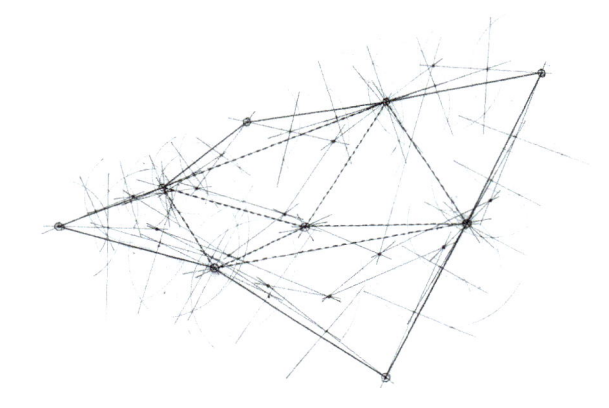

Figure 2.42:   Bézier surface, step two.

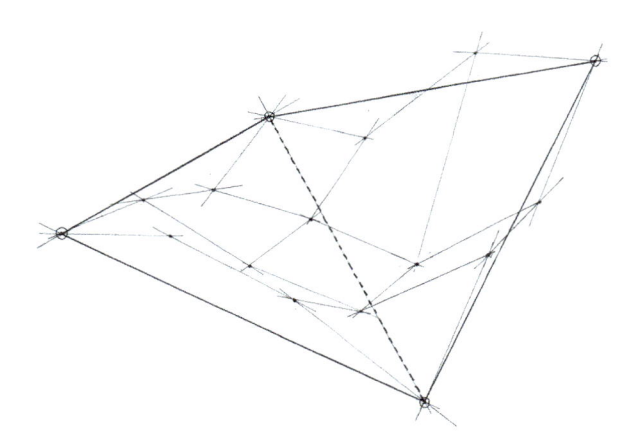

Figure 2.41:   Bézier surface, step one.

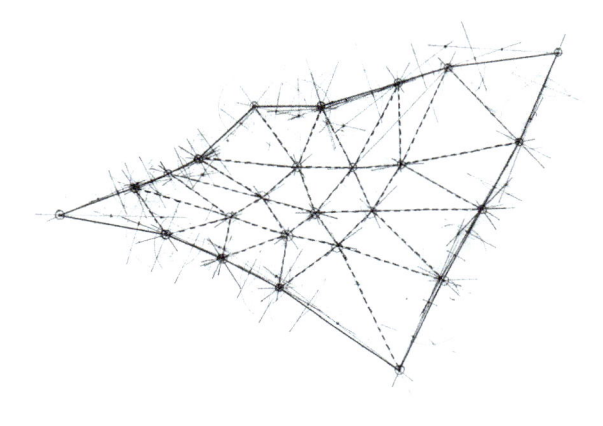

Figure 2.43:   Bézier surface, step three.

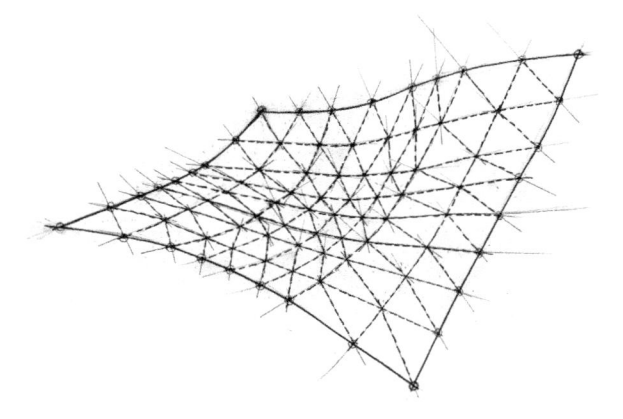

Figure 2.44: Bézier surface, step four.

Ed Catmull,[9] one of the founders of Pixar, developed a different freeform surface modeling technique that takes corner-cutting like Chaikin's algorithm into the third dimension. Called "subdivisional surfacing" (or just "Sub-D" for short), Catmull's methods have found widespread adoption in the visual effects and animation markets.

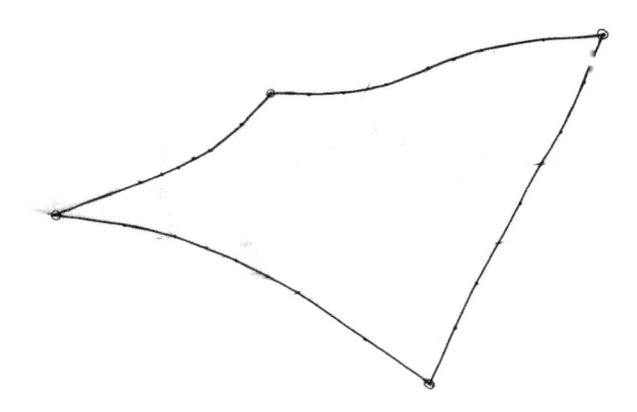

Figure 2.45: Bézier surface, step five.

---

[9]Edwin Catmull and Jim Clark co-authored what is considered the seminal paper on recursive subdivisional surface modeling in 1978 (Catmull and Clark 1978).

A detailed dive into the relative merits of NURBS vs. Sub-D would be well outside the scope of this book. Still, you should know there are many different freeform modeling techniques to choose from when you start looking for a modeling application to work on this kind of form.

I think it is wise to remember Chaikin's lesson: "Cutting corners always works." You don't have to be a mathematician to really appreciate how this kind of geometry works, especially not when you're sketching with it. The most critical distinction between drawing with Béziers, or NURBS or Sub-D surfaces, is that you are fundamentally defining a surface, similar in many respects to a simple flat Euclidean plane. Freeform surfaces have no thickness by themselves, only length and width. To make a form with them, you have to add more geometry.

Also, remember that all freeform curves are approximated with straight line segments. This is true when you draw them by hand, but it is also true when they are drawn to your computer screen or, even, when they are carved by a robot in an automated fabrication shop. To digital modeling systems, everything degrades for rendering to the simplest surface, the triangle. Even the most complex freeform surfaces are ultimately rendered as triangles on your computer's screen. Graphics display pipelines, no matter what their underlying geometry model may be, all render triangles to your screen. Freeform surfaces, with luxuriously smooth surface representations, are also drawn as triangles. They are just done so in a way that hides the edges so you can't tell.

When you are drawing by hand with pencil and paper, the surfaces in your sketches are implied by lines you draw at their boundaries. The physical presence of a surface is only implied by its bounding edges. You have now seen that the mathematical definition of a freeform surface is essentially the same. The curved lines at its edges define the surface.

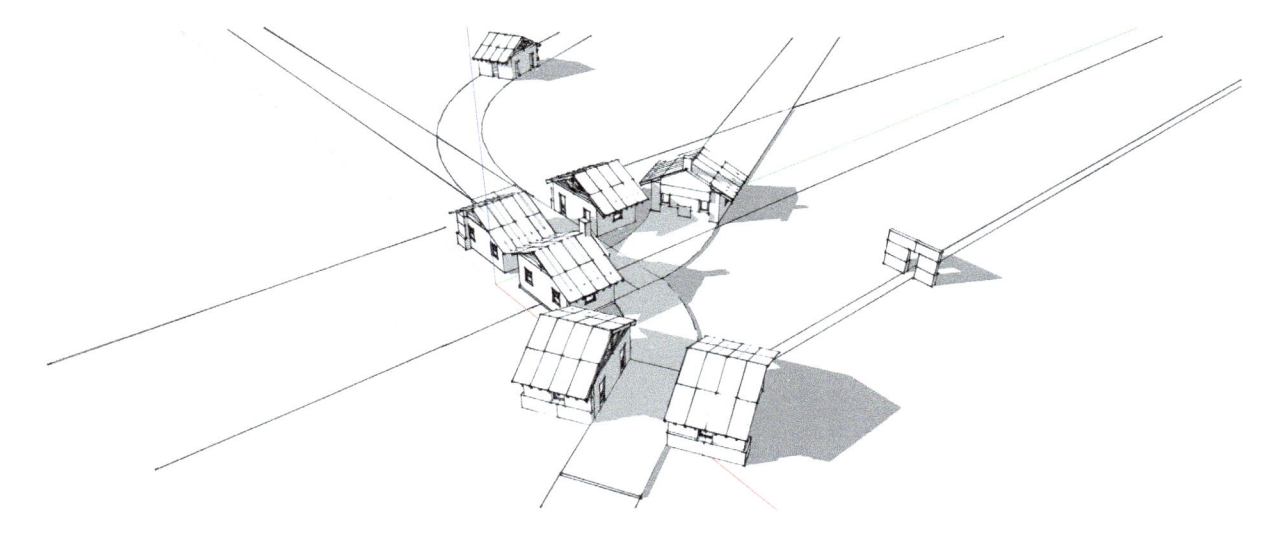

Figure 2.46: A plurality of architectonic forms.

Form

Where lines have one dimension, and planes have two, forms are measured in three dimensions. Architects, at heart, are form-makers, bringing new space into existence by shaping the material stuff that surrounds and defines it. Forms have topological insides and outsides, which can lead them, when rendered in some natural material, to have additional weight and presence. Buildings have a form and are themselves made of subassemblies of form. As Plato taught, we all of us live in a world of forms (Figure 2.46).[10]

Forms can be made in many different ways. In physical modeling, forms are made by cutting and shaping raw materials like wood or stone. When drawing with a pencil on paper, forms are defined by the arrangement of their boundary lines. Where there are no sharp geometric edges in the form, a combination of profile edges and or the play of shade and shadow can help to define the form. Form must be projected visually through some method to make it visible on the two-dimensional surface of a piece of paper or a computer screen.

Primitive Forms

As idealized by Plato, and later formalized by Euclid, there exist perfect mathematical forms that may be assembled together like children's blocks to create assemblages of greater and greater complexity. Classically, the Platonic solids include the tetrahedron, cube, octahedron, dodecahedron, and the icosahedron, shown in Figure 2.47. Each is composed of congruent regular polygonal faces, with all edges of equal length. Each of the five Platonic solids has a geometric relationship to one another. Each can be derived from the others, and all five nest perfectly together. It is no wonder the Greeks studied them with spiritual fervor.

---

[10] Plato wrote on what we now know as his "Theory of Forms" in many books, but notably so through his "Allegory of the Cave" in *The Republic,* Book VII (Plato n.d.).

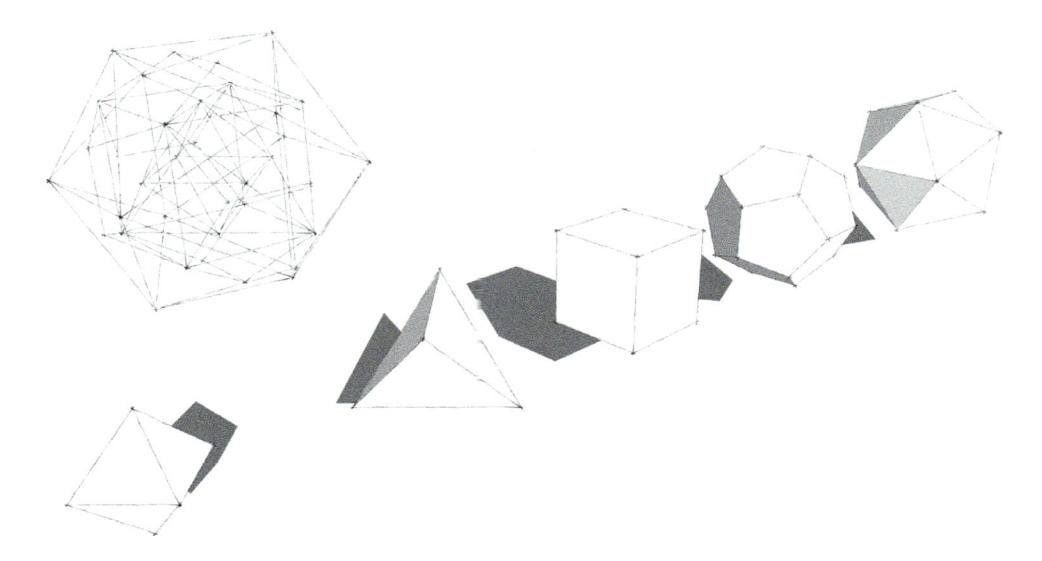

Figure 2.47: The five Platonic solids, individually and nested together.

Yet, in an architectural sense, there are few cases where the purity of the Platonic solids have much practical use to a designer. More commonly used primitives include cubes, cylinders, cones, and spheres, as well as their parametric permutations. The notion, however, that a more complex form might be built up from geometric primitives of any kind can be quite powerful, particularly when those "primitives" represent practical, commonly used higher-order building components like doors or windows, representing commercially manufactured objects. In computational terms, primitive-based solid modeling is traditionally referred to as onstructive solid geometry (CSG), shown in Figure 2.48.[11]

Even a limited palette of simplified construction blocks like cubes, cones, and spheres can be quite useful when sketching architectural forms. Quickly blocking and stacking individual forms, a building with the complexity of a gothic cathedral can be massed up quite efficiently. Adding only the simple Boolean[12] logical operations of union, subtraction, and intersection, a predefined palette of primitive forms can be combined together to create solid forms of great complexity, as shown in Figure 2.49.

Swept Forms

To Euclid, a surface swept through space also creates form. In simple sketching, projecting a plan into an oblique projection makes a quick massing

---

[11]Introduced by Mathematical Applications Group, Inc. (MAGI) in 1966, constructive solid geometry was used in an application called "SynthaVision", which modeled forms exclusively by assembling collections of primitive forms together. SynthaVision was used in the filming of the 1982 movie *Tron* (Peddie 2013)

[12]Boolean algebra, introduced by George Boole in *The Mathematical Analysis of Logic* (Boole 1847), provided a system of logic that underpins most computer programming today. In 3D modeling, "Booleans" have come to represent a powerful class of formal logical operations of addition, subtraction, splitting, and trimming operations, which are derived from Boole's logical mathematics.

SURFACE

Figure 2.48: some common architectonic solids.

study from a floor plan. In SketchUp, the "PushPull" tool is used to create form from any planar surface in the model. Just touch the surface and drag it in the direction you want to build. Surfaces need not be swept in a single direction. However, a form can be created by sweeping a surface along any path, as for example, to make pipes or railings (Figure 2.50).

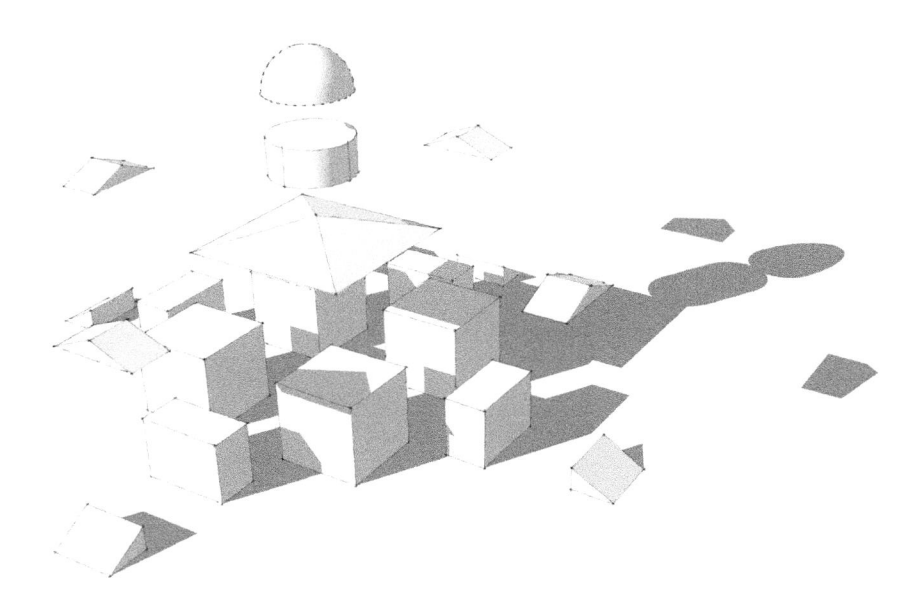

Figure 2.49: Architectural form from blocking and stacking.

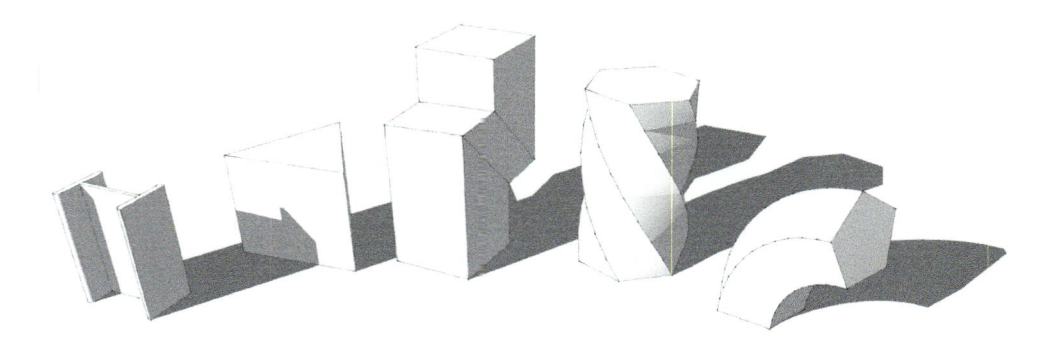

Figure 2.50: Simple extrusions.

A plane swept along a closed circular path creates a toroidal form. If the diameter of the circular path is zero, then a revolved surface is created. Surfaces of revolution are easily made on machine tools like a lathe, sweeping a profile equally around the axis of rotation (Figure 2.51).

Joined Surface Forms

Form can also be created by combining points, lines, and faces into larger objects. Multiple surfaces may also be joined together, much like gluing pieces of cardboard in a physical model, to assemble a form from component faces. Forms made of surfaces have, in addition to their points, lines, and surfaces, an overarching topology. "Solid" forms are topologically closed, with a distinct inside and outside. Alternately, the collection of joined surfaces may not be completely closed, with openings or gaps that prevent interpretation as a solid, just like modeling physically with cardboard, as shown in Figure 2.52.

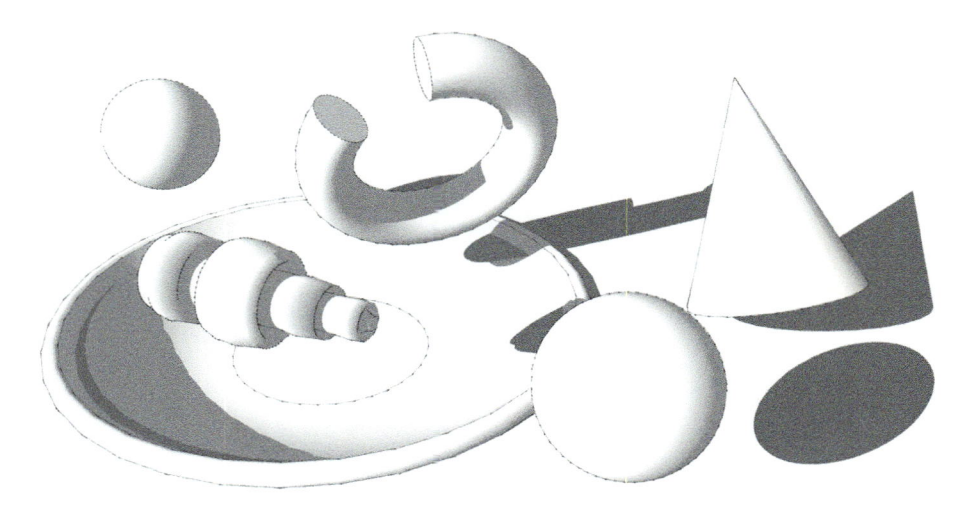

Figure 2.51: Revolved surfaces.

SURFACE

Figure 2.52: Joining cardboard pieces to make a house.

When sketch-modeling physically with cardboard, you must individually cut out every facet of your model, each wall, each plane of the roof, each layer of the site topography. While each of these individual pieces of cardboard may not by themselves represent a three-dimensional form, when assembled together, they can create a much more complex form than might be achievable with a simple composition of primitive forms (Figure 2.53).

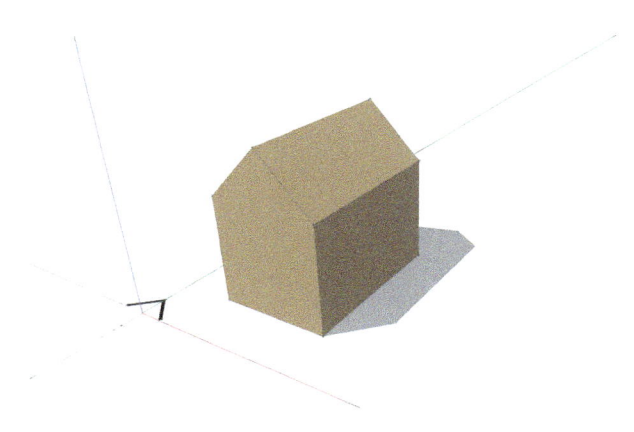

Figure 2.53: A computational surface model of the same house.

In computational terms, we also often refer to these kinds of form as *boundary representation* (BREP)[13] forms, and they are among the most common types of computational geometry used by commercial CAD applications today. The boundaries in this definition are formed by the points, lines, and surfaces that enclose and define form. There is no inherent materiality to a BREP form. However, when there are enough surfaces connected together to make a topologically closed form, they can be used to simulate physical objects with a material property like density or strength quite efficiently.

A BREP form that is not closed, and therefore not "solid," is easy to imagine and easy to create. Using cardboard to create a cube, for example, you might leave off one face so that you can easily look inside. In topological terms, the five remaining sides of the cube cannot enclose material; they are not, in a sense, "watertight" enough to be formally considered solid. But in architectural terms, a cube with an open face might well be precisely the sort of thing that you need to represent your concept, as shown in Figure 2.54.

There is nothing inherently degenerate about BREP forms that are not watertight, though there are certainly those who will make that claim. When you are machining parts from a CAD model, as with a 3D printer[14] or a computer numerical control (CNC)

---

[13] Boundary representation (BREP or B-rep) geometry was invented independently in the 1970's by Ian Braid (Cambridge) and Bruce Baumgart (Stanford). Their research eventually led to the development of commercial modeling kernels, notably ACIS and Parasolid, which are still in wide use today (Mäntylä 1988).

[14] Thanks to the innovative hacking of many contributors throughout the Maker community, 3D printing has become widely accessible for only a few hundred dollars. It remains, however, a tricky and fiddly technology to use practically. Models for 3D printing must be carefully prepared, and not every form that can be modeled is equally printable.

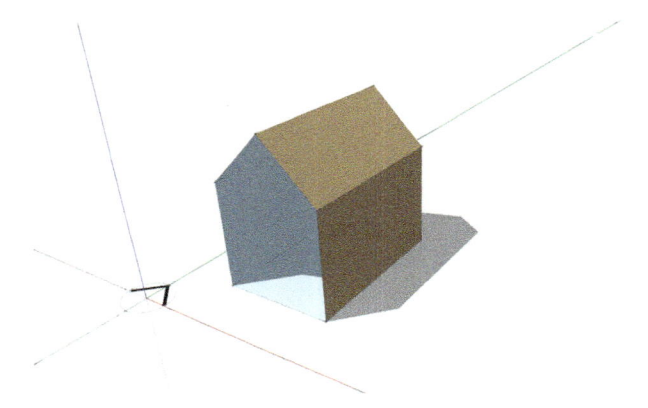

Figure 2.54: Removing a wall makes the space inside accessible.

router,[15] infinitely thin surfaces have no material property and are therefore difficult for a machine to understand. But when you are sketching, forms are really only defined by their boundary edges anyhow. The topology of a sketch model can and should be much looser, much freer of restrictions.

SketchUp uses what is called a winged-edge[16] data structure to store the geometry that you sketch with it. This structure stores vertices (points), edges (lines) and faces (surfaces), and the adjacency that they have to one another quickly and compactly in memory (Figure 2.55).

To efficiently model in a winged-edge data structure, you draw lines from point to point until a coplanar loop is closed, which then creates a surface automatically. Surfaces created in this way have a directionality, a front or back surface (called a "normal," indicating the normal vector direction perpendicular to the plane of the surface) that is derived from the order in which the boundary lines are stored in memory. By convention, winged-edge data models follow the "right-hand rule," which says that lines connected endpoint to endpoint in a counterclockwise direction have a face normal that points toward the viewer. SketchUp's ease-of-use is directly related to this data model (Figure 2.56).

In SketchUp, faces can have any number of sides. Additionally, faces can have as many holes in them as you might need. Surprisingly, these two characteristics of SketchUp geometry are still somewhat unusual in computer graphics. Though, of course, they feel quite natural to people accustomed to sketching by hand or modeling from real materials like cardboard.

Building Objects and Other Higher-Order Primitives

Construction teams don't build polygons; they build walls. Undoubtedly, the advantage of AEC-specific design software is that it helps designers get to the constructible level of detail faster and more efficiently. A building that is fully simulated before construction, down to the last bolt, is one that is quicker and more efficient to build. This is undoubtedly the direction our industry as a whole is moving. It is not, however, a complete solution by itself (Figure 2.57).

Building information modeling (BIM) is a process in which quantifiable objects are assembled into systems that together represent the building as a whole. There are established taxonomies of these objects and rules-based practices that describe how they may be used in great detail. This is all very useful in most cases. But before all of that, there

---

[15] Where 3D printers create models by building material up layer after layer on an empty build platform, computer numerical control (CNC) devices more commonly mill material away from a solid block. CNC routers are even more difficult to program than 3D printers, and they require extensive preprocessing of models before they are ready to be carved.

[16] Following his work on BREP modeling, Bruce Baumgart developed what he called the "Winged Edge Polyhedron Representation" to address some of the complexity of full BREP modeling kernels. The original documentation of this model can be found in an ARPA project memo (Baumgart 1972).

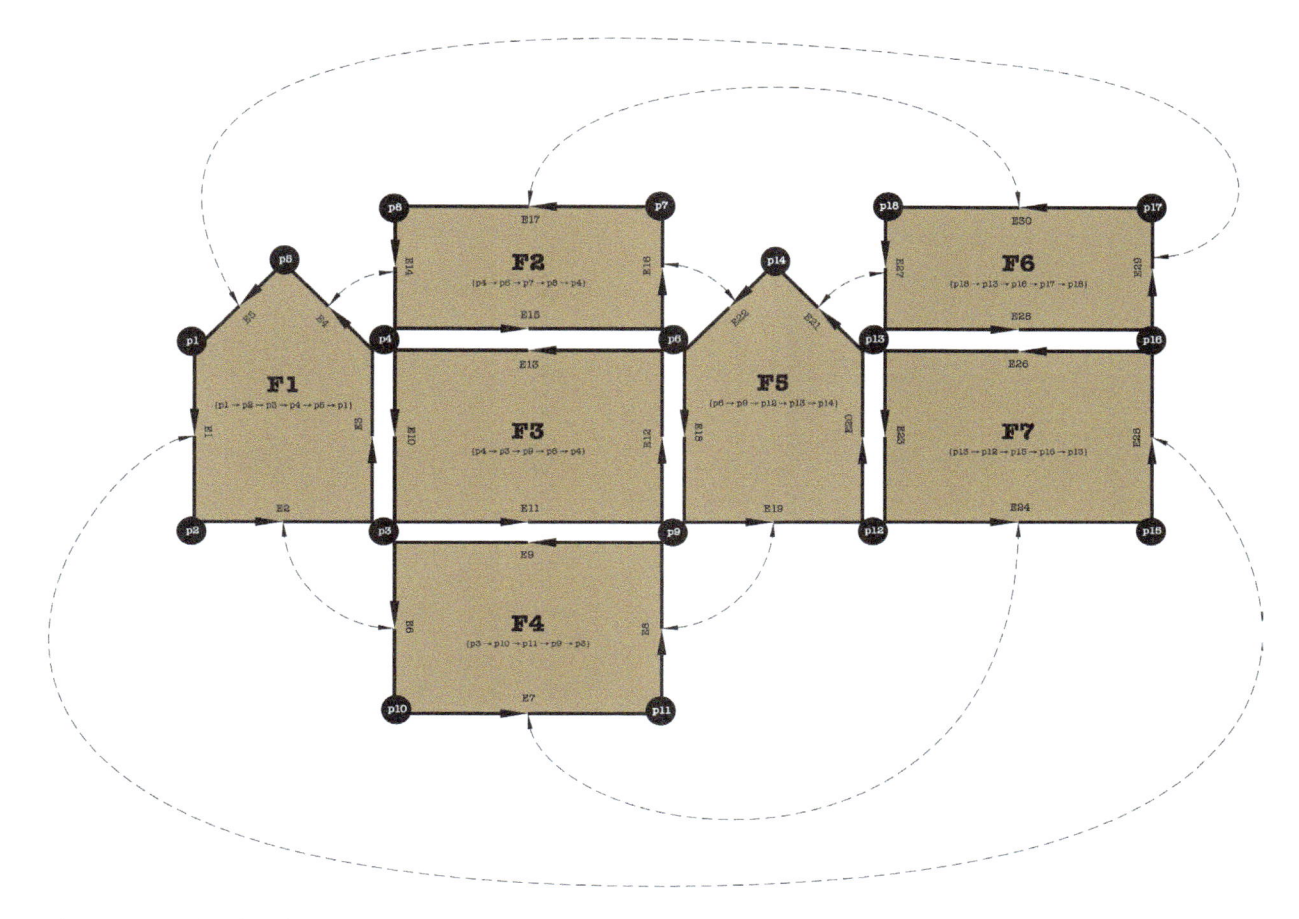

Figure 2.55: SketchUp's winged-edge data model.

is still a sketching phase. There is always a need to figure out, broadly speaking, what the building wants to become. And I fear as professionals, we forget how to do that part first.

If you can put aside your enthusiasm to jump to the highest level of detail right from your first (probably dumb) design idea, you'll design a better building. Premature detail makes your design process rigid and hard to change. But it need not be so. If you have a simple-to-use tool that automates the creation of detail with little extra work from you, then it should be just as easy to make a change as it is without the detail. But be cautioned – it is so tempting when you see a model that looks superficially like it is complete, to assume that there was more thought put into the creation than there may genuinely have been (Figure 2.58).

I so frequently hear from frustrated architects who, when reviewing construction document sets put together by newly minted experts in one of the monolithic[17] BIM applications, find that none

---

[17] Revit or ArchiCAD, mainly. These applications are quite similar in their approach to design modeling, leaning heavily on higher-order building objects and 2D drafting automation.

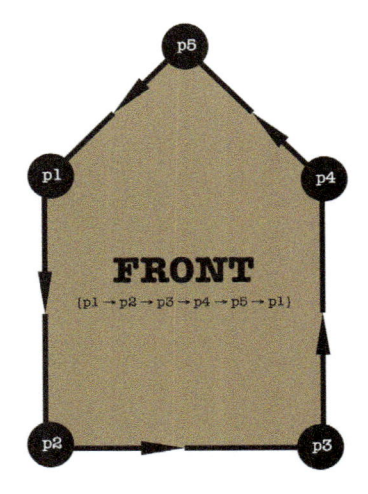

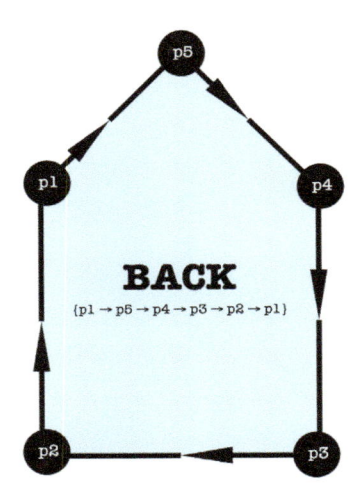

Figure 2.56: The "right-hand rule" in SketchUp defines the front and back sides of a face.

of the carefully presented details in the project actually work. The drawings look great, but the building is unbuildable. Too much detail committed too soon, and now it feels like there is a lot of rework to do.

The best description of a real thing is the thing itself. A brick is best represented by an actual brick (Figure 2.59). You can get everything you need to know from the brick. You can understand its color, its dimension, and how much it weighs. You can

Figure 2.57: A collection of chair models, available for SketchUp on 3D Warehouse.

Figure 2.58: "Ceci n'est. pas une pipe." With apologies to René Magritte, a 3D model of a p-trap from 3D Warehouse.

even know intangible things about it – how it smells, how it tastes. How hard it is to break and what it feels like if you drop it on your toe. You really can "know" the brick when you have it in your hand.

Everything other than the real brick in your hand is a projection, a simulacrum presented to you at some level of abstraction from some predefined frame of reference. Even the photo of a brick in Figure 2.59 is only an approximation of the real brick. The description of a brick must be consciously understood to represent "brickiness" from some point of view, in the service of some agenda. You might have a description of the brick that you can read on the manufacturer's website. Or maybe you have a verbal report from your client of a brick they saw once in a magazine. It could be that your contractor has a guy he always buys his bricks from, so he's sure that's what you'll want on this job too. Or, possibly, your digital design tool has a "standard" representation of the brick that makes it really easy to draw in your model. All of these alternatives may provide you the right information for your design when you need it. But all of them will be someone else's definition, not yours (Figure 2.60).

What kind of information do you need about the brick when you're sketching? What aspect of it is the one that you need to understand right at the beginning, before you know anything else about it or how you're going to use it in your project? You need the right information, at the right time, in the correct format, at the right level of detail, about the brick. Absent any other strategy, start as simple as

Figure 2.59: A photograph of a real brick.

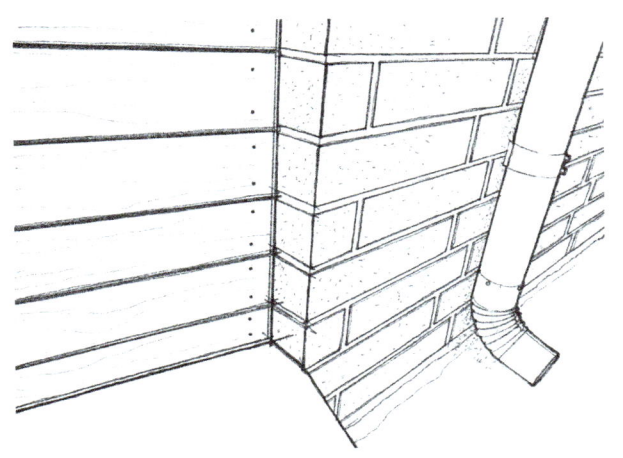

Figure 2.60: A brick sketched in context.

you can. In time, as decisions are made, alternatives considered, and a path forward clarified, specificity can and should enter your design work. But there has to be a logical progression, from sketch to geometry to quantifiable objects with richer attribute data attached for further production.

The simplest building objects to consider are those objects that are purchased from a manufacturer and installed in the project (e.g., a bathtub or a rooftop fan coil unit, anything that can be picked out of a catalog, purchased, delivered, and installed). Typically, there are few (if any) user-definable configurations of these kinds of objects available. Maybe you can pick a finish color or a fan size. But really, you are just choosing from among the manufacturer's available alternatives. Fixtures are the most natural things to place in a model without overthinking them.

Of course, you likely won't know which fan coil unit is going to be the right one when you're sketching. It might be helpful to know about how large the unit might be when it is eventually chosen by an engineer later on in the design project. As the designer, you need to design a screen, at least, to hide it from the neighbors. But if you know enough about the design of your building to know that rooftop fan coil units are going to be needed, you'll be better able to include a plan for their location early enough to integrate a solution into the building's overall design, rather than just glomming something on later.

You may want to use some attractive light fixtures in your design work at the beginning as well, and this leads to a different problem. If you use a beautiful mid-century modernist classic (like maybe those great artichoke lamps from Poulsen[18])

[18]You know the ones – they are found in an embarrassingly high number of renderings. And in a far smaller number of actual built projects.

to spice up a design presentation, it may make your design look better than it actually is. And then, when the inevitable cost-engineering stage replaces that lamp with something nice from IKEA, instead, your design might not look so great anymore. Design presentations always look better when they have some detail, and there are reasons why those great pieces of someone else's design look so great in every project. Adding them too early, when you probably have no idea if they can work with your client's budget, will be a crutch. And it will distract you from working on the parts of the design over which you have real agency.

Generic fixtures are the best when you're sketching (Figure 2.61). You don't want an object that looks too good, too specific, especially not something that your budget can never afford. You don't want to get distracted by detail, and you don't want to let that detail force you into making decisions you're not ready yet to make. You want to use objects that are detailed, but not so much that they distract you from looking at the parts of your project that you are actually responsible for designing from scratch.

Light fixtures deserve some special consideration if your design is one in which lighting, natural or artificial, is something you think of as a defining design characteristic. If you can simulate light while you're sketching, you are in a stronger position. When drawing by hand, some quick work with a white pencil or a bit of gauche can be enough to key your thinking about light. When working digitally, you have full-featured lighting simulations at your disposal in even the simplest of rendering engines. But just like any digital assist to design, you can easily find yourself mired in settings and configurations, tapping your toe impatiently as you wait for only

Figure 2.61: Generic model vs. specific model; which is the right level of abstraction?

one more rendering to finish. I will talk about designing with light simulations (rendering) in a later chapter, but for now, consider carefully how much lighting detail you really need when you're sketching.

Fixtures in a sketch can be indicated by something as simple as a point with a tag on it that shows broadly what class of object should be chosen for that location in the future. A rooftop fan unit can be just a box. A light can be a single polygon, maybe with a line indicating the direction light will be thrown. Any piece of furniture can be drawn effectively with fewer than a hundred polygons in a highly general way.

Giant repositories of building objects exist on the internet today, all of them full of detailed components you can collage together as you see a need to do so for your design. If you are using SketchUp (or even if you aren't), 3D Warehouse[19] is an excellent place to start your search. You may even want to pull together a collection of your favorites, the objects that most communicate to you, and yet are generic enough to prevent problems with your design later on in the process (Figure 2.62).

But always be wary of adding too much detail to your model too early. It will give you a false sense of just how complete your design really is. And if you're working digitally, you can really kill the performance of your modeling environment by adding unnecessary detail. But sometimes, a really great model of just the most essential thing to think about right now in your project can be incredibly helpful in moving your sketches forward.

---

[19] 3D Warehouse is a free-to-use repository of user-contributed 3D models in SketchUp format representing millions of different real-world (and imaginary) objects (Trimble Inc. 2020).

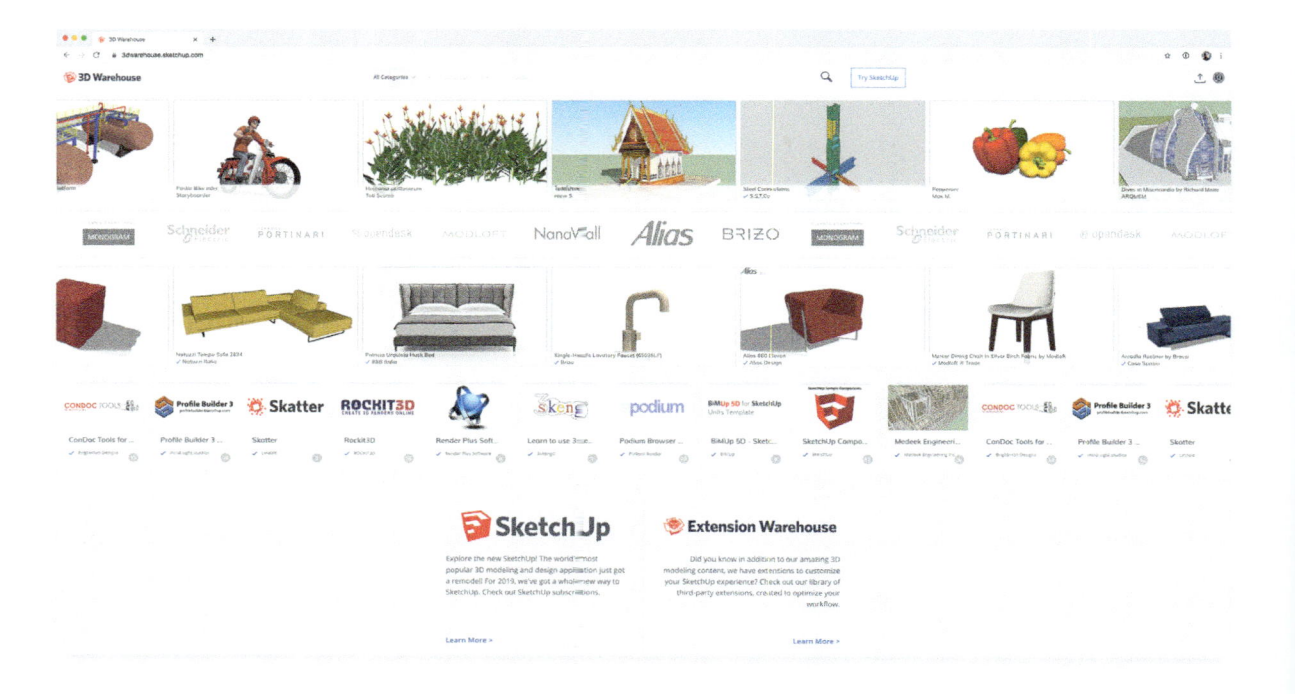

Figure 2.62: Trimble's 3D Warehouse.

chapter 3 Representations of Space

In concert with any basic formal grammar of shape and form must inevitably be considered the means of representation that allow you to both understand and explain your geometric concepts to yourself and others. All spatial design work, even when it exists only in the mind of the designer, is encoded and understood in some spatial projection system. This is so embedded in your design practice that you don't even think about it anymore. But it bears reconsideration in a new digital context.

The means of projection are not passive, even though they may not be fully visible to you. There is no single best and most natural system of representation. They all have some historical baggage, and they all bring some unique tactical affordances. Each also has its own problems, particularly concerning the distortions they impose on the viewer's point of view. You should learn each of them well enough that you can choose the right one for the particular sketching that you are doing. They are simply tools in your toolbox, like any other. But be sure you understand what they represent and how to make them work for you and your design goals.

While the various projection systems I'll discuss in this chapter are widely applied across many professions, architectural subjects demand some special considerations. For one thing, buildings have both insides and outsides – so your projection

Figure 3.1: The size of a building vs. the size of a piece of paper.

system needs to be able to accommodate that. Most importantly, however, you have to think about scale more than other professions might. And for a simple reason – architectural works are larger than all pieces of paper (or computer screens), which might be used to depict them. To represent a building, you are going to have to project it into a smaller space than the real world, as shown in Figure 3.1.

The notion that objects might be understood only through projection is an old one. Like many old ideas, it has become so embedded in our collective subconscious as to be basically invisible to ordinary practitioners who may even depend on being able to understand it and function in it every day. Your understanding of the graphical projections you use every day has become innate – but also, therefore, it may be something you have forgotten how to think about. Perhaps, however, we benefit from digging for a minute below the surface of this. To truly appreciate the benefits that come from working digitally, you need to know how to leverage the unique ways of seeing they depend on.

The earliest recognizable reference to a graphical system of projection that I know about is found in Plato's *Republic*, written around 375 BCE, in his allegory of the cave. He suggested that objects (or "Forms") in their purest, most complete, and truest existence might be best understood as though visible to humans only through the shadows they cast on a cave wall. There are prisoners in the cave, chained in a way that prevents them from turning their gaze anywhere but toward the wall in front of them. Behind them, a series of objects pass by, illuminated by fire and casting shadows on the wall that the prisoners can see. All they can ever see is the shadow of the objects, and all the conclusions they reach about them can come only from that information. Objective truth is made available to them only through projection (Figure 3.2).

Like the pure forms in Plato's cave, the ideal, perfect form of your work may also be hidden from your collaborators, tucked away in your imagination. You have a privileged point of view. You can

Figure 3.2: Plato's allegory of the Cave.

Figure 3.3: Hand shadows, From "Le Magasin Pittoresque," 1861 Le Magasin Pittoresque.

see the complete and perfect form with your mind's eye because it is yours to define. You know more about it than anyone else will ever know. You get to choose how it will be projected on the cave wall.

Imagine what a projection system that used only a single light source to describe form might include? How would it work? And what kinds of things might you learn from using it? How much information about a 3D form would be lost through a system that allowed you to see only the shadow cast on a wall? And at the same time, with a little strategic forethought, you can learn quite a lot about a form from only its shadow. Shadows alone can tell excellent stories (Figure 3.3).

Since at least the sixteenth century, geometers have known how to project the profile of an arbitrary object as a shadow on a plane. Albrecht Dürer demonstrated it in his 1525 *Painter's Manual*.[1] His technique, still graphically relevant today, defined shadow volumes by casting a ray from the light source through each of the critical points of the object to be described. This basic technique has many interesting applications in automatic 3D model extraction from photographs. It also forms the basis for the ray tracing found in all photorealistic rendering software. But for now, let's consider this precedent as an inspiration for many more sophisticated and capable systems that followed in history (Figure 3.4).[2]

Descriptive Geometry

While no longer regularly taught in architecture schools, the principles of *descriptive geometry*[3] underpin all of the more commonly used systems

---

[1] Dürer's *Painter's Manual* was printed twice in his lifetime, but has subsequently been reproduced countless times since. (Dürer 1525)

[2] As it happens, SketchUp's real-time shadow casting algorithm also derives from Dürer's seminal work.

[3] For a good general introduction to descriptive geometry, consider (Paré 1997). It was the textbook we used at Cooper Union in the early 1990s.

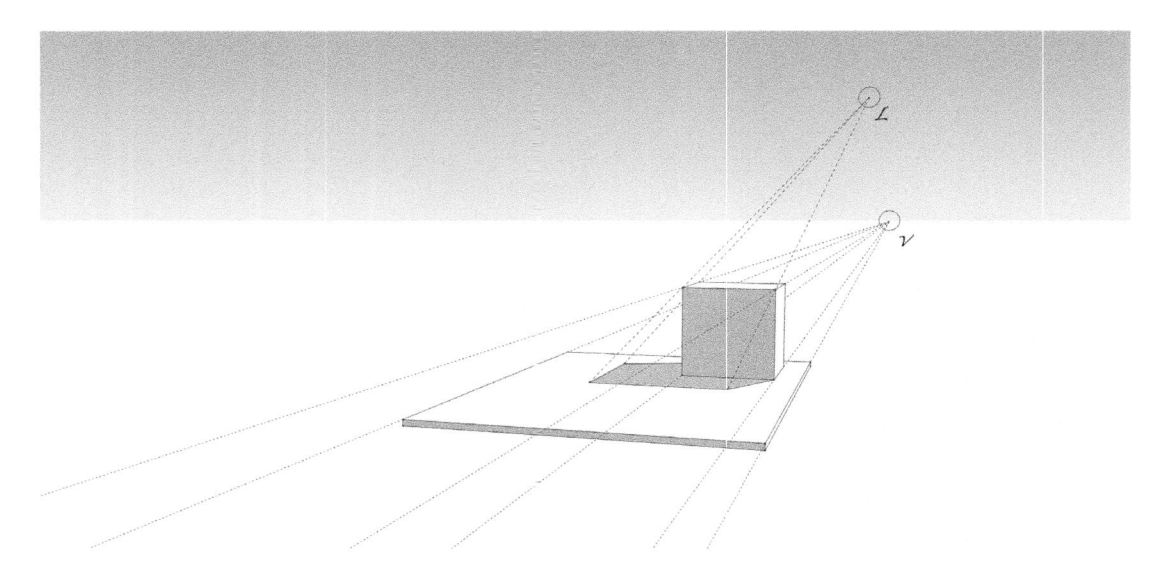

Figure 3.4: Albrecht Dürer's shadow casting algorithm, as recreated in SketchUp.

of projection that you are familiar with today. Descriptive geometry is a system of geometric constructions that allow a draftsperson to depict three-dimensional objects in two dimensions at true size with only a compass and straight edge in their toolbox. A 10 cm cube, for example, could be represented through descriptive geometry so that each edge in the cube might be represented on paper with a line that measures at 10 cm in length (Figure 3.5).

Descriptive geometry provides methods for rotating objects in space and preserving or (more importantly) displaying various features in their "true lengths." The system depends on the presence of a three-dimensional object and a two-dimensional plane of projection. The viewer in the descriptive geometric system is only able to see the two-dimensional projection and therefore, must be given the best and most complete version of it that is possible.

The methods of descriptive geometry can be wholly accomplished with only the simplest of drafting tools – a compass, pencil, and straight-edge. The techniques are purely graphical – no additional computation needs be performed on a piece of scratch paper beside the drafting table. Graphical transformations are accomplished by a series of spatial reflections across a plane perpendicular to the projection plane (Figure 3.6).

Descriptive geometry as a projection system is particularly useful when representing small, discrete objects (like machine parts), which are composed primarily if not exclusively from either flat surfaces or surfaces with continuous curvature like spares, cylinders of conic sections. More complexly curved surfaces like automobile bodies are difficult (though not impossible) to draft manually with descriptive geometric methods. For that sort of surface, you must look to different techniques. And while it can theoretically apply to objects of any scale, architectural scales where the point of view is inside an object can be tricky to visualize.

In descriptive geometry, the primary goal of the draftsperson is to remove any distortion from true

DESCRIPTIVE GEOMETRY

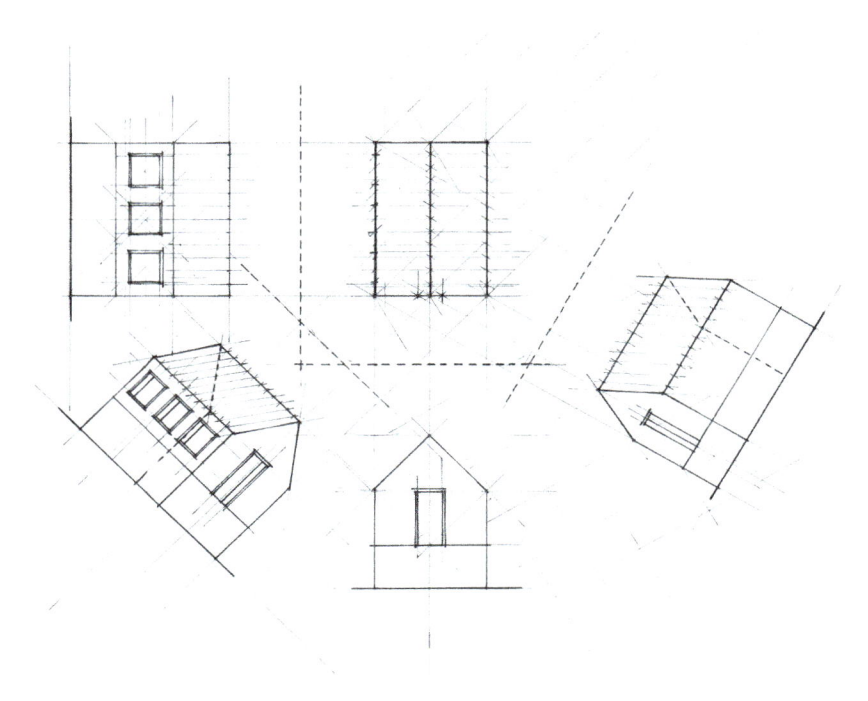

Figure 3.5: An example of descriptive geometry used to rotate an object.

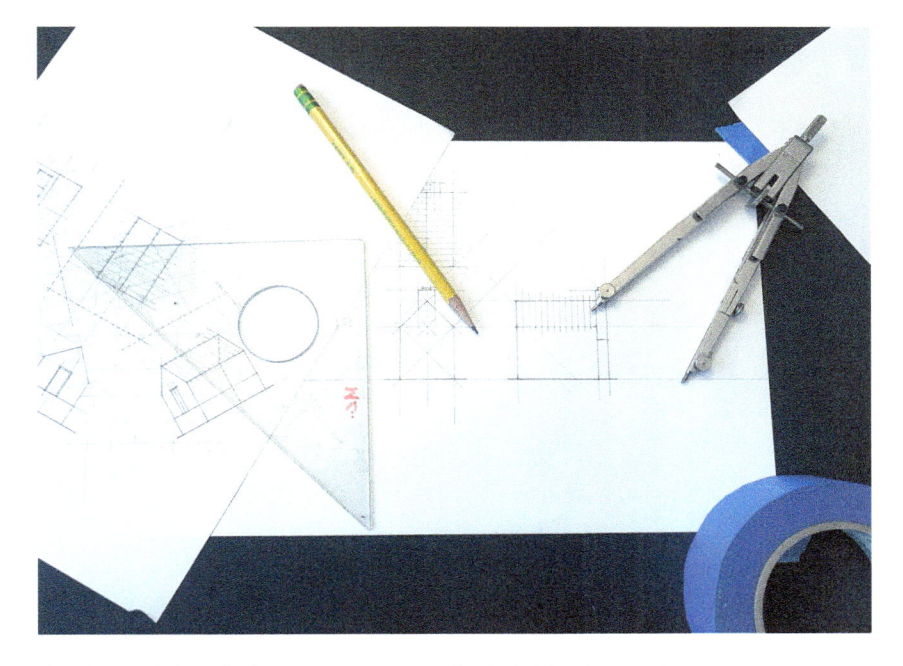

Figure 3.6: The technology of descriptive geometry: pencil, straight-edge, and compass.

length or true size, which might come from an edge displayed obliquely to the projection plane. Such edges are foreshortened and no longer reliably measurable by the viewer. For this reason, a series of standardized projections where the projection plane is parallel to a majority of surfaces on the represented object have entered into the architectural tradition, where they have become collectively known as the orthographic projections that you commonly refer to as "plan," "elevation," and "section" views (Figure 3.7).

But before we get there, let's give some more detailed consideration to the way that you use scale in any architectural drawing. Because by definition, descriptive geometry only works with lengths at true length. Alternately, you might think of descriptive geometry as being in a sense "scale-less" in that it only represents lengths in a 1:1 ratio with the lengths of edges in the projected model. Buildings have an annoying truth of being larger than any common projection surface (for example,

a piece of paper), and so we have to come up with a scale-able method to represent them.

Architectural Scale

When you are drawing a building by hand, as in your sketchbook, while you are enjoying a pleasant afternoon out on the Piazza del Duomo in Florence, it is likely not something you took the time to measure accurately before drawing it. More likely, you engaged in a complex set of decisions about proportion and adjacency that developed over time, improving with every new line you laid down. Of course, your sketch of the building is undoubtedly smaller than the original building (unless you have a really large sketchbook), but not in anything but a relative way. Many urban sketchers try to include people in their scenes as a kind of scale orientation. People generally understand how large a human figure is and, therefore, can judge the scale

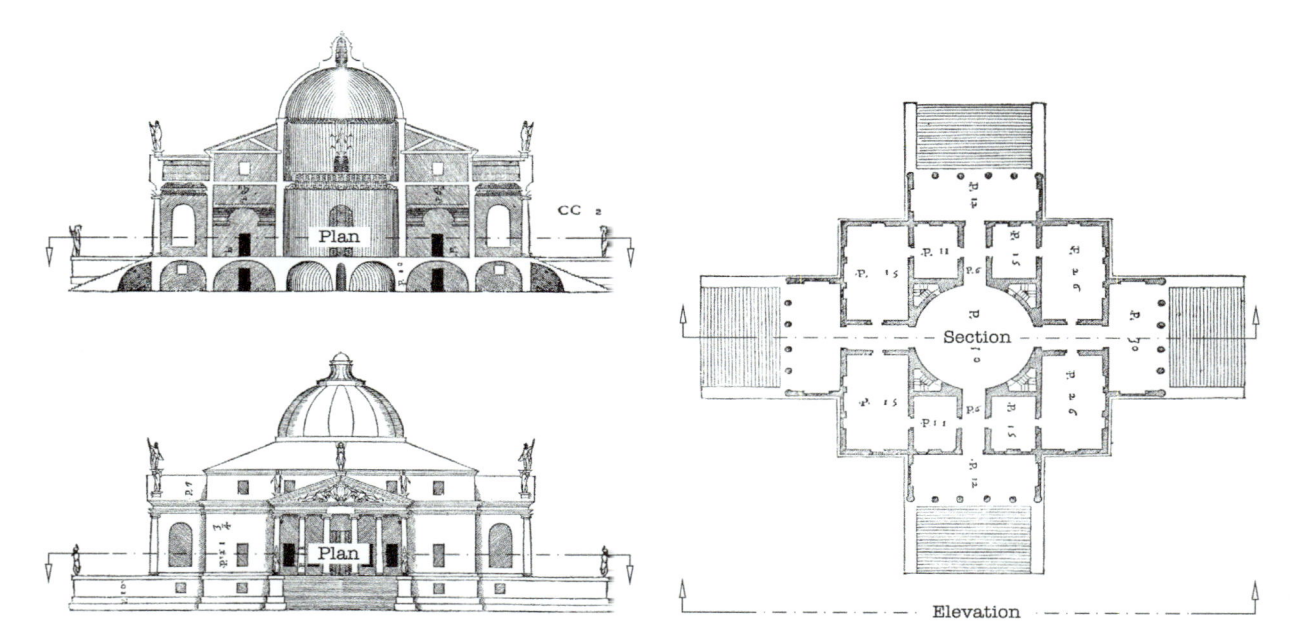

Figure 3.7: Plan, elevation, and section compared; Palladio "Villa Rotunda."

ARCHITECTURAL SCALE

Figure 3.8 Sketching at architectural scale.

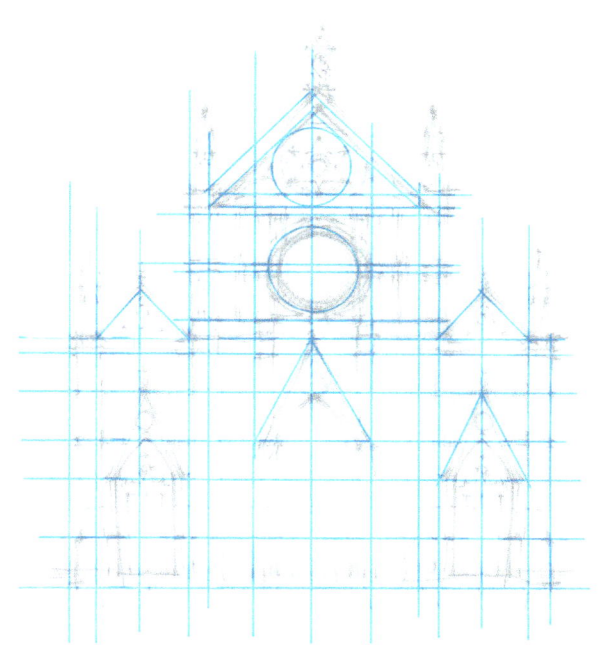

Figure 3.9: Understanding the proportions.

of other things in the scene relative to that reference (Figure 3.8).

But when you start to work on more detailed plans with more critical dimensionality, you need to start thinking about things in true scale, not just in proportion. Architectural scales are formal ratios between the true length of a feature in the real world and the representation of that feature on your smaller sheet of paper. Given a trusted scale reference on paper, you can infer the true size of other adjacent features base on their relative proportion (Figure 3.9).

You might think about this in terms of degrees of freedom. In a traditional sketch, in pencil on paper, the scale of the drawing is fixed. You can't stretch the paper to make it bigger or smaller. In a digital system, however, you can zoom to any scale you want. Scale doesn't mean as much in a CAD system as it does on paper. You are perhaps accustomed to managing this with the odd abstractions of "model space" and "paper space." Scale makes even less sense in a 3D modeling viewport. Tellingly, many of the earliest CAD systems were, in fact, completely unitless, relying on the user to define whatever measurement system they wished.

The freedom of dynamically zooming in, out, and panning around in a drawing is not available when you are working on a piece of paper. You can look closer, of course, if you need to see some detail in the drawing, and this is the experience that the original designers of CAD systems were trying to simulate. But it can become quite challenging to retain your inner sense of the scale of things as you zoom around so be very careful about this while you're sketching digitally.

Traditional architectural scales are something that likely was baked in your subconscious as a student. You know, if it is how you were trained, that

Figure 3.10: A graphical scale reference.

a drawing at 1/8" = 1' scale has specific properties that help you judge size and distance in it without needing to resort to a measuring device. You become accustomed to judging distance intuitively at scales that you know. When you're sketching, this is critically important to understand. You need to be able to tell if there is enough room in your plan sketches, for example, to fit a staircase, or a doorway, or a bathroom. You need to be able to judge this without resorting to a ruler or some other measuring tool (Figure 3.10).

Architectural scales also help you to manage the level of detail effectively (Figure 3.11). A 1/8" = 1' scale drawing, for example, cannot show any feature which is smaller than the scaled thickness of a line in the drawing. You would never show where flashing details go in a wall section at that scale. On the other hand, a detail view of a wall section at 1" = 1' scale would show all the components of the wall clearly. But you would never draw the whole building at that scale, so you will naturally tend toward drawings that show only a vignette.

Users of CAD systems that allow infinite panning and zooming easily forget these two fundamental principles of abstraction in drawings. First, they have difficulty judging the true scale of objects they draw, particularly at the beginning of a design when there is little surrounding context. Ideas may feel like they work proportionally, but the designer may later find they aren't scaled in any reasonable way. I have had this problem before – that I have begun sketching freely in the seemingly infinite void of a digital modeling environment only to discover later that the carefully detailed representation of a building I was making was actually over a thousand feet long. . . or, in another case, less than an inch. If you have no scale reference when you're sketching, scale may feel irrelevant. Until it isn't.

The more difficult challenge comes from a failure to understand the appropriate level of detail for the work at hand. It becomes possible, even probable, that the designer will get lost in details while sketching digitally. If there is no sense of scale in the drawing, then there's no penalty associated with getting too caught up defining the little details. Of course, there's nothing wrong with working out details early in the project if that is important to your overall concept, but it can be a problem if, in doing that detailed work, you forget to figure out

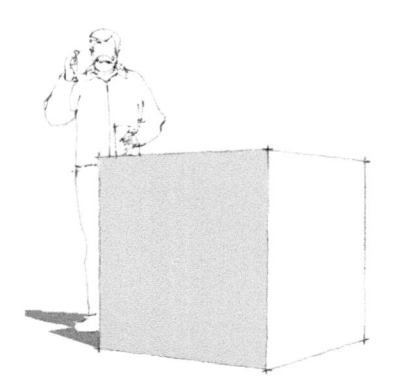

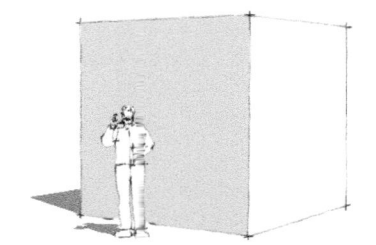

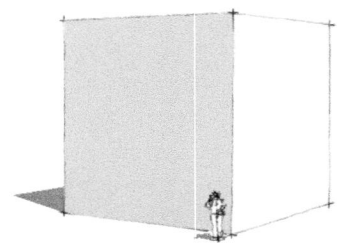

Figure 3.11: How big is this cube?

the big stuff. Or if you become unable to reverse detailed designs that you committed to before the rest of the design was ready.

Scale is an important design tool, and it is one that, while intuitively apparent in traditional sketches, is easy to lose track of it in digital sketching environments. Unless you are designing at 1:1 scale, you should never lose track of the size of your work. Always use some scale that makes sense to you, something that fits in your professional milieu. Be aware, of course, that traditions vary regionally – the 1/8″ = 1′ scale that makes intuitive sense to me won't help your colleagues in Europe, with their 1:100 scale drawings. So keep that in mind.

If you don't have an effective way to manage scale in your digital drawings, then you should always work at full scale. Give yourself a scale reference, some object that has a size you intuitively understand. In SketchUp, we give you a scale figure, standing next to the coordinate origin in every default modeling template. Before you have drawn anything else in the model, you can still judge the scale of your scene by judging proportionally against the size of a real human being.

Cutting Planes

Architectural forms are also unique in that their insides and outsides are equally essential to visualize. Cutting planes expose the interior when representing three-dimensional forms that include both an inside and an outside. Designers of mechanical parts that are never experienced from the inside don't typically have to worry about such things. But for sketches of buildings, a cutting plane of some orientation is almost always used. Plan views are derived from a horizontal cutting plane and sections from a vertical one (Figure 3.12). Even roof plans and exterior elevations, which show only the

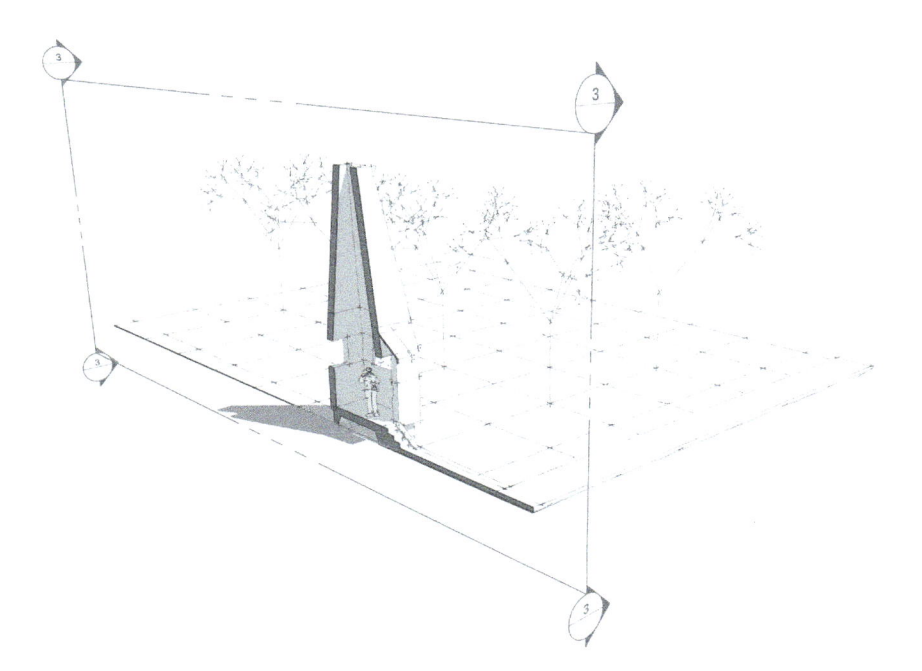

Figure 3.12: Standard cutting planes.

outside of a building, are actually derived from cutting planes.

Cutting planes work from an abstract notion that the world is made of some material that can be cut through cleanly to expose its internal properties. Some materials are rendered transparently, like air or glass. Others are rendered with opacity, like structural materials of wood, steel, or concrete. Some sectional views consider only a difference between solid and void, filling the inside of cut solids with a single fill color. When greater detail about constructability matters, sectional views may show solid material types with representational hatch patterns.

Cutting planes are a powerful design tool, valuable from your earliest sketch to most precise construction details. They give you a supernatural power to understand both the interior and exterior of your design at the same time. Penetrations from inside to outside, as, for example, doors or windows, give your building's users a glimpse at this special point of view, but without the privileged oversight that you have with a full sectional view.

Orthographic Projection

Orthographic projections of space are idealized abstractions that preserve true dimensions, in scale, while minimizing perspectival distortion. This is accomplished by passing a cutting plane through the space of the project parallel to and projecting plane, projecting project details beyond the cut directly onto the picture plane. In an orthographic projection, all dimensions parallel to the cutting plane are drawn at true scale (Figure 3.13).

Encompassing plan, section, and elevation views, orthographic projections are a natural abstraction of space that predates the invention of perspective by thousands of years. Orthographic projections can be found in drawings throughout ancient Egypt.[4] Possibly the earliest identifiable architectural drawings, carved in clay tablets in ancient Mesopotamia, are orthographic in nature (Figure 3.14).

If you are sketching by hand today, you are likely sketching orthographically. With a roll of trace and an architect's scale by your side, sketches of plans and elevations are among the fastest ways to churn through the essential spatial components of any architectural project. In plan, a single fat line drawn with a black marker can quickly progress a simple network diagram to a representation of a credible structure with walls and floors. Sketching in orthographic projection is easy, fast, and natural.

Plans

Plan views of a building represent the three-dimensional space of a building as projected to a plane parallel to the floor, cutting through the vertical walls of the building and showing them in their true length. By convention, plans implement a cutting plane through the walls somewhere around waist height, showing both doors and windows in sectional views. If plan sections are taken through a part of a building with built-in furniture (cabinetry, for example), that is conventionally shown in as though it is below the section cut (Figure 3.15).

Plan views are a useful abstraction when thinking about paths of movement through a building. While a plan view offers you a god-like overview of the building as a whole, your users will experience the plan only by moving through the space you have designed. To them, the building's space is experienced by moving around in it. Perspectives might be closer to something they can understand,

---

[4] For example, Egyptian carpenters used basic orthographic projection in documents relating to their clients' commissioning of custom furniture for fabrication (Killen 2017).

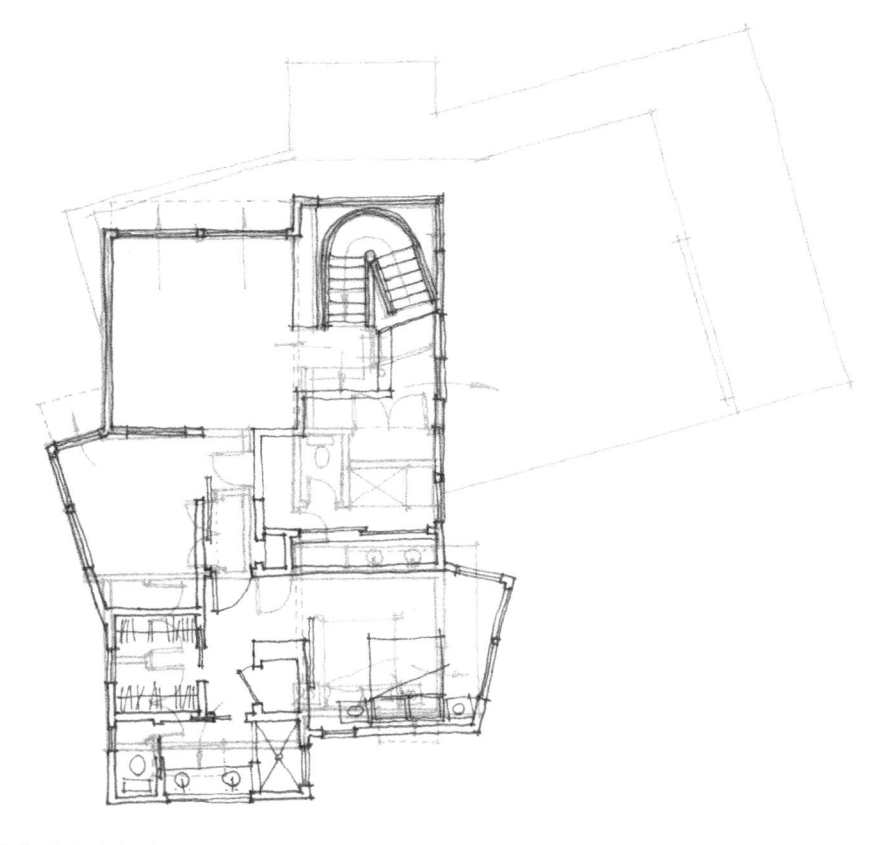

Figure 3.13: Detailed sketch plan.

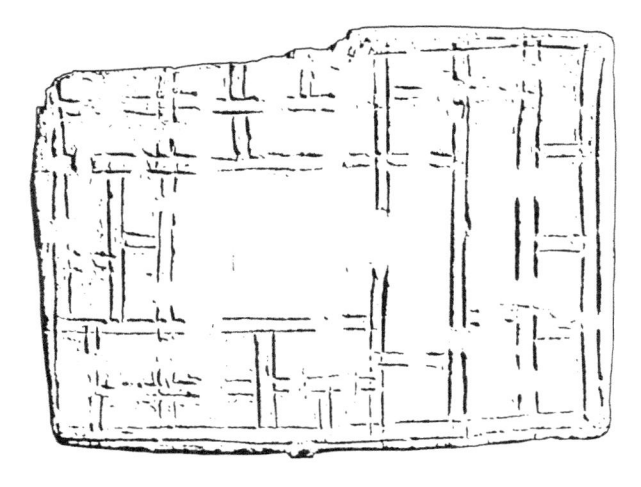

Figure 3.14: "Plan d'un sanctuaire ou d'une maison privée," AO 338 Musée du Louvre.

though they have problems in other ways that we will discuss in the next section.

Plans are tough for the uninitiated to understand, and must typically be accompanied by some other representation when explaining the unbuilt project proposal to its future residents. For an architect, the plan is often the most useful way to think about the building. They will go to a plan view before almost anything else. But for most people, a plan view is too much of an abstraction. They just can't convert from a plan to a habitable vision of what the project will become when it is ready for them to move in. Regardless, there is no better projection to help you understand the sequence of experiences in a building. The experience of

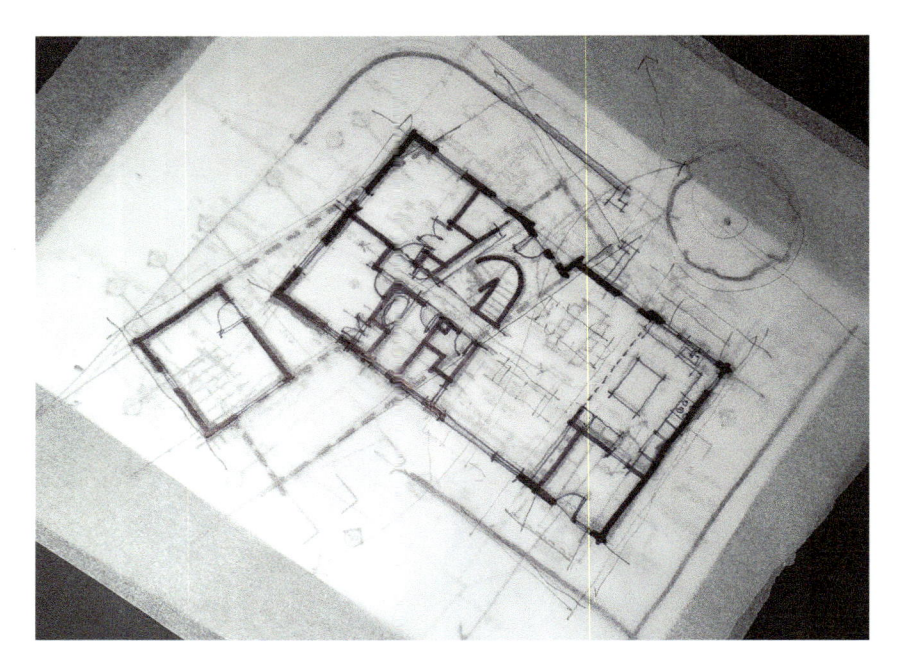

Figure 3.15: A pile of plan sketches.

space is more natural to inhabit with a plan than almost any other projection you can imagine.

If you don't know where to begin on a new project, sketching in plan is never a bad idea. Many design projects start with a diagram exploring the forces that shape the site and the relationship that programmatic elements have with one another. More often than not, these diagrammatic relationships can be "hardened" into a reasonable first approximation of a plan for the building with only a few quick strokes of a pen.

You may think of sketching in plan as being "only" 2D problem-solving. Working in a full 3D model does make sense in many cases, and I'm certainly an advocate of using modeling to sketch early in your design process. But even when modeling in a rich 3D environment, you still need to be prepared to think primarily in plan projection. To elevate the walls of a model, you'll likely need to draw them in plan first.

Elevations and Vertical Sections

Elevation views of a building represent the three-dimensional space of the building projected onto a cutting plane perpendicular to the plan. Where plan views are oriented according to gravity (typically looking down; up only in the case of a reflected ceiling plan), elevations can be oriented in any direction. Gravity typically points down in an elevation, with all vertical dimensions represented at true scale. Typically, elevations are also oriented parallel to a particular wall system so that lengths can also be shown in true length, avoiding wherever possible distortion due to foreshortening (Figure 3.16).

You may choose to make elevation sketches when you are trying to work out built-in casework (bookshelves, cabinets, etc.) or when designing a building's facade system, with window or curtain-wall designs to be figured out. Elevation views are

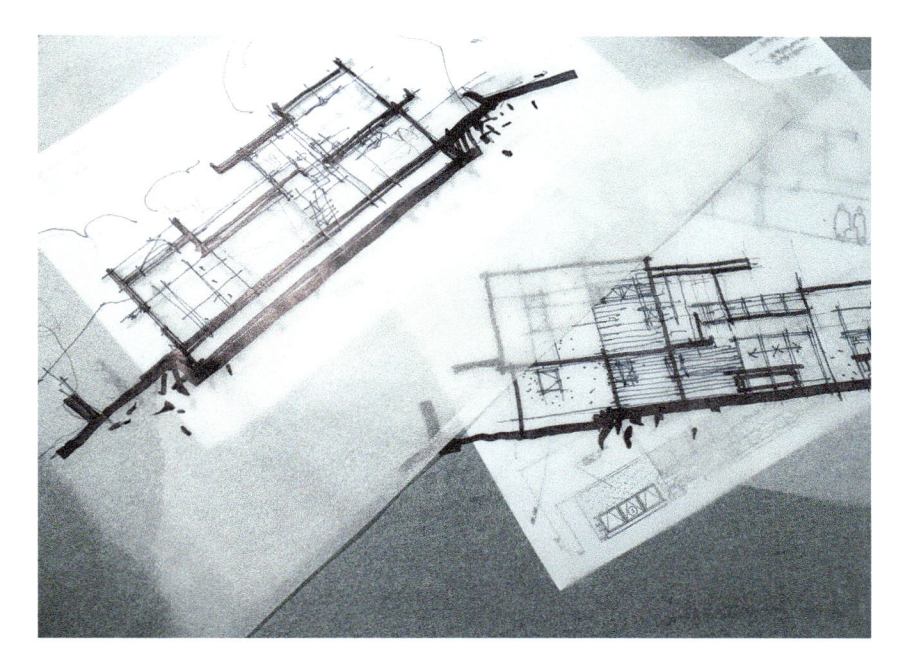

Figure 3.16: A pile of sections and elevations sketches.

perfectly suited to exploring complex rooflines or the vaults inside them.

For parts of your building where vertical circulation from floor to floor needs to be worked out, a section is invaluable. Elevations and sections are different only in the location of their cutting plane. With a cutting plane that passes through a stair tower or elevator core, you can quickly come to understand the vertical relationships between things. Similarly, if you need to understand how the structural frame for your building will work, a judiciously chosen section view can show you exactly where the lines of force will web down to your building's foundation.

If plans allow you to sketch through the spatial experience of circulation in the building, through an exploration of the sequence of programmatic function from one connected space to the next, then elevations are closer to the visual experience of what the space will look like while circulating. Elevations are perpendicular to the viewer's natural gaze. It is not hard for most people to imagine they understand what the space will look like based on a set of interior elevations. Exterior elevations are similarly easy for most people to understand as *"what the building will look like."*

Oblique Projection

Before the invention of perspective by Brunelleschi and his circle in fifteenth century Florence,[5] the most naturalistic representations of three-dimensional space were without the

---

[5] According to Vasari's *The Lives of the Artists* (Vasari 1550), Brunelleschi demonstrated work in linear perspective as early as 1415. He did not publish results directly, but his results were documented by Leon Battista Alberti (Alberti 1435) in his book *De Pictura* (*On Painting*).

Figure 3.17: A curious vignette from "The Kanxi Emperor's Tour of the South."

concept of a vanishing point. Parallel edges in space were drawn in parallel on paper, permitting a moving point of view, which encouraged the viewer to explore the scene more freely. Where perspective projections favor a single point of view – the point in space from which the perspective is constructed, axonometric projections allow the viewer to roam a bit more freely through the space of the projection. There is no singularly privileged point of view to which things can be either showcased or hidden from view.

Prior to the introduction of perspective from European missionaries in the sixteenth century, oblique projections were widely used to depict space by artists throughout Asia. Nowhere is this more evident than in the work of seventeenth century Chinese painter Wang Hui, whose 72-foot-long documentation of the Kangxi Emperor's Tour of the South is delightfully filled with the most complex oblique representations of space.

David Hockney (1988), in his short film *A Day on the Grand Canal with the Emperor of China*, picks out several of these key moments (Figure 3.17).

Unlike perspective projections and the physical cameras they rationalize mathematically, oblique projections are often more experiential in nature, perhaps more illustrative than rationalized in common usage. That said, in some formal projections ("cabinet," the French "cavalier," or the top-down 90 oblique popularized by the New York Five[6] architects, for example), they can represent edge lengths in true scale, and in this way are also more analytical than a perspective (Figure 3.18).

---

[6] As named in Drexler (Frampton 1975), The New York Five included Peter Eisenman, Michael Graves, Charles Gwathmey, John Hejduk, and Richard Meier. They favored axonometric projections in the documentation of their work, especially (in Hejduk's case) the 90° projection.

OBLIQUE PROJECTION

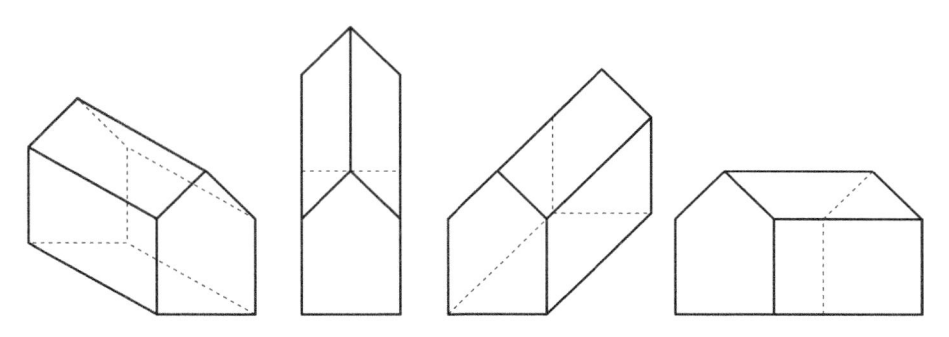

Figure 3.18: Formalized axonometric projections.

In the simplest definition, an oblique projection is one in which parallel edges remain drawn parallel to one another in projection. Perpendicular edges, however, may vary in angle relative to one another from style to style. In the most useful oblique projections for sketching, at least one face in the object is drawn in true dimension in at least two axes. If your first sketch was a plan view, then an oblique projection is quick to accomplish by rotating the plan to some comfortable angle and projecting walls straight up on the page (Figure 3.19).

Oblique projections may, in some respects, be an unreliable depiction of the true dimension of all aspects of the projected subject matter. Still, they are much more delightfully capable of depicting space in a way that is easy to mentally inhabit. There are fewer privileged moments in these drawings, and therefore it is easier to imagine oneself wandering freely throughout the depicted space (Figure 3.20).

Among my favorite moments in Wang Hui's scroll is a depiction of the formal entry gate to the city of Suzhou, where the emperor will be spending the night. Arrayed in front of the gate, ready to greet the emperor, is a crowd of people. Because of the oblique nature of the projection, it is simultaneously possible to peer over the heavily fortified walls of the city to glimpse the daily life of those within.

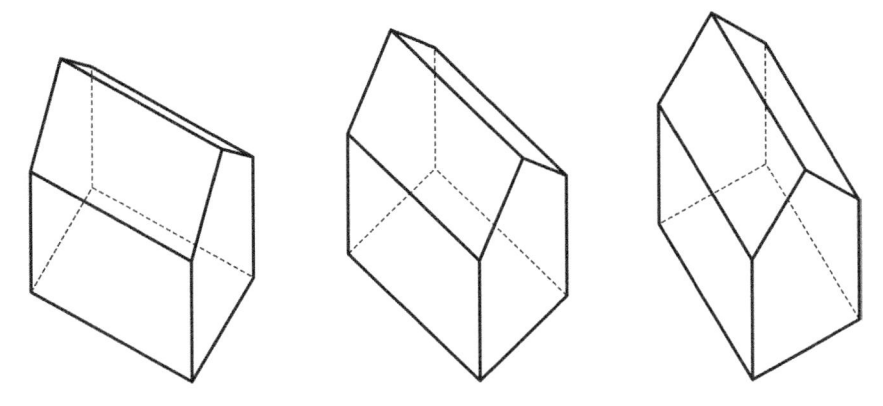

Figure 3.19: Military projection.

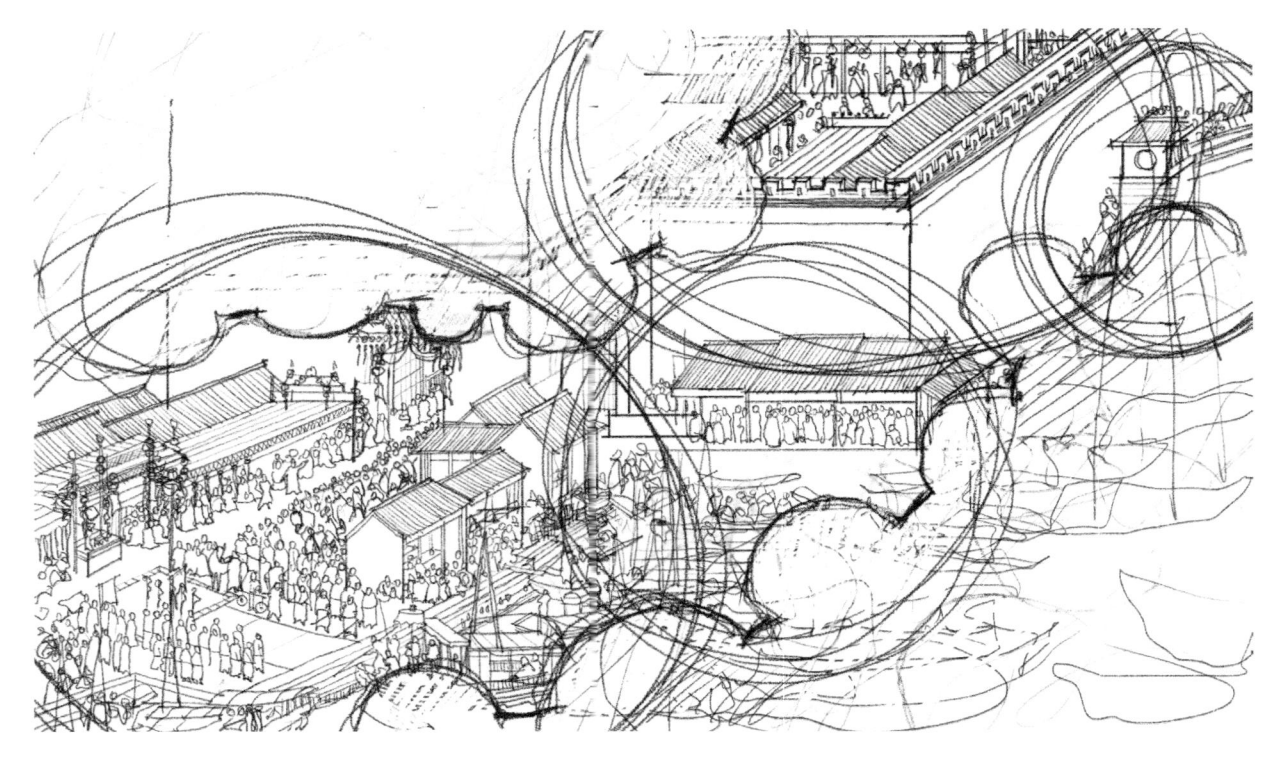

Figure 3.20: Sketching vignettes in the emperor's arrival.

In a traditional perspective view of the same scene, it would have been impossible to see over the walls in quite the same way. The viewer can simultaneously understand both the pomp and ceremony of the emperor's arrival and the everyday work of the city, making its preparations behind the scenes (Figure 3.21).

Similar representations can be seen in the early renaissance work of thirteenth-century Italian artist Fra Angelico, notably his "Scenes from the Lives of the Desert Fathers" (or "Thebaid") (Mugello 1420 c.). Fra Angelico recognized that while in visual observation, parallel lines could be seen to converge as they receded into the distance, they did not need to all converge together into a single vanishing point in the depicted scene. Instead, by allowing for multiple points of convergence, a more vibrant narrative was possible in the scene – one in which the viewer was rewarded for allowing their attention to wander.

Rather than a single idealized perspective, paintings like Angelico's "Thebaid" contain a network of discrete vignettes, each independently consistent as geometric projections, with architectural forms staging a scene, acted out by figures enclosed within the projection. A roaming gaze across the entirety of the painting is rewarded with a series of individual scenes playing out. But taken as a whole, there is another geometric order that becomes apparent, with some fascinating geometric projection in the connective spaces between the vignettes.

You should allow yourself similar degrees of freedom when sketching. Soon enough, the cold realities

Figure 3.21: Detail sketched from "Theabid," Fra Angelico.

of physical construction will assert themselves onto your process. Still, it is essential to remember that you are designing not just a physical space but also the experience of the space and its effect on the future occupants. You are equal parts engineer and storyteller for your projects, and a skillfully executed axonometric sketch can capture that in a way that no other projection can match.

When sketching by hand, oblique projections are a natural extension from orthographic plans, sections, and elevations. By picking a third angle of projection from any orthographic, a quick spatial representation can be constructed. You can do this quickly and without losing properties of either scale or proportion. In a presentation, an oblique projection is more natural for an unsophisticated viewer to understand and inhabit.

Axonometric projections are common in digital 3D graphics systems, and they are straightforward for graphics systems to rationalize in software. While you may naturally gravitate toward perspective projection, I challenge you to consider the alternatives. All 3D modeling environments with even rudimentary control available for their camera simulation are capable of removing perspective distortion completely, leaving a catalog of oblique projections available to you.

Perspective Projection

The history of world architecture since the Renaissance has been written by those for whom the perspective projection best represented the human experience of space. Audiences in sixteenth-century Florence marveled at Brunelleschi's demonstration of perspective in front of the Baptistry; never before had such a realistic representation of space been seen (Vasari 1550). Audiences were awed by the precision of representation, indistinguishable from the actual scene in front of them.

Even today, four centuries later, we still live in a world dominated by perspective projections (Figure 3.22).

There are some disadvantages in perspective. Perspective projections have a unique quality in that they have a fixed relationship between the viewer and the vanishing point. As the viewer's location changes in space, the perspective projection must be recast to accommodate. In the simplest perspective case, a one-point projection cast at the end of an endless road, the viewer may travel forever and never reach the destination.

Perspective projections can also feel somewhat more constricting in presentation. The point of view, the viewer's location, is only fixed in space; static even. Where a viewer can allow their experience of a space to freely roam through an orthographic projection (being as there is no special presence afforded any singular vantage point), viewers of

Figure 3.22: Albrecht Dürer's explanation of perspective.

PERSPECTIVE PROJECTION

a perspective projection may feel there is something unique and special about their point of view. If you are designing a building with long axial vistas, perspective projections tell a great story. If you are designing a more modern free plan, they may serve your needs less well.

If the Renaissance invented perspective, it was the invention of cheap photographic cameras that cemented perspective into the modern designer's gestalt. With a simple press of a button (OK, it is never quite that simple), any designer can capture a realistic perspective representation of the world in front of them. Once captured, the perspective can be extended, edited, and enhanced just by tracing over it. Easy. And *easy* means you can do it frequently, experiment with it, and learn from the experiments. If you are thinking in perspective for your design work, a camera, even the one in your smartphone, should never be far from your side (Figure 3.23).

There are many different kinds of perspective projection, and it is worth exploring some of the most popular variants in a bit more detail. Begin with the linear perspective constructions of the Renaissance, which are still some of the easiest and most common projections, and are likely the ones you learned to do by hand at an early age. In these simple linear perspective cases, you must establish a horizon line (by convention, it is flat and horizontal, located at the eye height of your viewer). On this horizon, you place a vanishing point (Figure 3.24).

One-point perspectives are great because they are easy to draw. Also, they are not uncommon in typical architectural design contexts – as down a hallway, or through a window to a scene outside. One-point perspectives are the quickest kind of perspective to sketch by hand. But they are also among the most artificial perspectives, and the most likely to exhibit the projection's more problematic qualities.

Figure 3.23: Tracing a photo in perspective, digitally, on an iPad in Procreate.

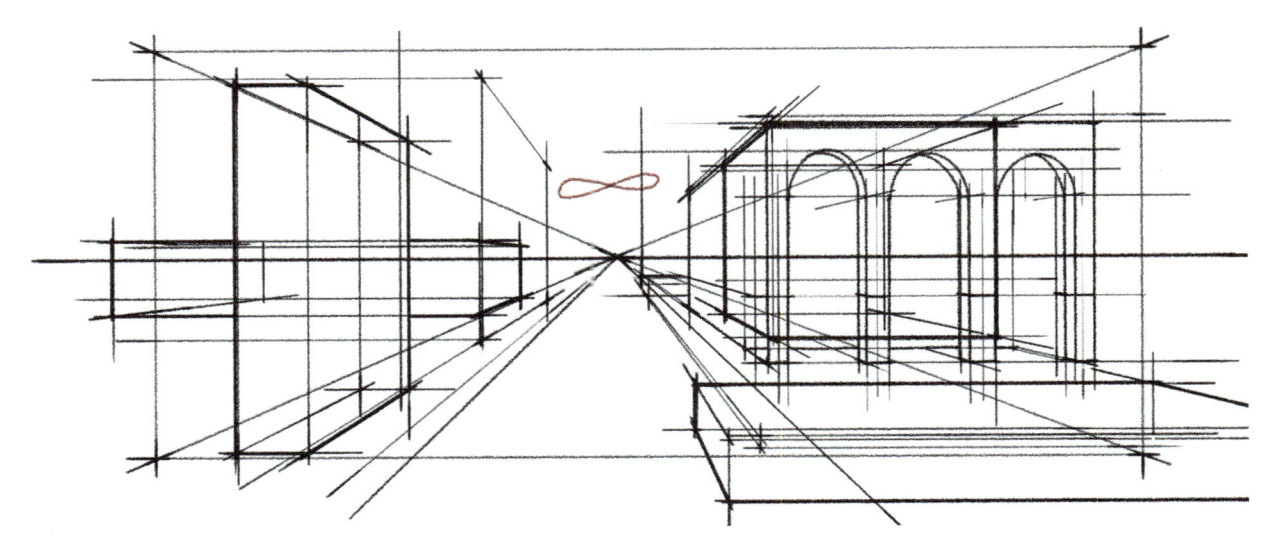

Figure 3.24: One-point perspective, simple case.

One-point perspectives are only really accurate when the viewer's gaze is on an axis exactly parallel to the axis of the majority of parallel lines in the scene. And they really start to look funny if the vanishing point is located anywhere but the geometric center of the projection plane. Any divergence from that will necessitate the addition of a second vanishing point in a two-point perspective projection. The "two points" in the name of this projection refer to two vanishing points, both located on the horizon. In a two-point perspective, your viewer's point of view no longer needs to be located strictly in the center of the page – you have a bit more freedom to experiment.

There are still problems with this kind of perspective, however. Your point of view can move more freely, but it is still assumed to be fixed to the horizon. This is fine if you are representing a scene in a way where the viewer might naturally be able to see the horizon. But if you move their gaze above or below the horizon, you must add an additional vanishing point to preserve the illusion. Three-point perspectives include three vanishing points, one representing each of the Cartesian coordinate directions.

There is a common misconception that human eyes operate identically to cameras – and that by studying cameras, we can build effective rationalizations of human vision. This is increasingly understood in scientific circles as more than a little naive.[7] The camera-like characteristics of your eyes are only a small part of your complete vision system. Vision is tempered by memory, by experience, and by expectation. Perspectives sketches alone cannot capture this complexity. But your eyes are similar enough to cameras that a perspective traced over a photo will remain convincing to anyone who looks at it (Figure 3.25).

Human eyes, like camera lenses, actually capture what is referred to as a spherical perspective, with infinite numbers of vanishing points distributed circularly around the center of your field of view.

---

[7] For an exhaustive analysis of this, Erwin Panofsky's "Perspective as Symbolic Form" is a good starting point (Panofsky and Wood 1997).

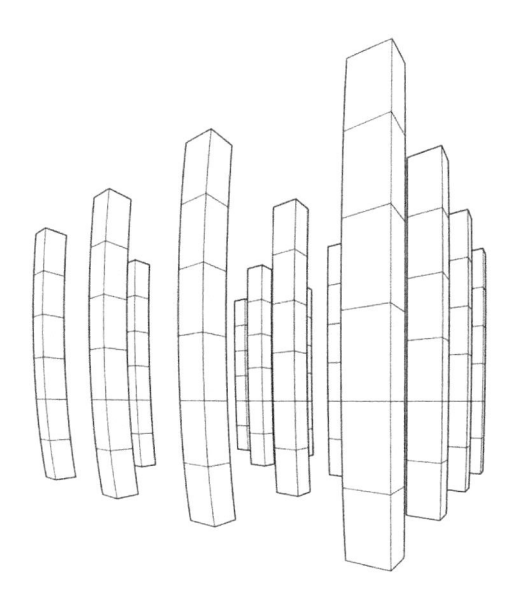

Figure 3.25: Constructed perspective (two-point) compared to camera lens perspective distortions on the right. (Adapted from Panofsky and Wood 1997).

With a camera, this is easy to model. It is the center of focus for the camera's lens system, and by convention, it is found at the center of the camera's rectangular sensor. Human optics become complex when you realize that they are not projecting light on a planar sensor; rather, they are projecting it on the hemispherical surface that is your retina (Panofsky and Wood 1997, especially Section 1, p. 33).

Computer graphics systems, starting with Silicon Graphics' IrisGL[8] included an easy-to-implement perspective system that simplified the projection of the abstracted 3D model in the system's memory to the screen so that it could be appreciated by the system's users. In doing so, favoring this over other possible systems that might have been less rationally implemented, the die was cast, and (I would argue) designers began to shift inexorably into design languages that more naturally fit into perspective projection (Figure 3.26).

These hesitations aside, perspective projections are immensely useful in a design context, perhaps even more in their digital forms than when sketched by hand. Perspective projections are laborious to lay out by hand. They are trivially easy to access digitally. In fact, they are the default projection in most 3D modeling applications.

It is hard to find an architectural illustrator today, even among masters of traditional art like Steve Oles (1979), Mike Doyle (2007), and Jim Leggitt (2010), who would still think of laying out a perspective by hand. All of them use digital modeling techniques today to block out their renderings before committing to a hand rendering. This is true throughout the entertainment industry as well, where concept artists are habitually doing the same thing. OpenGL makes it trivially easy to experiment with perspective views of a 3D model,

---

[8] IRIS GL (Integrated Raster Imaging System Graphics Library) was the predecessor to OpenGL, running only on Silicon Graphics computers and the IRIX operating system.

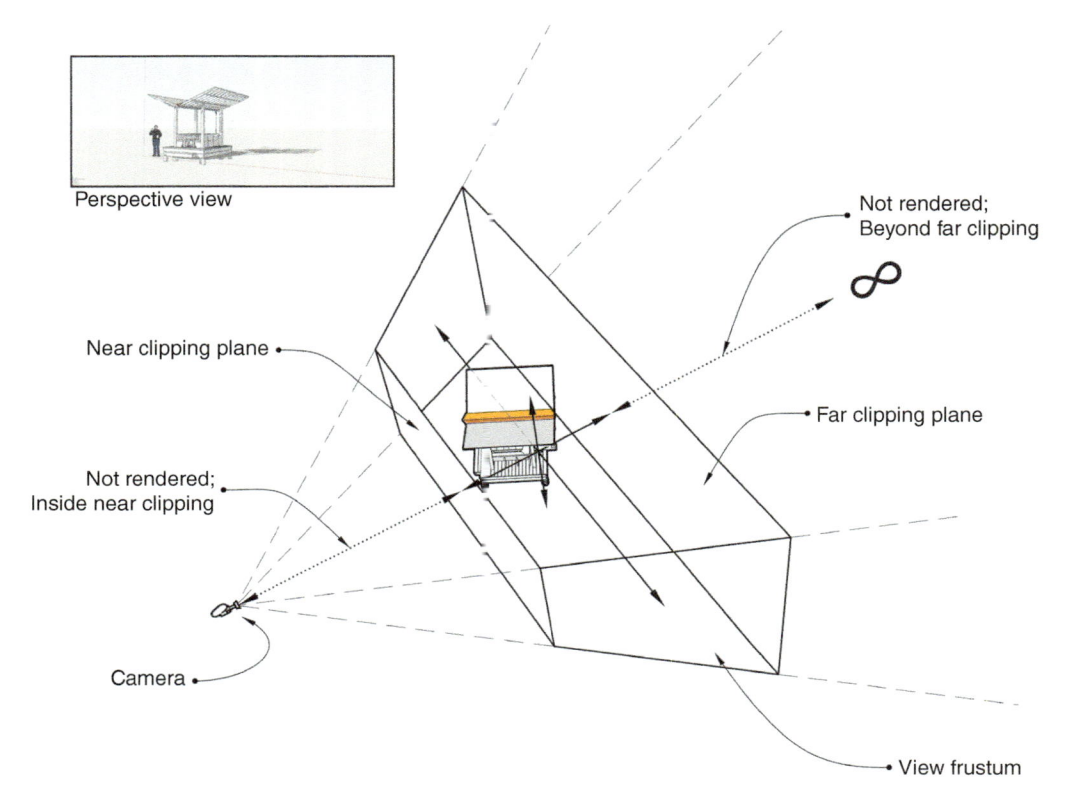

Figure 3.26: A typical OpenGL perspective construction.

fluidly moving the point of view until just the right composition is discovered.

It is this ability that I think allows digital perspective to transcend the inventions of the Renaissance. By automating the construction of perspective and allowing the viewer to change their point of view fluidly, we have invented a new kind of perspective, a dynamic perspective that is easier to inhabit and easier to explore for untrained viewers.

Cinematic Perspective

If a perspective projection is like a still camera, then cinematic perspective is different only in that the camera is allowed to move on a prescribed path. Just as cranes and dollies freed Hollywood's physical cameras to begin moving through the sets they were filming in the 1920s, architects are free to create animations simulating the cinematic phenomenon of traveling through a building before construction.

Cinematic perspectives are scripted, and they retain the notion of a primary point of view, which the viewer cannot change. The only difference is that the designer can move that point of view through time, and by doing so can open up the viewer's understanding of the space more than they might form a single fixed point. The experience of architecture, considered cinematically, is demonstrably different from the experience from a sequence of static views. A moving camera is one that acknowledges time as an independent dimension.

Russian film director Andrey Tarkovsky (1986), in his book *Sculpting in Time,* wrote, *"...the infinite cannot be made into matter, but it is possible to create an illusion of the infinite: the image."*

It is difficult to imagine an architect taking the time to assemble this sort of representation with drawings created by hand. Even the cell-animation powerhouses (with fleets of skilled animators) were loath to tackle animation sequences where the background changed radically from frame to frame. There is just too much drawing to do to really consider doing it by hand.

With digital assistance, however, it becomes not only easier to assemble cinematic perspectives, but in fact, it is easy enough that a designer might rationally include this kind of projection in their sketching phase. For example, knowing that a particular approach to the building is a given (driving down a long entry drive, for example) before knowing exactly what the sequence of geometric forms along the way (gates, sign pylons, framing structures, etc.) might become. A standard path could be established, then you could play freely with everything else in the scene (Figure 3.27).

Cinematic perspectives, "walkthroughs" taken from a human perspective, or "fly-throughs" if imagined from the air, are powerful storytelling tools for a design presentation, and might, therefore, be worth spending some of your precious design time to bake out just right before the big presentation.

Free Perspective

Where cinematic perspectives rely on a predetermined path for their moving point of view and perhaps also a scripted sequence and pacing for that movement, free perspective projections do not.

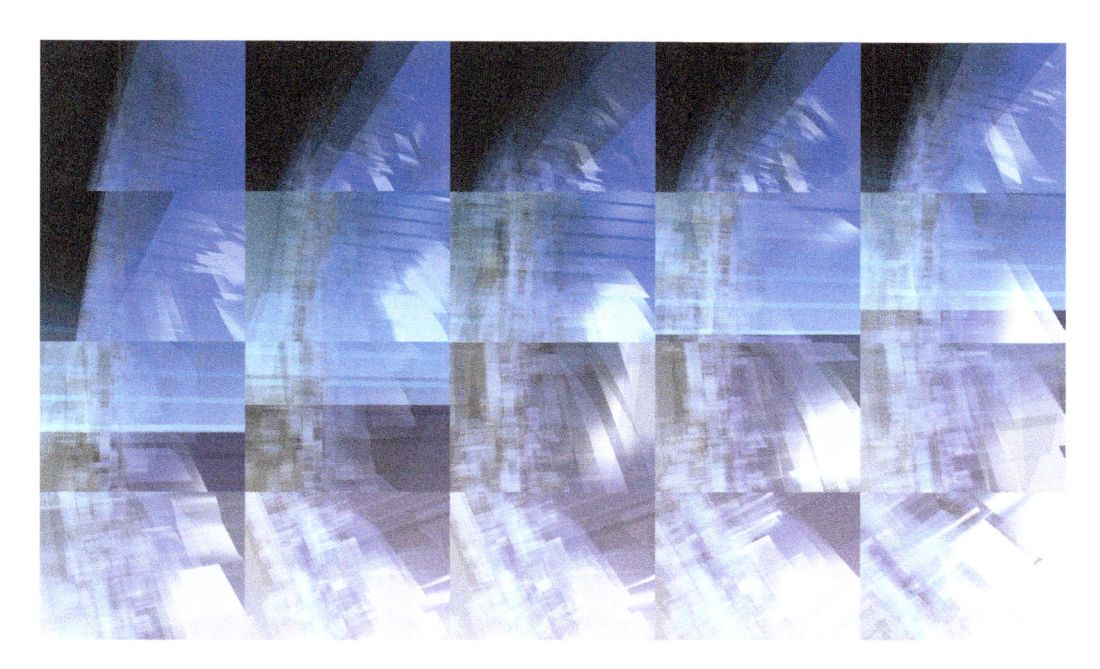

Figure 3.27: Frames from a walkthrough animation (Miles Davis Archive project).

They put the viewer in control of their gaze, freely allowing the viewer to control where they are looking in what feels to them like a fluidly continuous space of exploration. Free perspectives are only practically possible to present in a digital system. A manual draftsperson couldn't hope to be able to keep up with the demands of a viewer who was really exploring.

An individual designer, however, sketching through a particular sequence of design questions about a project, might well assemble a collection of perspective views. One view might lead to a question about what is around the next corner, about where the design might develop on the other side of a door or through a window. This kind of linked spatial consideration, while sketching, can be quite effective. And when considered as a set, this sequence of perspective sketches might provide a cohesive and freely explorable story of the project for a viewer.

Contemporary digital modeling environments depend on free perspective visualization. In fact, they include predefined interactive tools that may support many different ways of "navigating" the model, matching a variety of experiences that match prior experiences navigating the physical world, because in the physical world, a free perspective is strongly similar to the way we, as human beings, navigate through real space. We are able to walk from place to place, to look up and around. We can pick up objects and tumble them around to examine them from different angles. Our natural visual senses are, in fact, attached to mobile platforms – our bodies are capable of hauling our eyes around from place to place as we freely explore our environment (Figure 3.28).

The simplest and most common digital analog for this kind of experience is a freely orientable 3D perspective projection onto the flat projection

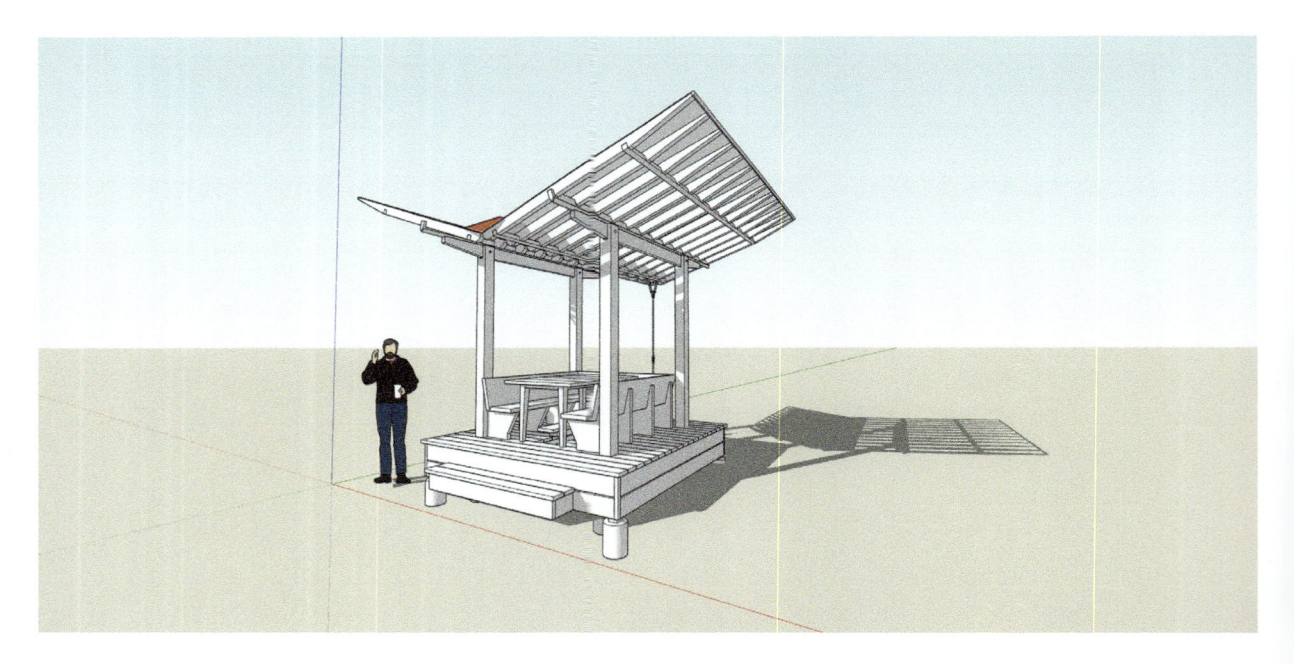

Figure 3.28: A typical 3D modeling viewport, free perspective.

plane that is your computer's monitor. With a collection of view manipulation tools (commonly cast as "camera" tools), you are free to explore the virtual space of your model. The system can smoothly interpolate frame by frame the changing perspective projection as your virtual camera moves through space. The system is still behaving in an essentially cinematic way – with the computer calculating a sequence of frames and blowing them onto your screen as fast as your hardware will allow.

Stereoscopic Perspective

With a free perspective, in its most capable digital expression in a 3D modeling environment, you are still stuck to a flat projection plane, as though observing the digital world through a glass window. Your computer's monitor allows the view to change dynamically at or beyond the visual fusion rate for your eyes in a way that makes the illusion of 3D space behind the screen quite convincing. But you are still interacting with your design through a single picture plane. It is easier to mentally inhabit the space of your sketches than it might be in a static perspective or some other more experiential oblique projection, but there are still limits to be overcome.

The human visual system is binocular in nature. We have two eyes offset from one another widely enough to allow, through parallax, for our minds to infer depth in space through direct physical experience. A projection system that matches human binocular gaze, particularly one that completely wraps the full human field of view, could be quite compelling.

With a stereo perspective that matches the optical geometry of our vision, it will feel as though the viewer is physically embodied in the projection. Everywhere you look, you will see your project.

Imagine what it would be like to interact directly with it, moving walls and windows around freely as though you were actually inside the space itself. This will, one day, be the most powerful possible way to imagine architectural space.

Stereoscopic viewers have been available to designers for decades (Figure 3.29). In fact, they were quite popular as parlor entertainment early as the 1850s. In principle, stereoscopic projection requires only a matched pair of perspective images rendered from two points of view separated horizontally by a few inches. If the images are focused at the same depth of field and in all other ways matched, it is possible to fool the eye into experiencing depth in an image.

There are a variety of devices that help to make the experience more comfortable for untrained viewers, but it is possible to experience a stereo image pair without any additional equipment. I'll print an example in this book for you to try. Focus your gaze on the black dots below the image in Figure 3.30 and gradually cross your eyes until you see the two dots first double, then converge on a new dot in the middle of your vision. Once

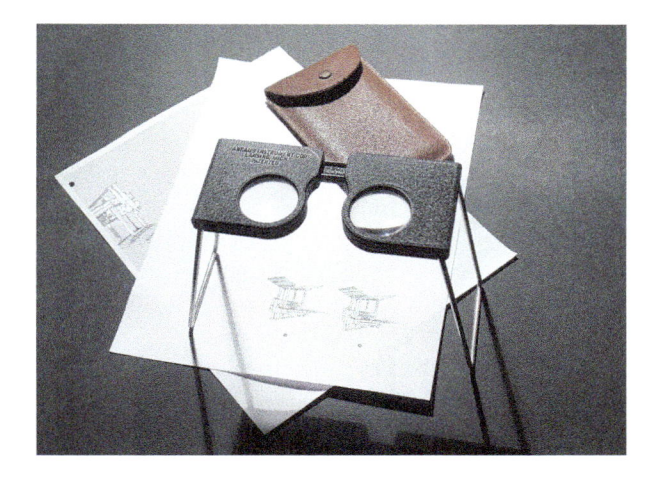

Figure 3.29: A mechanical stereoscopic viewer.

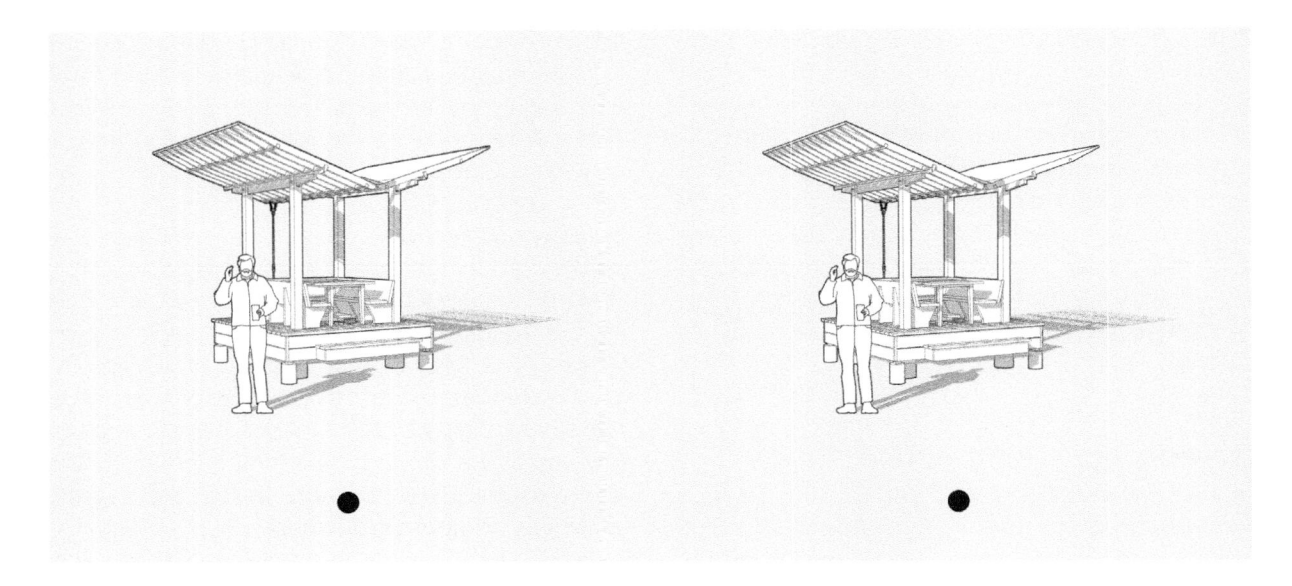

Figure 3.30:   A cross-eyed perspective projection, generated with SketchUp.

you have converged the dots, shift your gaze up to look at the model. With practice, you will see it pop into 3D.

With a stereo viewer, it is easier to force each eye to see only the appropriate image for it, and as a consequence, it becomes easier for casual viewers to experience the 3D effect. But even the simplest of technical solutions, even a hand-drawn sketch on paper with pencil, can be used to make a stereoscopic perspective projection.

But these images are, in the end, still just static perspective images. They are no more effective in a sketching practice, really than a parlor trick. Where stereoscopic perspectives begin to get really interesting is when they are paired with a fast digital modeling system that frees the viewer's point of view and allows them to explore the space of a project dynamically. There are, however, some technological problems still to solve.

Head-mounted virtual reality displays have had a new renaissance in the last 10 years,[9] but their conceptual design has been known for 50 years or more. The first stereoscopic digital displays were pioneered by Ivan Sutherland at Lincoln Labs in 1968, proximate to his team's development of the first CAD systems. Nicknamed the "Sword of Damocles" by those who used the prototype in deference to the nest of wires and sensors that hung above the viewer's head while using it, Sutherland's "Ultimate Display" overlaid a wireframe model of 3D objects in space optically so that it appeared to hover in front of the user's eyes wherever they looked (Figure 3.31). Coupled, however, with a collection of sensors that measured head position,

---

[9] Since the successful funding of their Kickstarter project (and subsequent purchase by Facebook), the Oculus Rift opened the door for a range of new Head-mounted Displays (HMDs) that are increasing in capability very rapidly. Virtual Reality is within the reach of most computer users for the first time in history.

STEREOSCOPIC PERSPECTIVE

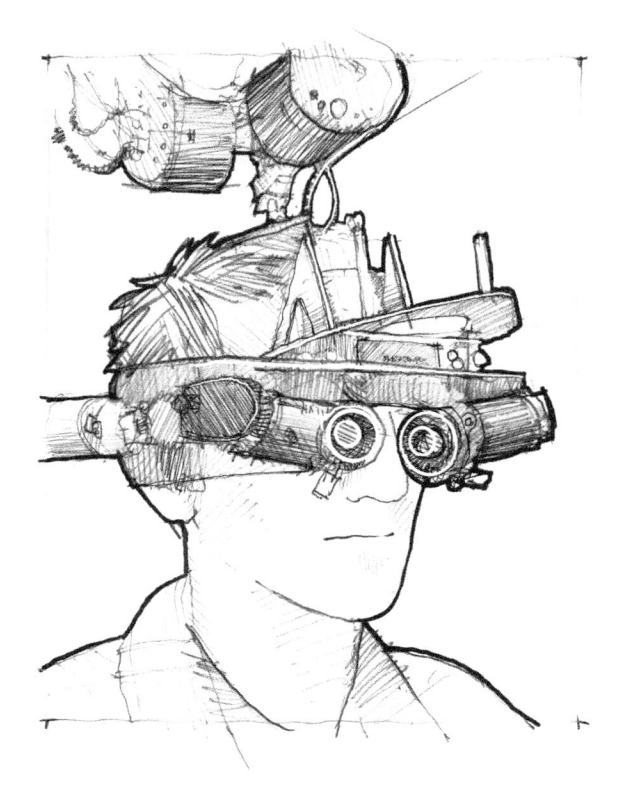

Figure 3.31: The first demonstrated augmented reality (AR) headset; from Ivan Sutherland at MIT Lincoln Labs in 1968.

the system could track its digital rendering with the viewer's gaze. Sutherland (1968) wrote, "Thus displayed material can be made either to hang disembodied in space or to coincide with maps, desktops, walls or the keys of a typewriter." This system, which we now call "augmented reality," is manifest today in devices like the MagicLeap One and Microsoft's HoloLens.

Fully immersive head-mounted displays were popularized in the late 1980s, notably through the efforts of Jaron Lanier and his company Visual Programming Lab (VPL), where some of the first commercially available head-mounted displays and other related physical interface devices were developed. Lanier is often credited with coining the term *virtual reality* (VR) to describe the fully immersive experience of devices that completely overlaid the human senses of sight and sound with digitally produced analogs

Access by designers to a fully (or partially) immersive experience of perspective is now becoming commercially accessible for almost anyone with a computer. Since the 2010 launch of the first Oculus Rift device, virtual reality applications have become cheaper, and faster, and they have gained increasing degrees of visual fidelity. There is still a lot of work to do before we can give ourselves over to a completely believable digital experience of reality, but the path ahead is clear.

Some obstacles remain. Chief among them may be the simple result of human physiology's reaction to having its sensors hijacked incompletely. VR systems are capable of convincing the visual cortex to believe that the digital world being presented is a real one, but our inner ear is left out of the simulation. For those who are physically sensitive to scenarios where visually apparent motion is unmatched by the rest of their body's positional awareness, VR systems can give you motion sickness. I suffer from this, unfortunately.

I think there is an important lesson in this, however. Our technology, from the invention of orthographic projection to perspective through cinema to the slickest of fully immersive VR, has finally become so good that we are actually able to fool our bodies at a physiological level. Immersive virtual reality is no longer just a parlor trick – instead, it is (almost) a physical enhancement to the human visual cortex. Given only a little bit more development, technologically, designers will have the most powerful tools for sketching that have ever been imagined.

Unfolded Projections

To be sure, perspective and orthographic projections are the most common methods you'll use to represent spatial ideas in your sketching work. Many centuries of thought have gone into their development, and the conventions that guide their use are deep, complex, and widely understood. But rigorous exploration and invention around the means of projection did not come to completion in the renaissance – in fact, there have been some key developments since then that haven't found as ready a home in most digital sketching systems.

I am fascinated by the visual projection systems employed by cubist painters and sculptors of the first half of the twentieth century. In their work, the viewer's point of view was free to move around the subject, allowing it to be simultaneously viewed from many different angles. Cubists like Georges Braque and Pablo Picasso freed themselves from the rigid sequence of cinematic perspective to present multiple points of view in a single instant. Futurists like Boccioni extended and perfected the art, adding motion and even states of mind to the multiplicity of points of view (Figure 3.32).

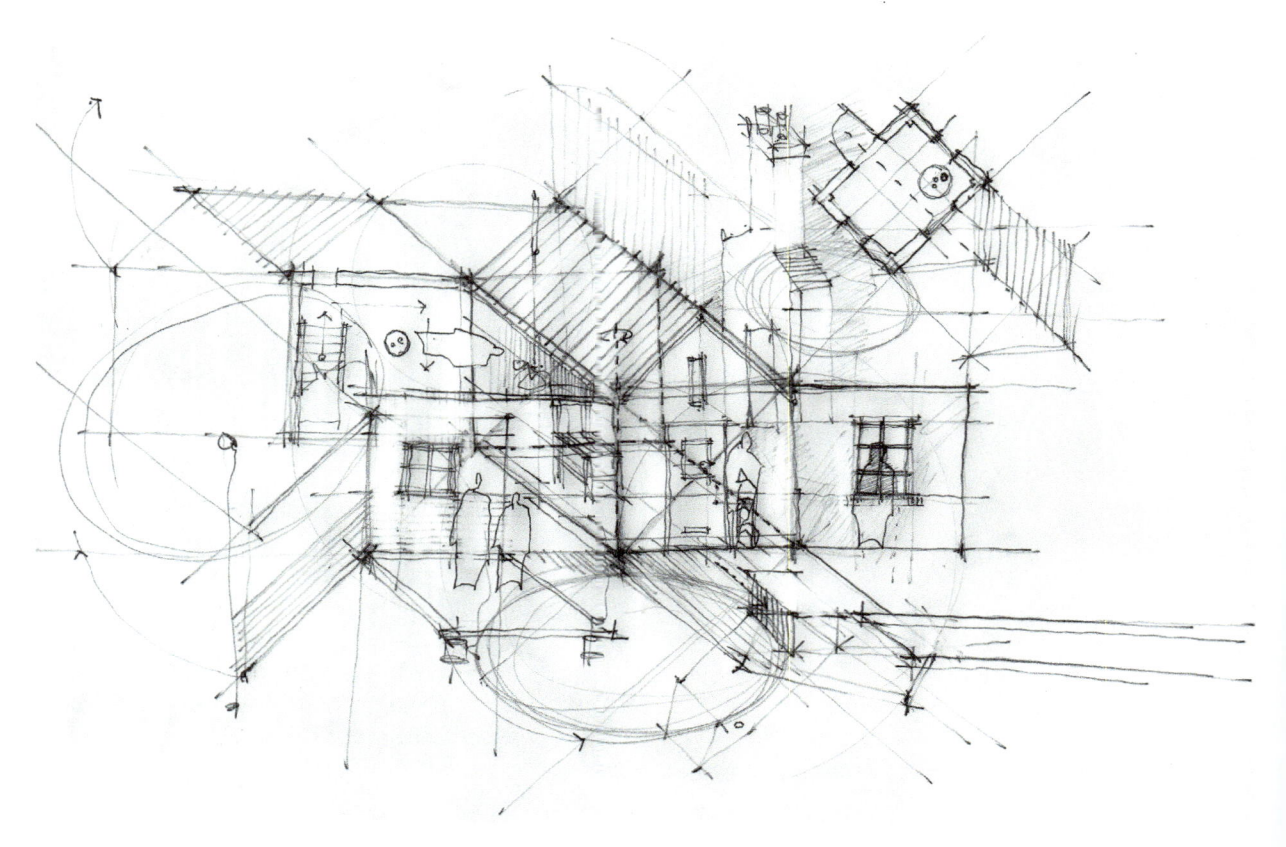

Figure 3.32: An unfolded sketch.

UNFOLDED PROJECTIONS

In a rational sense, a standard sheet of architectural drawings includes multiple points of view in a single diagrammatic representation. Still, there are other, more straightforward ways to consider this in your sketching practice as well. If you find yourself wondering what is around the next corner when you are sketching on some concept, you don't need a new piece of paper to explore it. Just draw right on top of whatever you were working on last. Multiple images lead to a discontinuity of experience. If you are interested in the flow of one space into the next, you can't just draw the endpoints – you need to find a way to draw the flow between them.

I have never seen a useful representation of cubism in a digital sketching tool, but I think it would be a compelling way to work. In a way, 3D modeling systems with multiple viewports onto an object are also somewhat cubist in nature, but they only offer four highly analytical views. What would a system be like if it allowed the designer to sketch simultaneously in 20 viewports? Can you imagine what you might do with such a system?

For now, cubist projections of space can perhaps only be drawn expertly by hand, though there are lessons from descriptive geometry that can help to lend more rigor to their construction. I would encourage you to experiment with this to see what kind of rationality you can invent. Cubism offers perhaps a glimpse through pure ocular rationalism into something beyond, something more relevant to a poetic interpretation of space and the experience of it.

Possible entry into other projections systems that present a multitude of points of view in a single projection might be found by unfolding the subject. Similar to Vesalius' anatomical studies, in an unfolded projection, connected surfaces are separately viewable in true scale – unfolded and flattened onto the projection plane (Figure 3.33).

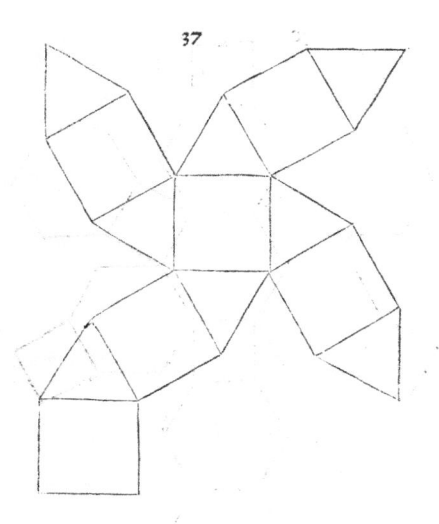

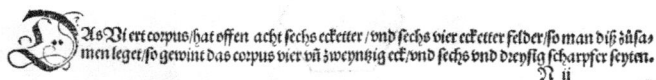

Figure 3.33: Dürer's polyhedral unfolding technique.

If the resulting projection were printed on a piece of paper and cut out with a hobby knife, it might be folded back up to form a three-dimensional model. Unfolded projections are useful, of course, for physical model making, but they have many other uses as well.

Unfolded projections are commonly used by geometers who study polyhedral theory – to them, the unfolding of a form is the best way to understand the true shape of the individual faces of the form they are studying. They might as well use a network diagram to understand which faces are connected to one another in such a way that adjacency can be quantifiably maintained on a flat projection. Albrecht Dürer showed a similar projection system in his *Painter's Manual* (Dürer 1525).

Unfolded projections are a good way to think about a sequence of adjacent elevations in a model, as for example when sketching out the arrangement of cabinets in a kitchen, or perhaps a complex sequence of wall treatments or facade details. Unfolding, rather than elevating each wall separately, allows you to consider the continuity of visual lines as they wrap around corners.

There are some much deeper opportunities to consider, as well. Consider, for example, the anatomy of a biological organism. It is quite challenging to understand this through traditional drawings. Architectural form is primarily made up of flat planes and discrete geometric forms, forms well described with orthographic and perspective projections. The organic forms of anatomy, however, are not so easily described with the same tools. Especially as they shift and deform against one another, wrapped in layer after layer of different systems. A section cut through a body can tell many things, but it isn't particularly useful at describing the interconnectedness of systems (Figure 3.34) A different way of drawing is required.

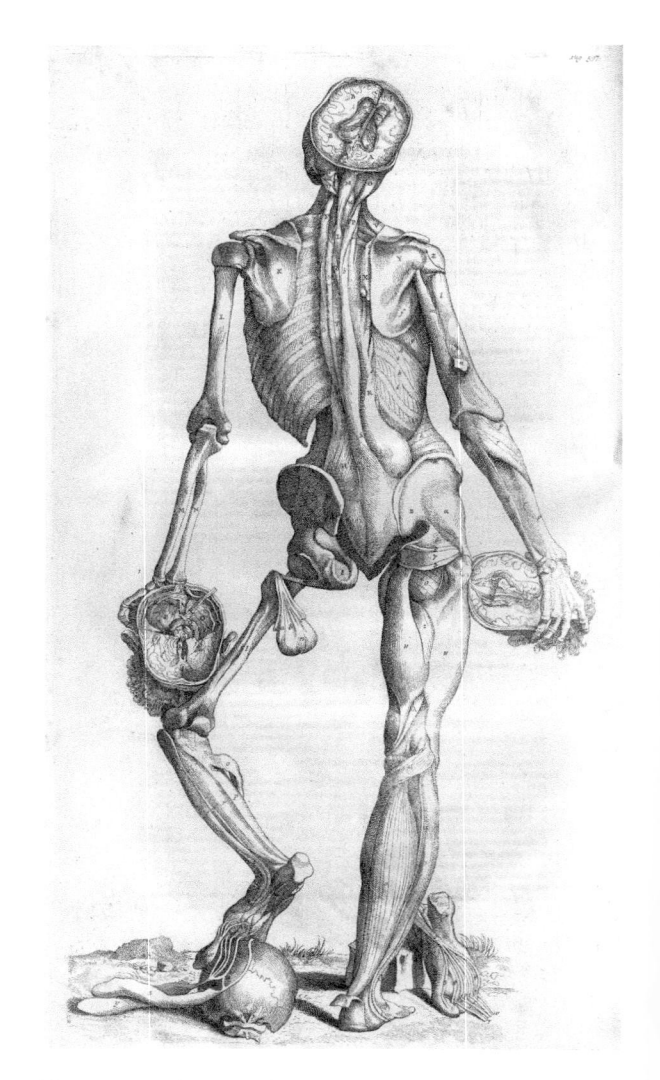

Figure 3.34: From *De humani corporis fabrica* (Vesalius 1543). Suzan Oschmann/Shutterstock.com.

The practices of traditional life drawing are very effective at capturing organic forms on paper. Andreas Vesalius, the sixteenth-century Flemish anatomist, created some of my favorite early examples of the unfolded projection (Vesalius 1543). And it is through his work that I think the real value of this kind of projection can be best understood. Sometimes referred to as an

SKETCHING IN DIAGRAMS

écorché projection, Vesalius sketched anatomy as though capturing a physical dissection. Systems were flayed from one another, splayed out for inspection. Critical connections were preserved, allowing the viewer to understand concepts of connection and critical adjacency without occluding other related systems, which might usually be hidden behind. Vesalius' projections tell the story of the systems they expose in a highly narrative way – and through them, it is easy for a viewer to understand what would otherwise be a very convoluted visual mess.

Sketching in Diagrams

There is a special place in my process for a kind of spatial projection which is purely diagrammatic in nature. Before there is space in a design, there is a concept for the building found in the relationship between the requirements and ideas of its conception. It isn't space in the traditional sense, but it can inform future spatial interpretation and their projection profoundly.

The most straightforward diagrams to sketch for a work of architecture are network diagrams that allow you to explore the experiential relationships between named functions in the building's program. The kitchen should be near the place where you will eat; also, to capture the morning light, it should be located so that it points toward the sunrise. Your site might have other conditions to consider, like a prevailing wind direction from the nearby mountains or adjacency to a neighborhood park.

Diagrams that document the network of programmatic relationships can quickly capture a project's most salient opportunities and can provide you a quick visual language for exploring variations on a theme that help you to optimize for as many of them as possible at one time. You will quickly find that there are relationships in your design that are mutually exclusive, or that cannot be resolved in a singularly "best" way. Which way is the best one for your project? Being able to try many different solutions quickly is essential.

On paper, it is easy to create iteration after iteration of a diagram with layers of cheap tracing paper and a fat marker. Working digitally, however, you can bring other optimization schemes into play as well. If there are many choices to consider, why not let the computer try to find an optimal solution for you?

Imagine the nodes in your diagram are planets in a kind of solar system. Each is asserting a gravitational force on the others, attracting in some cases, repelling in others. The "gravity" between two nodes in your diagram might indicate a desired spatial adjacency – the kitchen to the dining room, for example. By telling the computer which adjacencies you want and which you do not, you can set up a physical simulation that the computer can jostle around until a reasonable solution is found.

Imagine the nodes in your diagram are planets in some kind of solar system. Each is asserting a gravitational force on the others, attracting in some cases, repelling in others. The "gravity" between two nodes in your diagram might indicate a desired spatial adjacency – the kitchen to the dining room, for example. By telling the computer exactly which adjacencies you desire and which you do not, you can set up a physical simulation that the computer can jostle around until a reasonable solution is found.

Here's an example of a traditional Craftsman-style home plan, described in terms of desired adjacency.

```
            "Street" -> "Porch", "Porch" -> "Hall",
"Hall" -> "Closet", "Hall" -> "Bed Room", "Hall"
-> "Living Room", "Living Room" -> "Dining
Room", "Bed Room" -> "Bath Room", "Bed Room"
-> "Bed Room Closet", "Master Bed Room" ->
"Bath Room", "Master Bed Room" -> "Master Bed
Room Closet", "Kitchen" -> "Master Bed Room",
"Kitchen" -> "Screen Porch", "Dining Room" ->
"Kitchen", "Screen Porch" -> "Back Yard", "Back
Yard" -> "Screen Porch", "Master Bed Room" ->
"Kitchen", "Bath Room" -> "Master Bed Room",
"Screen Porch" -> "Kitchen", "Kitchen" ->
"Dining Room", "Dining Room" -> "Living Room",
"Living Room" -> "Hall", "Hall" -> "Porch",
"Porch" -> "Street", "Bed Room" -> "Hall"
```

The arrows between named spaces indicate the desired adjacency, which will be interpreted as a gravity vector that I'll ask my computer to try to resolve. When you see an arrow, think, "This should be close to that." The algorithm I'll use will find an optimal solution to the set of all these gravities. I've also included connections to the street in front of the house, as well as to the house's back yard.

Here's what the algorithm made of my network of requirements (Figure 3.35).

I added a few extra elements to the algorithm to start to get some sense of the spaces between nodes in my graph. Still, it is already apparent that there is a kind of "racetrack" of connections the forms the dominant circulation in the house. This is more or less correct to the way a traditional Craftsman plan works. So my algorithm did end up predicting something that I know to be apparent in the original house plan.

I could use the same basic algorithm to try all different kinds of programmatic adjacency. They don't all have to be named spaces in a house plan. I might drop in an adjacency to represent the desire for a need to work from home to be isolated from the noise and activity associated with preparing dinner in another part of the house. Or maybe, for my client, there is a desire to have those two activities be extra close together. I might add a different kind of "gravity"

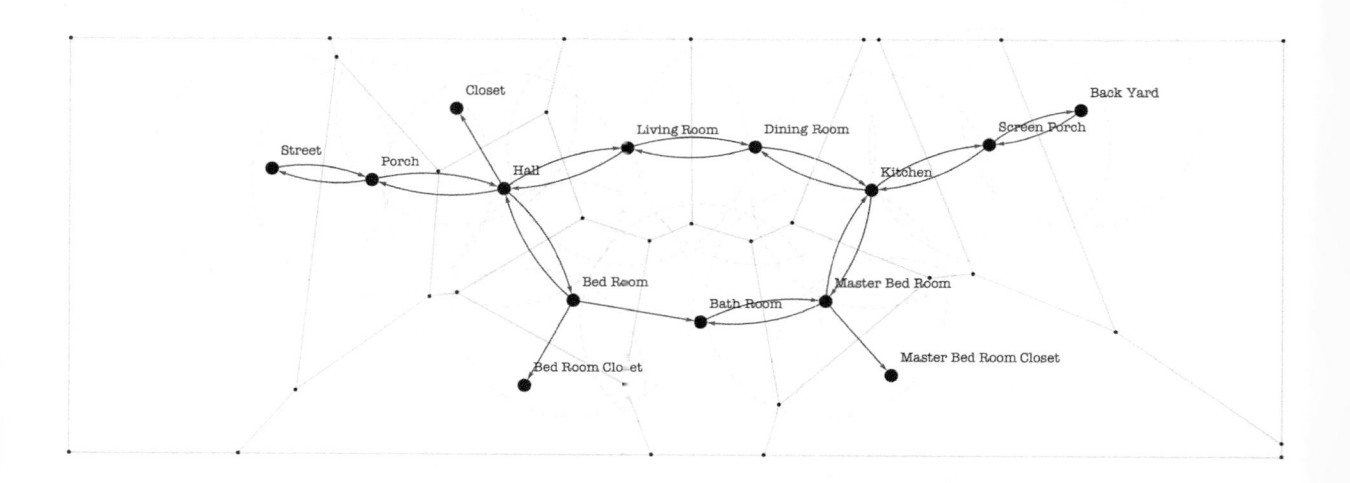

Figure 3.35: An optimal diagram of the adjacencies in a house, digitally assisted.

SKETCHING IN DIAGRAMS

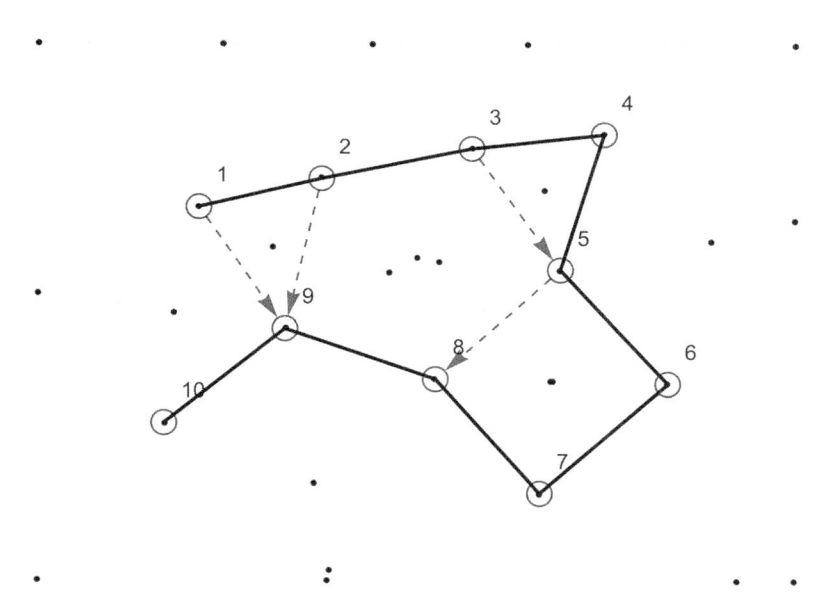

Figure 3.36: A path through space, folded by desired adjacency.

to accommodate these more ephemeral requirements than I might use for two spaces intended to be physically connected to one another. What might the abstract relationships, with the actual names of spaces stripped away, begin to look like? (Figure 3.36)

Once I have an algorithm in place that seems to be giving interesting results, sketching digitally allows me to immediately try many variations at the same time. How many different variations on this basic theme might be available? Let's try five different configurations, perhaps something interesting will turn up in them (Figure 3.37).

These examples are, of course, all shown in 2D, as though projected flat on the ground and perhaps most useful for imagining idealized plan views for my design. With digital assistance, however, I can also almost as easily explore the same basic diagram in 3D just by freeing the line to fold up and down as well as in the ground plane (Figure 3.38).

The space I defined digitally to help me resolve these diagrams has a kind of physical property, and I am depending on basic conventions of spatial projection to read into them. The space of a diagram isn't exactly nonspace, nor is it yet fully

Figure 3.37: Variations on a theme.

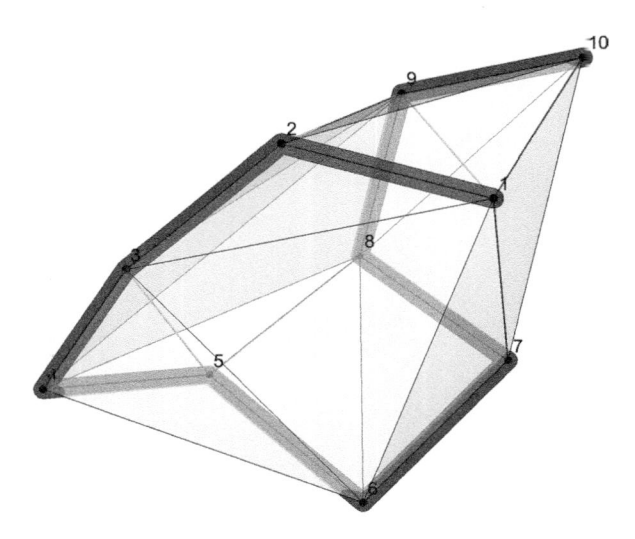

Figure 3.38: A three-dimensional path through space folded by the same desired adjacency.

measurable in Cartesian terms, either. It is a space of ideas, a space of opportunity, and affordance.

It is, to the computer, a purely numerical space that can be iterated over and over until an optimal solution is found to the set of simultaneous equations that together represent the network I defined. The computer is better at solving such things than I am, and just maybe it will come up with an optimization that I might otherwise have missed. As a designer, I can use the computer's solution as yet another way to think about the overall planning for the project. The computer, in a sense, is a partner to my process. It helps me think of things I might not have done by myself.

Rather than confuse matters prematurely, I won't share the code for these simulations specifically. I built them in Mathematica, using its powerful built-in graph visualization functions in only a few lines of code. But let's save that for Chapter 6, "Sketching in Code."

When you are just at the beginning of a project, and you know little more than the basic requirements supplied by your client, these sorts of diagrams can be the best way to get the conversation started. You don't know much at the very beginning, and you'll only begin to learn where the real challenges and opportunities lie with a new project once you have the basic stuff documented and out of the way. But a basic network diagram can quickly become a spatial study of the ultimate form and character of a building with only a couple of quick moves. The path between adjacencies in a network diagram becomes a doorway, or a hall, or (in 3D) a stair. The nodes in the diagram become discrete rooms or maybe just some kind of polarity in an open plan. Regardless of where your sketching takes you next, you will always be grounded by this kind of work.

chapter 4 Sketching in 2D

Architects are taught to think with their hands, usually by drawing. The kinesthetic experience of a hand moving over paper with a pencil provides space for thought and encoding of formal ideation that cannot be duplicated easily through other means. Other professions are taught to think in different ways, through writing, tabular calculation, through verbal debate or quiet observation and contemplation. But architects, like other designers of physical things that are destined for physical manufacturing or on-site construction, work best when they think by drawing.

Your Sketchbook

Any architect worth their salt is never far from a sketchbook and a pencil. Or a pen. Whatever you like best. There's nothing that really defines an architect more than this, with the possible exception of their obsessively mannered glasses and the broken-down Saab in the driveway. Most designers come to have a special relationship with a particular kind of paper, a specific manufacturer's notebooks, and whatever mark-making tool works best for them (Figure 4.1).

Bruce Chatwin, in his book *The Songlines,* writes reverently about his *carnets moleskines,* travel notebooks that he carried with him as he traveled the earth on walkabout for more than 20 years (Chatwin 1987). He purchased them from a small papeterie in Paris, buying as many as he could every time he traveled through. Surely there were other pieces of paper that Chatwin could have used, other notebooks that would have satisfied his needs. But he knew what he liked, and they helped him work.

The modern moleskine notebook is a good enough analog to Chatwin's favorite. However, the family business that made them in his time has long since closed, relicensing the brand to the modern note-booking behemoth that you know of today (Moleskine 2020). I still like their notebooks and typically carry one or more of them in my shoulder bag every day. I like plain, unlined paper, and a Dixon Ticonderoga No. 2 pencil. Lately, I've been buying the small "Cahier" notebooks and slipping them into a well-worn leather journal cover. I carry a flex-nibbed fountain pen as well, which is helpful in other ways. It's what works for me, anyway.

For you, it might be Italian vellum from Florence and a Montblanc fountain pen with organic Japanese squid ink. Whatever suits you, whatever makes it work for you. But you should pick something you like and stick with it as long as you can. Creative professionals of all kinds need a trusted place to capture in-progress thoughts and ideas, in

Figure 4.1: One of my old sketchbooks, a sketch from Piazza Navona in Rome.

whatever form they may present themselves in the moment. The important thing is to be ready and to practice sketching at every opportunity.

It might also be that you have adopted a digital sketchbook, either as a companion to your physical book or maybe even a complete replacement. After many years of development, tablet computing, notably with Apple's iPad devices, Microsoft's Surface tablets, and many other equally competent alternatives, has finally matured. I carry, in addition to my sketchbook, an iPad and an Apple Pencil. They are pretty great together, too. Maybe, I'm learning to love them.

Digital sketching is, for me, still more laborious than sketching with a pencil on paper. I'll talk more about that later, but for now, I think it is essential to understand that when I'm sketching on the iPad, I'm using most of the same parts of my brain that I am using when drawing on paper. "Digital" is just another medium, to me, and there are times and places where it is much more appropriate. What I find most interesting about the various sketching apps that I have used over the years is just how uncertain they all seem to be about where they should fit in the designer's toolbox. And they are all very different, in my experience.

Sketching with Purpose

Sketchbooks are typically found hand-in-hand inside an architect's office with rolls of cheap tracing paper that is treated with disposable disdain. Some even refer to it as "trash-paper" or "bumwad" to further denigrate its long-term importance in their process. But a roll of tracing paper is as crucial to the conceptual design process as any other technology. What makes it special and unique from work in a sketchbook is its ability to be layered up, sheet on sheet, to develop an idea iteratively.

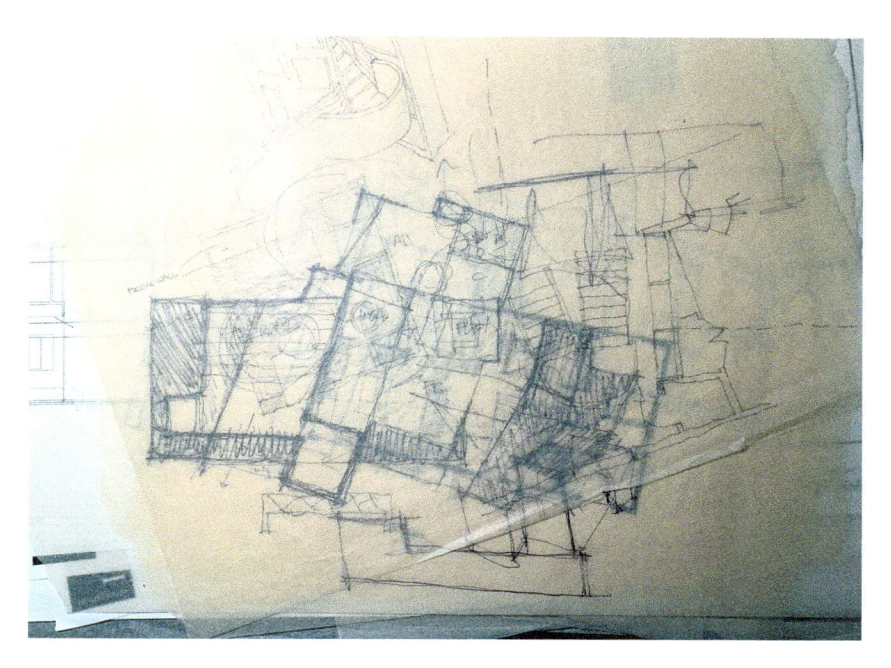

Figure 4.2: A pile of sketches on tracing paper.

By stacking a fresh layer of trace on top of the last drawing you did, the best intentions can be carried forward to the next drawing. And the worst can be forgotten (Figure 4.2).

Sketches are essential both in their creation and in their destruction. A sketch is an encoding of an idea on the way to the next idea, overlaid, torn, taped, and crumpled up and forgotten as quickly as necessary. Among the greatest sins, a designer can commit during the conceptual phase is to treat any single drawing as too "precious" to them. Sketches are like ideas – some good, some bad, and most of them fleeting. And, rarely, preciously, some of them persist long enough to change the world.

Let us imagine a sketching session – how does it work? A first approximate sketch might be made, possibly overlaying an as-built drawing, a photo or a sketch from some other source. Maybe, you might trace from a rendering of a digital 3D model.

Quick lines lay out the bones of the new idea, and the designer stops to consider the implications. Problems are discovered, and perhaps some new opportunities. The original design problem is reframed. A new sheet of trace is laid on top, and the idea is iterated. A new idea emerges. And again, and again, and again.

Layer after layer of trace builds up good ideas gradually over time. The oldest thoughts, if there is nothing memorable about them to preserve, disappear into the building opacity of the stack of newer iterations – literally fading away into irrelevance. Good ideas may rise again and again as the designer remembers them, or consciously pops them back to the top of the stack again and again. "Undo" operations are possible by merely poking back through the pile to some prior state. Forks in the design's development are possible by slipping a good idea out of the stack and starting it on a new parallel stack of its own. Many previously divergent ideas may be merged back together again simply by slipping their piles back together.

Collaborative sketching with trace is similarly straightforward – separate designers working across the table from one another may pass sketches back and forth for consideration, revision, or adaptation. Each might take distinct responsibilities – one working, say, on the site plan while the other works on the interior plan. A natural division of labor can occur simply by sharing single drawings back and forth between stacks.

Important calibration points must be agreed to at the beginning, as well as some kind of shared language of projection and representation. If two designers are working on plans together, they'll find it easier to collaborate if they are working at the same scale. That way, when it comes time to reintegrate their work, previously separate drawings can simply be taped together. A primitive but highly effective merging technique.

Donald Schön, in his anthropological analysis of a conceptual design critique between a student and teacher in an architecture studio, describes a "language game" in design thinking that combines both verbal utterances and lines drawn on trace layered one on top of the other (Schön 2017). Taken separately and out of context, neither the speech nor the sketching would make much sense. But in the context of design, working together to solve a commonly held problem, they are a highly efficient method of communication.

Sketching, then, is about communication, about sharing ideas for review and critique. But it is also about forming those ideas, just as any verbal communication. Thought, it seems, is made of language. To "think" a thing into existence is to be able to describe it in some structured way, either to yourself or to others. Sketching is a visual language, equal to any written or verbal-linguistic counterpart. Once expressed, thoughts can be shared, can be considered, and can be critiqued and reframed.

In Schön's work on sketching, he identified a universal component of professional practice that he called "reflection-in-action," and it bears deeper consideration. There is a consistent process of action, reflection, and reframing that seems constant across all design disciplines. The designer makes a first approximation of a solution, encoded in a sketch or some other durable form. The sketch represents the designer's current best thinking on a solution to the problem at hand. That sketch also likely exposes some critical flaws or tantalizing opportunities, but the designer does not know how to proceed. Through reflecting on the sketch, the designer recognizes that they cannot take the idea further – they are "stuck" in the current way of thinking. Upon reflection, the designer discovers there is a way to reframe the original problem, and in doing so, they become unstuck and can proceed forward in the design's development. At least, until they get stuck again and the cycle repeats.

This process of action, reflection, reframing, and acting again happens so quickly in the mind of the designer that it is almost invisible to external observers. This is what Schön refers to as "reflection-in-action." In other words, it is a state of being where the designer is continuously flowing from action to action, reflecting, critiquing, and adjusting so fast that it appears to be a continuous, laminar stream of activity. This is also what designers often refer to as their *flow state*.

But sometimes, a design gets really stuck. So stuck that the designer needs help from an external voice – a desk critique or something like it. All good designers know this is going to happen, and they know how to set aside their ego long enough to ask for help. They know that external voices from team collaborators are the most valuable kind of input possible, in fact. Design problems are always full of self-contradictions that resolve themselves only through compromise and optimization. Only by working together with others can we consider every facet of the problem.

It is easy to tell when you are around a designer who is really comfortable sketching. They are the ones with a desk covered with the detritus of design. Sketches everywhere carefully tacked up or crumpled in a ball under their desk. Pens, pencils, markers, and photos scattered across the desk and stain rings from spilled coffee. Active conceptual design is a messy, non-linear process. Like a crime scene investigator, a good designer will follow the design inquiry where it leads, and that can be (and typically is) an unruly process. I'm always suspicious of clean desks in a design firm. I wonder to myself where they are keeping the creatives on staff if everything looks spic and span when I visit. Or, worse, I wonder if they actually have any?

I love visiting Frank Gehry's studio, and I was privileged to have been able to do so on varying occasions several years ago. What struck me immediately was just how creative it all felt. And I mean that in two ways. First, it looks like someone bashed a blimp-sized piñata over it, full of bits of paper, wire, wood, plastic, and tiny scale models of trees and people. It is covered with sketches – on every horizontal surface, and underfoot everywhere. Also, and this is particularly striking, I noticed that it was full of different things every time I visited. Those guys are really working, burning the candle, and getting things done. I am always energized by my visits there.

I feel the same thing when I visit architecture studios in the best schools of architecture. Cooper Union, which was my home for six years in the early 1990s, felt the best. The place literally hummed with creativity. Rigorous creativity, far from oxymoronic, is, in fact, the most intellectually stimulating thing in the world. There are sketches everywhere, like Gehry's office. And physical models of all kinds covering every horizontal surface. You could see at a glance that people were fighting with one another over space to work, space to make the next sketch. It exuded a sense of discipline, of rigor in process. Fundamentally academic; it knows the history, can cite more precedent than you can, and (if you ask it to) you'll find the math all checks out as well. These students know how to be rigorously creative.

And in it all, as well, you'll find a light-heartedness. A whimsy that lightens the heaviness of the kinds of design problems that architects are regularly invited to confront. Great designers work *so* hard, stay up *so* late, and keep *such an* unprecedented connection to the culture around them. Every now and again, you have to let off the pressure a bit. When you are up late, and everyone else has gone home to their families, what might creep into your sketches?

When I talk to people who do not have any experience of their own with sketching as a form of thinking, they typically assume that what I'm talking about some kind of fine art process. They think of sketches as something decorative in nature, something appropriate to hang on the wall if it complements the color of the couch. And to be sure, some drawings do end up looking beautiful. Especially the good ones – you know, the ones where the designer really broke through on something new. It is funny how, like in a piece of fine art, people can feel the ideas that are encoded in only a few strong marks on a piece of paper. Even if they cannot understand the full story of either the making or the thinking that led to it, they can tell when a sketch has something to say to them. But there is more to a sketch than this. Sketches serve ideas about something else.

It is a persistent mystery in design activity that neither the designer nor their audience may be able to fully explain the why or how of it. And yet, reproducibly, they all know when the design is good and recognize that something valuable has been done. In fact, that is the best justification that I know for preserving the messy, irrational, and nonlinear process of design through the constant rationalization that is happening everywhere else in the construction industry. That is, even though none of us can explain why, we all know when it is excellent – and also, when it is not.

Let's Get Sketching. . .

It is clear by now that I think sketching by hand with traditional media is still among the best ways to get yourself into the conceptual design process. This book is, of course, not exclusively about that, however. We need to spend some time thinking through the ways that physical sketching with traditional media can be either enhanced or (maybe) replaced by digital methods. I'll explore some alternative positions on this, and share some of my personal experiences with it as well.

And in the final analysis, I think you'll agree that the digitalization of your sketching process is something important to do. It will help you to be a better and more effective designer. You can be more integrated with larger and more complex production and construction teams. And you will be much more likely to have your design ideas heard and internalized by contractors who just want to get the building built as fast as they possibly can. Time is money, folks. It is up to you to make that worth it.

Time is also history and legacy. You're leaving a mark on the landscape with every building that gets built. Never forget that you, as the designer, are the one most responsible for making that mark a good one that enhances lives for generations.

Sketching is not something to be casual about. It is an earnest business, and you have a responsibility to get it done right.

Above all, I'll focus on practice, not product. I want you to feel comfortable performing your design work on demand, to find ways to enter into that reflection-in-action state quickly and effortlessly with any medium, physical, or digital. Let us see how that might work out.

Physical Sketching

Drawing: The most ancient, modern, difficult and the cheapest form of expression in the world.
—Barcelona Manifesto 1979[1]

Sketching by hand, with a pen or pencil on a piece of paper, is still one of the best ways to design. It may be an unexpected position on my part – after all, I make a living by selling much more advanced computational design tools to architects. But there are good reasons why physical sketching is still the best. It is tough to beat physical drawing for speed, for expressiveness, and for convenience.

It is trivially easy to get set up to sketch with traditional media. A pencil and a piece of paper, all you need to make a sketch when an idea strikes, are ubiquitously available. You are never far from the means of making a sketch. You do not have to wait for a computer to boot up or for an application to load. The tools are easy to learn, at least the mark-making parts of them. They are cheap enough that you can have lots of them tucked away everywhere

_____

[1] Rix Jennings, one of my old drawing teachers, shared this Manifesto with me. He remembers it having been informally shared around art schools in the early 1980's, but I was unable to find further reference to it. Still, it is available online at Rix's website (Sanchez 1979)

that you might find yourself needing to jot down a quick thought – in your office, at home, on the bus or in the car. You can carry the essential tools of sketching around with you in your pocket all day.

Paper and pencil are also a remarkably durable and versatile medium of expression. Drawings made with charcoal on rock walls are among our earliest recognizable design acts as a species and have lasted tens of thousands of years. More recently in history, marks made by architects who we know by name with ink on paper have lasted hundreds of years, displaying the highest degree of rigor and visual sophistication. There are sketches absolutely everywhere. There are even sketches scratched into train windows and spray-painted on the underside of bridges. These methods of working have withstood the test of time through centuries of continuous use, and they are still good and useful today.

Pencils and paper are also just about the most expressive medium imaginable for design thinking. Your pencil has no a priori notion of what you will draw with it beyond the basic rules of mark-making. The pencil does not know the marks you make with it have anything to do with anything, and as a consequence, the only meaning in a sketch (from a design perspective, anyway) is found in the intention you bring to it. The sketch means what you make it mean in the context in which you present it. Nothing more, but also nothing less.

I think that every single designer should learn how to draw. By hand. Without exception. Digital sketching grows from physical sketching in every way. To lazy students who believe they can get away without learning to draw by hand, I suggest you are deluding yourself. You are going to have to learn to draw at some point. Your digital skills are dependent on lessons learned by drawing by hand. And to those of us who are years past finishing our education, I challenge you to remember why you learned to sketch in the first place. It is one of the most valuable, if mysterious, skills you have as a practicing architect.

In the first year of my design education at Cooper Union, I spent as much time learning to draw, primarily life drawing from the human figure, as I did in my first year architectonics studio. There was a recognition from Cooper's design faculty that students were not worth much in the studio until they had gained some basic skills and confidence drawing by hand. Desk critiques were pointless if students could not engage in the rapid back and forth of drawing and speaking that characterizes studio-based pedagogy. Not knowing how to draw was treated as being like not knowing how to write or how to speak. How could you hope to think critically about architecture if you could not draw?

Like any creative task, drawing improves through practice. Great drawing skills are hard-won through consistent practice, not innately gifted to you by your DNA. And I have never met a person who truly could not learn how to draw with the help of practice and competent instruction in a studio environment. If you are telling yourself you do not have the talent required to draw competently, you are unnecessarily beating yourself up. Stop it and learn to draw. It is just another language. You can learn it to at least a conversational level if you put some time in.

I'll emphasize that point – because it is one where I think professionals lose a lot of time and self-confidence. Anyone can learn how to draw, at least how to draw well enough to communicate an idea to others in the form of a simple sketch. It is incredible to me how mysterious it is to those who do not (or will not) draw when I produce a quick sketch of some concept on the spot in a conversation. Even a quick diagram, just illustrating a concept, has a mystical power over some people. I'm not above using that to my advantage on occasion, to influence people to listen more closely to an idea that I have to offer.

"How do you learn to draw – I could never do that," is a constant refrain in meetings with folks who did not go to architecture school. But for those of us who did, drawing is a superpower that should never be discounted. Cultivate it, refine it and use it every chance you get. Practice it every chance you get until you can do it without thinking, just like riding a bicycle.

In life, my most influential teachers have always been drawing teachers. The lessons I learned in the drawing studio are the most durable, the most insightful and the most lasting of my entire education. David Hallaka, Rix Jennings, Anthony Candido, Sue Gussow, and Basilios Paulos, even 30 years later, have their little voices in my head that I hear almost every day. On the off chance that any of them are reading this book, know that I owe each of you a deep debt of gratitude for what you taught. Truly, I'm thankful for that.

Drawing is not really much dependent on learning how to lay down a nice-looking line on a piece of paper. That makes a difference, of course, but it is more of a knack than a skill. It is instead almost entirely about learning how to see with clarity. How to really understand the nature of the subject you are trying to capture in a sketch. If you cannot draw what you are trying to draw, ask yourself how well you really understand what it is that you are drawing. Look closer, think harder. Take the time, it is the most important thing.

Drawing from life, from real observation, is the most valuable way to start. Eventually, you want enough confidence with your skills that you can draw anything that pops into your mind quickly and clearly. But at the beginning, drawing from life will provide you a great source of discipline and practice. Life drawing, from a live model, is the apex drawing practice for anyone serious about learning to draw, but you can start absolutely anywhere. Look around you right now, at the space where you are reading this book. You can draw that.

I'm fascinated by the growth of the "Urban Sketching"[2] movement, springing up in cities around the world. As a traveler, I like to find some time to sit in a new place I'm visiting to capture it in a quick sketch. I find I see it better, more completely, and with more compassion than I might through any other means. Urban sketching offers a way to justify the same kind of attention, of careful consideration and reflection on the world around you at any time. As an architect, your trade is mostly in the creation of buildings, and principally (not entirely of course) exercised in cities. Take some time to sit and draw your city, and I guarantee you'll find you have learned something new about it.

It is not hard to find formal instruction in drawing, but if you cannot, there are other great options available today as well. The internet is a seemingly endless source of learning materials. My daughter, who draws beautifully (you'll see some of her work in this book), is almost entirely self-taught. YouTube videos from other artists provided her with most of the support she needed. Sharing drawings for support and critique is easily accomplished via Instagram (or whatever social network happens to be in vogue when you are reading this book.)

Drawing, particularly from observation, is a mysterious process whereby you convert your spatial perceptions of the world around you into representational marks on a 2D surface. To accomplish that, your brain has a lot of work to do. Your eyes have to perceive the light bouncing off the subject you are drawing and convert those signals into a mental

---

[2] Like all great movements, they have a manifesto (Urban Sketchers 2019).

model. Not everyone does this the same way (isn't that marvelous!), and there is no singularly "right" way to understand space. What your brain makes of the signals your eyes pass to it makes you uniquely you.

Drawing well begins with seeing clearly. Always spend some time really looking before you begin drawing from life. And then when you start laying marks on paper, anywhere is fine. Nobody gets precisely the right line drawn with their first try. Drawing is a process of progressive refinement. Start from the first approximation of the right line, test it against your observations, then refine it. Do not scrub madly at it with an eraser. Instead, learn from your mistakes, adjust the line gradually. Does it need to move over a bit? Is it curved the right way? Is it dark enough? Too dark?

Your first line sets the context for your next line, and so on and so forth. With every new line you lay on the paper, your understanding of where the drawing is going improves. Proportional relationships emerge. Is your line too long? Too short? Is it at the right angle from the line next to it (Figure 4.3)?

Life drawing is wonderful because you rapidly learn not to hang on a particular area in your drawing for too long. The physical act of posing is exhausting work for the model, regardless of how languid yours may appear. They are going to need frequent breaks to stretch and get the blood flowing again. And when they return from those breaks, the pose will be different. Which means you really do not have time to obsess over details. Embrace that, and use it to drive you forward quickly. Start with the most prominent features, the essential elements of the pose. And then leave the obvious bits unexplored if you run out of time.

Like all design activities of any kind, drawing must consider constraints. The rectangular bounds of a piece of paper are the first true representation of an

Figure 4.3: Drawing live from a model.

architectural site for me. You cannot draw past the edge of the paper, so you'd better plan ahead. It is funny to watch from the outside, but novice draftspeople often distort their subject when they start crowding up against the edge of the page. Learn to treat the edge as a constraint and draw past it if you have to. Never let it warp what you are seeing.

Similarly, novices often end up distorting their work because they are looking at their paper at an angle. Things near the bottom of the page (physically close to them) end up drawn smaller than things farther away from them at the top of the page. The physical, spatial qualities of the way you are drawing can have a surprising impact on your work. Imagine that.

Get yourself some space to spread out at the beginning. Tony Candido, my first-year drawing professor at Cooper Union, frequently told those of us in his class to "Get a bigger pad," when we got "precious" with a drawing. More paper to cover in the same amount of time meant you had to work faster, bigger, and with more energy. Beginners, maybe because they are afraid of exposure, perhaps just because they do not know how to really get moving yet, tend to make tiny little drawings floating cock-eyed in the middle of the page. Drawing is a whole-body experience. When I'm drawing, I move from the hips, not from the wrist. I like to really throw my whole body into it. Almost like a dance form or martial art. It takes more space than you think. Do not be ashamed of that in any way.

Do not be precious, either. Most of the sketches you make are going to suck. So what – that's nothing to obsess about. Sketches are for you, to help you think. Maybe one day you'll make a final polished "rendering" of something by hand, but that is not the goal of conceptual design sketching. Sketching in conceptual design is about thinking, about seeing and about learning what the building wants to become. And maybe that quick sketch that you burned out without thought will be the one that turns your design around and makes it into something transcendently great that nobody would have expected at all.

Materials for Physical Sketching

Start with simple materials. I like plain old Dixon Ticonderoga #2 cedar pencils. The kind you can buy at any office supply store for next to nothing. The feel of graphite dragging on a piece of cheap white bond paper has always made me immensely comfortable. I can get set up to draw with almost no cost (Figure 4.4).

At the same time, have some self-respect as well. Spending a little bit more on some quality art supplies (the good stuff can be expensive) will make

Figure 4.4: Some of my favorite drawing materials.

a difference in the long run. You never want your choice of materials to inhibit your work. Higher-quality materials leave you more headroom to work before you start to struggle with their limits. You will not appreciate them at first, but in time you will see why they are worth the money.

I recently decided I wanted to bring some color back into my work. I am lucky that the sandstone deserts of Utah and New Mexico are close to my home, and I have been finding ways to spend more time out in them lately. The colors are amazing, and I had to try to capture them. I had an old field box of watercolors from the late 1980s that I pulled out again and an old block of Arches cold-pressed paper. Pretty good, but I wasn't getting the vibrancy I saw in nature. Then I picked up a new mini pan of QoR watercolors[3] ($60! So expensive, I thought...) to try and BLAMMO. I did not know you could get color that saturated out of a pan of paint. My heart leaped when my brush hit the paper with the first blob of paint. There was so much pigment in the colors that they transformed what I saw in nature.

Excellent materials will not make you a better artist, but they will keep you from getting frustrated. A pen that skips, a pencil that breaks, or paint that gives just "meh" colors can seem to confirm your worst suspicions about your abilities. Good materials, on the other hand, will let your growing skills show proudly. Pay particular attention to the paper you are drawing on. Its roughness, resilience, and color will all impact the work you are doing. Cheap paper will not take ink well or will pucker when you hit it with watercolor. Take the time to experiment and find what really works for you.

There is a powerful allure to drawing materials that were favored by some artist or architect that did work you admired. There's no shame in some hero-worship, and one of the best ways to feel what it was like for that person to do the work they did is to try out their preferred tools. Like any tool, though, you should never assume that you will be able to create the same results that your hero did just by adopting their tools. Great work comes after lots of bad work – after lots of practice. Whatever means you choose when you are learning to draw, commit to the ones that feel right for you. You'll know it when it happens – it will feel like falling in love. You might not know why that particular brand of Grumbacher vine charcoal is the one, but it feels just right for you. Buy a big box of whatever it is and commit to it.

Carry a Sketchbook, Really

I made this claim before, but I'll emphasize it to drive home the point. Because you cannot always be sure you'll be at your desk when inspiration strikes, you should always carry a sketchbook. Also, you need the practice, no matter how good you think you are. It does not have to be anything nice, and you never have to show the sketches you are doing in it to anyone else, but know that it is always with you if you need to jot something down.

I have years' worth of old sketchbooks tucked away on a bookshelf in my house. They are nothing special, mostly, and I rarely look back through them. Sometimes, when I do, I cringe a little. But keeping them around reminds me that I have been sketching for a pretty long time. It is an excellent sense of history for me. And evidence of practice.

As a design professional, you are not going to be able to restrict your design thinking to typical office working hours. Design does not work like that. Keeping a sketchbook means you can capture some thought or idea you have about a project you are working on whenever and wherever it happens

---
[3] QoR makes a small field box of watercolors called "QoR mini" that includes 12 half-pan cake colors that are just breathtaking.

to strike you. There does not have to be much work done to capture the thought, but the more practiced you are, the easier it will be to capture those thoughts. Some design ideas, especially the ones you need for the most thorny of design problems, can be distressingly fleeting – like when you wake from a dream and try to remember what happened in it. Your brain stores design ideas in a way that is almost more like a dream than a tactical problem solution.

But specifically, in the context of physical sketching, you become better with practice. A sketchbook is also a personal challenge and a traveling reminder of the work you have committed yourself to do. Sketch something every single day, whether or not you are being paid for it. Be that mysterious person on the train who is drawing the old lady sitting across from her. Be the artsy-fartsy guy sitting in front of the public library, sketching the elaborate beaux-arts facade. Always be sketching something, and take your time at it. You can give up half an hour a couple times a week to practicing your drawing. It is worth it.

Tools for Digital Sketching

Tablet computers, with a stylus and a good piece of software, can be as useful as traditional sketching. In some respects, they can actually be far better. Also, in many cases, they can be far worse. Mainly, they are just different, with a different set of trade-offs, advantages, and distractions. With time and practice, you can learn to use them as effectively as paper and pencil.

Like the art supply choices you have when sketching with traditional media, there are countless apps for mobile tablet and phone devices today that offer artists a complete digital analog to a box full of art supplies. You can readily find them in your friendly neighborhood App Store, rather than the friendly neighborhood art supply store. In some respects, digital sketching applications are just another drawing medium. You'll resonate with some of them and not with others. And you'll probably have to try a lot of them to find the one you really like. Luckily, they are all very affordable, even free, to try. I have some favorites, but they might not match yours.

If you go back to the very origins of design computation, you'll find the stylus before any other graphical input device. Ivan Sutherland's "Sketchpad" included a light pen for input, along with a dedicated button pad. The now-ubiquitous mouse as a pointing device came far later in the development of graphical computing. Most software development projects look back to some kind of prior experience among their hoped-for future users that will (in theory) make it easier for them to pick up the new digital paradigm as smoothly as possible. A stylus is a natural choice for architects who have already learned how to draw on paper.

There have been digital analogs to the architect's sketchbook for many years, particularly on contemporary tablet computers. Historically, and in my opinion, they have been kind of awful for real drawing. They are getting much better. But they have been universally fantastic for their integration with cloud storage, cameras, and all the fantastically useful ways that we have to send one another pictures and blocks of text over the internet. A marginal drawing experience may be worth adopting entirely into your work only because it is just so connected (Figure 4.5).

I do not want to fall into nostalgia about paper, but I think digital sketching tools that try to simulate traditional media aren't all that much more useful than those that do not. If I wanted the feel of paper and pencil, I would stick to drawing in a sketchbook. I do not see much value in a digital

TOOLS FOR DIGITAL SKETCHING

Figure 4.5: My iPad and Apple Pencil, ready for some digital sketching.

simulacrum for sketching. I seem to spend an inordinate amount of time messing about with settings trying to design a digital pencil in the latest sketching app that I like drawing with as much as a real pencil.

I personally find it a little hard to draw on tablets. I have been telling myself for years (since the very first Wacom Cintiq combined a screen and a pressure-sensitive stylus in one unit) that all I really needed to do was get more time in and practice. This is probably true, but I have an entire lifetime of experience with pencil and paper that is hard to leave behind. Change is hard, especially when you are pretty satisfied with the results you get from the old way of doing things.

I've given this a lot of thought over the years, and I've tried dozens of different digital drawing devices. In many respects, I like all of the tools I have used... but I just cannot get myself over the hump to full-on adoption. I think these are the reasons why:

- The stylus on the glass is too slippery for me. I'm most comfortable with a Dixon Ticonderoga on a sheet of plain white bond paper. I like the friction-y feel of the pencil dragging on the paper and the slight give as the pencil digs into the soft surface of a beautiful sheet of paper. I have trouble drawing with pen and ink for the same reason, so this is probably something I can get over in time.

- The distance between the tip of the stylus and the image (because of layers of glass) causes a parallax/offset that is disconcerting. As tablets get thinner and thinner, this problem is going away. Apple's latest iPad Pro tablets have a laminated screen that puts the pixels so close (apparently, anyhow) to the surface that I cannot really tell the difference.

- More complex brushes in the kind of natural media painting programs I want to use (Procreate,[4] for example) have some lag between the stylus gesture and the line that is left behind in the drawing. Performance is improving all the time, though, so I expect this to eventually disappear.

- Pressure, tilt, and rotation gesture support is helpful. I use the heck out of those gestures when I'm drawing with a real pencil without really thinking about them. But I spend endless amounts of time tuning them in digital painting apps to try to simulate the results I'm expecting. Settings twiddling is anathema to my creative flow. Better brush presets would resolve this problem. Better for me personally, of course, is pretty hard for someone else to predict.

- I like the feeling of my pencil digging into the thickness of a sheet of paper. In fact, when I'm drawing, I want to have a couple of layers of paper under my drawing as additional padding. Tablet screens are hard and unforgiving. It's like drawing with a pencil on glass, which is, of course, what you are doing.

---

[4] (Savage n.d.)

- Even the largest (and most expensive) tablet computers are still smaller than a tabloid (or A4) sheet of paper. They are nowhere nearly large enough to take into a figure drawing class (Figure 4.6).

At the end of the day, though, people who have grown up drawing on tablets are capable of genuinely excellent drawing with them. Probably, they would have trouble backing down to the traditional media ("...where's the undo?!") that I prefer. I'm in a transitional generation, anchored in my prior experience and change-averse to anything different – even when the benefits are rationally obvious. I recognize the benefit of moving to something digital.

As a sketchbook, I think I could love working on an iPad someday. Imagine never worrying about losing your sketchbook, because your work is always backed up to the cloud. Or simply never running out of pages – digital devices are limited only by their storage capacity, which is now effectively infinitely large. And the freedom to share a quick sketch has such universal appeal as well.

There is plenty to like about the idea of a digital sketchbook.

The real challenge for me is that I have not put in the time to honestly know these new tools. I've played with them endlessly, but only recently really made some drawings on them that make me proud. Artists depend on knowing the tools of their art (whether a pencil, a tuba, or a typewriter) in a completely transparent way. They need to be able to think through the tool in a completely seamless, laminar way, to "forget" that the device is there and focus entirely on their actual work. Any tool in the hands of any artist can reach that point, but it takes years of practice to reach true virtuosity.

I think my ability to focus may be the biggest hurdle ahead before I can completely replace my physical sketchbook. It may be the diversity of things that I can do on my tablet computer that ultimately hurts its viability. There are just too many distractions. Too many notifications, too many temptations. Focused attention is hard to give. These days, it may actually be my most valuable asset. I have enough money, and I have a fixed quantity of time left in my life. But the things I choose to pay attention to are precious.

Practically speaking, drawing has more in common with playing a musical instrument than most people immediately recognize. Great musicians aren't born with their virtuosity. They may have some genetic gifts that get them up and going faster, but behind any virtuoso is a ton of hard work and practice. Drawing is precisely the same. To be really great at drawing, you just have to do a lot of drawing. There's no better way, and there's no other way. Software tools for sketching cannot make up for practice time any more than a software instrument can make you a great pianist.

Figure 4.6: Tablet computers, compared in size to common drawing pads.

But let us be realistic about the considerable benefits that drawing on a tablet brings you. If you are just starting out, you should consider starting your sketching practice in a purely digital form. If instead, like me, you have decades of experience working with traditional media, they may well be worth the time it takes you to adjust:

- *Undo:* My Dixon Ticonderoga pencil has an eraser on the tail end, which I use the heck out of. But it is not a complete undo that simply wipes out the mistake you just made. All digital sketching apps include Undo.
- *Layers:* Layers make it so much easier to build up intricate drawings from sketches, inking, and color, all tidily managed separately. Layers are a quick way to build up complexity in your drawing gradually over time, strategically flattening them together only when you are ready to fully commit. Layers are also great for simple version control.
- *Archiving:* If you are willing to save to a cloud sharing site (you absolutely should be doing this), you'll never lose another sketch. Everything can be transparently saved and backed up forever. There's no practical way to match this with traditional media.
- *Sharing:* Sending someone a work in progress for comment and review has never been easier to accomplish. Or maybe you just want to share your work with an audience on Instagram or something. Boom, easy.

Tablet computers aren't new by any means, but I think they are really maturing as a tool for sketching. Screens are higher resolution, brighter, and have far better color representation. Backlighting is more brilliant, and batteries last longer. The devices, as a whole, are lighter and more durable than ever. Laminated screens are removing the parallax problem between stylus and image – Apple's 2019 iPad Pro has an almost imperceptible distance between the stylus and the image. It is close enough, anyway, that I do not notice it anymore.

Actually, the resolution of the 2019 iPad Pro display is nothing short of magical. Apple's iPad devices were among the first consumer devices on the market with "Retina" display resolutions in the vicinity of 300 dpi. You just cannot see the pixels on these things, and that makes such a difference when you are drawing. I'm sure this will improve with time, but this technology has reached the "who cares" moment where I have more or less everything I need from it.

I'm also close to in love with Apple's Pencil device, and I use one when I'm drawing. Of all the digital styli I have used in my time (and there have been so many that I have tried[5]), Apple's is undoubtedly the best. But I'd still like to see it improve a little bit more before I can say without hesitation that I think it is the one everybody should use. A stylus is, I suspect, a very personal choice. Personally, I like this one. Maybe it would be a little better if I could dig it into the display a little more, but we are close enough that I can adapt.

I remain unimpressed with the experience of drawing on a smooth glass surface. It is just too slippery to me – I feel like I'm still not really in control of where the pencil is going when I draw. I added a slightly toothy screen protector to my iPad several months ago, and it made a shockingly immediate improvement for me. I went from thinking that the iPad might still be a drawing toy for people who had not learned to draw properly somewhere else to thinking that maybe, just maybe, this thing could one day replace my sketchbook. It really made a big difference for me.

---

[5] I think I have tried, at one time or another, just about every single pen-based computer that has shipped in the last 20 years. I even tried sketching on an Apple Newton.

I think there are still many improvements to come to the experience of digital sketching on a tablet computer, and I have not really gone all that deeply into the current options here. There is one new class of device that I'm quite curious about, keen to learn more, and to try one out for an extended test. There are now devices based not on traditional LED display technology (with a battery-draining backlight) but instead with a passive "e-Ink" display that looks more like paper – visible in bright sunlight, as for example, outdoors on a construction site. Some of these devices now include the ability to draw on them interactively, and I think that is really intriguing. More development to come on that, I'm sure. Maybe if I make some money from this book, I'll buy one to play with.

I'm basically optimistic that digital sketching hardware and software will both continue to improve dramatically as their core technologies continue to develop. I think we are beginning to appreciate outside the specialized professional domain of architecture just how valuable it is to be able to sketch an idea out by hand to help communicate with others. Also, research continues to mount, suggesting that writing notes out by hand causes them to be stored differently in our memory than when we type them. Notes taken by hand seem to promote better learning (Mueller and Oppenheimer 2014). So maybe things that architects have known for generations are beginning to crosstalk into other professions as well. It is a good sign, anyway.

Physical and Digital Together

I had the pleasure of working closely with several really amazing traditional architectural illustrators in my time, notable among them was Mike Doyle at Communication Arts. We worked together for almost six years, on many different projects and in many different contexts. Whereas I hung my professional hat on computational design tools, Mike was an unashamed traditional draftsman. His book *Color Drawing* (now in its third edition) was a biblically scaled resource for all of us interested in design communication at the time. Mike is a master of visual communication (Doyle 2006). If we wanted to sell a problematic design proposal to a client, we turned to Mike. He could bring out the charm, the life, and the grace of anything we put in front of him. Including some (I'm not ashamed to admit) really ugly projects.

One of the last projects I worked on with Mike was a new building on the University of Colorado campus to house the "Alliance for Technology, Learning, and Society," also known as the ATLAS building. At the time, I was primarily running an application called Lightwave (still available; (Lightwave 2020)) on a Silicon Graphics O2 workstation (sadly, no longer available).

Mike and I had a really good collaborative process going. I was pretty inexperienced as a designer at the time, but Mike was a gracious mentor and carried me along very kindly. Where Mike did beautiful, emotional, engaging renderings, I could just bang out massing studies on the computer all day long. I made a dozen different mass models for every drawing Mike took all the way to completion. Mike knew which ones were the "good" ones for reasons I did not fully understand. I'm not sure he knew how to tell why the ones he liked were better, but it was impossible, for whatever reason, to disagree with him.

What I learned from this process was that manual sketching, with traditional media, goes much faster when there's a rendering from a computer model to sketch over. Mike did not have to do the mundane drawing tasks of laying out the perspective (always a laborious and time-consuming process) because my computer was able to do that easily for him. And because of that, he was able to spin through dozens more design variations than he might have when working only by hand with his preferred traditional media (Chartpak markers and Prismacolor pencils). The combination of computer modeling and traditional sketching proved a very effective mix – we won the project and built the building. In fact, I teach a class every spring in that very same building today.

Mike's book changed in exciting ways from its first to second and, ultimately, third editions. In the first edition, published in 1992, there is no mention of anything beyond traditional hand-drawing techniques. By the second edition, he had added a photograph of my Silicon Graphics computer to the Preface. I think he recognized there were ways to use the computer in design practice that augmented and improved design work without losing really anything about the way that a sketch communicates. Rather than thinking of the computer – particularly of photorealistic rendering from a detailed 3D model – as a replacement for traditional design communication, the computer helped to automate the drudgery of perspective projection and freed the designer to go deeper into the design, to try more options and (ultimately) deliver a better answer to the client. Mike's third edition has computers in it everywhere.

In a sense, this is nothing particularly new. In addition to Mike, I also had the pleasure of meeting (though not working directly with) Jim Leggitt, another of Colorado's expert architectural illustrators and design communicators. In the first edition of Jim's book, *Drawing Shortcuts*, he suggested that one of the best ways to save yourself time when sketching is to shoot a quick photo of the site where you are going to make a design proposal and draw over that (Leggitt 2009). Sketching over a photograph saves you the time and effort of formally and manually blocking out the perspective projection matrix for your sketch.

Jim later (following the publication of his book in 2002) took to a drawing strategy very much like the one Mike and I used when we worked together. He calls it "tradigital" drawing, a portmanteau of "traditional" and "digital" methods that remains one of the most evocative methods of its kind today. Jim is an accomplished painter, in addition to his work as an architect and illustrator – so I know he knows what he is doing. If he sees an opportunity to take a shortcut without damaging the design process, I trust him to do it with confidence.

In the second edition of Jim's book, he replaces photography with 3D modeling, tracing over digitally generated wireframe perspective renderings. He still sketches over the digital renderings by hand, preferring the warmth of a physical sketch over the precision of digital rendering. And his sketches do communicate very differently to his audience. A hand-drawn sketch feels more authentic, more personal, and more a part of a continuum of design decisions. Clients respond more openly to a sketch than to a hard-lined digital drawing. The level of abstraction is just right; they feel included in the design and are more open to talking about it.

Both Mike and Jim were students of W. Kirby Lockard at the University of Arizona in the late 1970s, and it is from him that many of their most enduring lessons about drawing came. In the end, it does not matter what medium you choose to help you sketch. The formal rules of sketching apply, both for communicating with yourself in your own reflection-in-action flow state and when you use drawing to communicate your ideas to others. Kirby Lockard taught that design drawing is a unique form of communication, distinct from drafting, separate from fine art. Drawing is, as he wrote, *". . . a means to architecture"* (Lockard 1965).

chapter 5 Sketching in 3D

As important as any drawing, architects also sketch by making models. For some, a sketch model is even more useful than a drawing. Models are abstractions of a project, just like a drawing. Models can be detailed or highly abstract. They can attempt a naturalistic representation of the project or show a simple poetic about material and form. But whatever their subject, models bring new clarity to any sketch, with real gravity, materiality, and substance. Modeling may take a little more time than drawing, but the results are always worth the extra effort.

There Is No Hiding in Models

When I was a young designer working at Communication Arts, I worked on designs for a lot of shopping malls. Our reputation depended on being able to add life and interest to spaces which, without our attention, were otherwise just cavernous empty boxes. Inevitably, the spaces we designed in perspective sketches were crowded with happy, excited people having fun – surrounded by all kinds of exciting design features. Just adding people to the sketches made them feel alive. They looked inspiring and fun to occupy.

When I was asked to translate those sketches into models, I was always amazed at how barren they looked. If you start with a cavernous airplane-hanger of a space and add a couple of benches, banners, and specialty light fixtures. . . it is still basically a hollow box. With the people, the pop and excitement removed (or rather, the non-physical sketch elements that signaled these things to our clients), the cold reality of the space we were working to improve quickly reasserted itself. There was just no hiding in a model. We could soon see in the model exactly how much more work was going to have to be done to get our clients the result they needed.

Benefits of Sketch Modeling

You should move quickly from sketches on paper to sketches in 3D, adding modeling to your paper and pencil sketching but never replacing it altogether. Sketching on paper is quick, easy, and disposable. Physical modeling is more laborious and slower but very useful when you're thinking about materiality. Digital modeling is much faster, probably than either physical sketching or modeling. It is not always better to go more quickly. Sometimes you need to slow down to let your thinking catch up with the sketches you've already made. I encourage you to try to work both into your process, and I'll spend the remainder of this chapter working through the basic principles that should govern when and how you bring them in.

Never forget that architects think with their hands. Thinking by doing, thinking by making – these are the hallmarks of an architect's reflection-in-action design process. The kinesthetic thinking that you do when drawing is multiplied when instead of the simple tools of paper and pencil, you have the harder materials of wood, plastic, and metal at your disposal. Materials like these have weight, presence, and they force you to confront their base materiality. A keyboard and mouse, even really great ones that activate all your kinesthetic triggers, are anemic substitutes for a bandsaw and a belt sander.

As with any sketching medium, there is a danger when modeling that you will add too much detail to your work too early. If it is equally easy to spend your next hour designing a decorative handrail or a site massing study, you can quickly become distracted in ways that will cause your overall design to suffer. And with every extra minute you spend on some small part of your model, you increase the risk that your models will become too precious to change. You may find that you hang on to your first ideas too long at the expense of more significant opportunities that might only become apparent later.

When modeling physically, this will manifest as a feeling that the model you are working on is just too pretty to modify. Maybe you put an hour into cutting out a sophisticated façade, which, now that you've glued it in place, may never be questioned again. Or perhaps you spent an afternoon building some kind of parametric form-finding algorithm into a digital model that was hard enough to program that you're afraid to touch it again. If you find yourself doing something like this, you have to immediately tear it out of your model. How can you be sure that you've made the right decision? Don't confuse the effort expended or the time spent with evidence of design quality. You have to let it go. This is just a sketch. The energy you spent in making it is in the way of forwarding progress on your overall design.

Sketching, done well, is a way of testing many ideas quickly, of exposing the good ones and carrying them farther while letting the bad ones drop gracefully to the wayside. Never forget that you are at the beginning of a design process and that your ultimate goal for the design is something much bigger and more complex than whatever thing you are making right now. As an architect, you are ultimately responsible for making buildings, not drawings or models of buildings. If all you want to do is make drawings or models, maybe you should have gone to art school.

Sketch models, like sketches on paper, should be taken cheaply. If they are easy to create, they are also easy to discard. But an idea that you are sure is a good one may deserve a better, more detailed, and more precious construction. The very same impulse that prevents you from discarding a bad idea that you spent a lot of time to document can help you to remember which ideas were the good ones – the ones you don't want to lose quickly. A carefully handcrafted element in a model is evidence of your love for what it represents. Maybe, in some cases, that one element is the only thing you want to preserve. Perhaps what you really need to do is tear away all the rest of the sketch model, instead. Remember to *". . .exaggerate the essential, and leave the obvious unclear"* (Jansen 2013).

Admittedly, it is harder to treat physical models abstractly than it is to do the same with a sketch on paper. Models require you to think about gravity, about connections, and about the challenges of cutting and shaping the material they are made from. There are no abstractly thin surfaces in a model; everything you build has some kind of thickness. And there are no plastically curving organic surfaces in your model, either – curves have to be developed up from their material first principles by folding, cutting, scoring... or sculpting from some abstract plastic material that you'll probably have trouble rationalizing for construction later.

Most architects at some point in their education spend a few late nights (or maybe weeks of them) cutting and gluing together little pieces of paperboard and wood in the run-up to their final end of the year studio critique. In their professional lives to follow, they may find themselves engaged in the same kind of work in the run-up to an important client meeting or a presentation to a city planning department. Physical models present very well, and most ordinary people are fascinated by them. A presentation model of a client's building is like a gift – it is a beautiful, gem-like miniature of the project to come that captivates any audience. Perhaps, this is precisely the result you want – your client may be so dazzled with the model that they entirely forget to question the design it represents. Beware of this – it can help you get an approval, but if your client hasn't really understood the proposal you're making, there might well be big problems later when changes are harder to make.

The digital analogy to this sort of physical presentation model is the photorealistic rendering. Digital model-making is, in many ways, the perfect space for sketch modeling. The models are quick to make, and it is easy to think of them as temporary and disposable. Digital tools actually make modeling faster in many respects than sketching on paper, and this is a significant advantage for your process. You can build a dozen massing models in the time it takes to draw a single detailed perspective sketch. But just like physical modeling, it is easy to get caught in details too early – details that will lead you to fall in love with a model that beautifully represents poor design decisions. Rendering engines, employed too soon, can be a real distraction.

For me, a photorealistic rendering, peppered with trees, people, and other realistic entourage, often does nothing for the design. There is undoubtedly a place for rendering in a sketching practice, for "sketch-rendering." If your design exploration has taken you in a direction where the physical presence of material needs to be considered, a digital rendering can help. Similarly, if your design depends on the play of natural light along a roughened wall, you should add renderings to your toolbox. But beware, as with any sketching tool, the allure of too much detail too soon. With digital model making, it is often easier to pull a photorealistic rendering than it is to create some more sketchy-looking analog. But the means of representation do matter, especially to less visually sophisticated viewers.

Fundamentally, this chapter is not about presentation models. Every sketch you make should, of course, be available to you when you're putting together a formal presentation. Still, for every presentation, there are likely dozens, maybe hundreds, of sketch assets that might only make sense to you. I want you to think about the models you make that you never plan to show to anyone – the models you make just to help you solve a problem or figure out a particularly complex situation in your design scheme. Models with the rough and tumble status of a sketch. Maybe you're having trouble figuring out how your roofline is going to come together, or perhaps you're working up a detail for an ornamental handrail on a staircase, and you don't know what materials will work the best. Sketching, in 3D, either physically or digitally, may be just the thing you need to do.

And just as when drawing, there is no single best place to begin when modeling. If you don't know where to start, anywhere will work. Doing something, even if you know it is wrong, will break through the static inertia of doing nothing. You have to throw the first thing out before you can understand what the second might be. And just as when drawing, you can't know if an idea is good or bad until you've tried it out. Good ideas need context to be understood. If the idea is good, then it is so only in comparison to something that isn't. You have to start somewhere.

And in modeling, there are so many different ways to get started. You might carefully construct a chipboard analog from a detailed set of plans and elevations, carefully cutting out every door and window, meticulously crafting every step of every staircase. Or, you might just find an interesting rock on a walk down by the river that makes you think differently about how your client's parking lot should be arranged. Sketch models, like any sketch, can begin with anything.

And I'm not joking about the rock. Models made by assembling bits and pieces you have found around you in the world may well be the right way to get something new under consideration in your design. I like to have piles of stuff around me when I'm working on a tough idea. I find that I work better and faster from scrap material than anything else. I know that I am not going to have to mill something down for half an hour before it is ready for use. I can just collage bits and pieces together quickly to see what sticks. And, as scrap, I know there's no waste from me making a mistake. Scrap material starts with no value, and it is only through work with it that anything worth saving emerges.

There is no single silver bullet that will be the right sketching medium for you every time. Every project is different, just like every practice. You may not know in any rational way where your sketches will take the design you're working on right now. Just like sketching with paper and pencil, sketch modeling can be an answer to a design question, or it may simply open new questions for you to consider further. I like to use modeling as a way to help me reframe a design problem when I've gotten stuck.

Because every designer gets stuck sometimes, actually you probably get stuck all the time.

Free-wheeling though your exploration may be when you're in that reflection-in-action flow state that we all love so much, you will run into problems in your design as often as you find solutions. You might not be able to see your way clear of in a particular problem when you are standing in the middle of it. To move forward, you have to reframe. Maybe you can reframe by thinking about the original requirements differently, or by taking a look at the problem from a different point of view.

I have found the best way to reframe any design problem is to switch media. If I was sketching with a pencil on opaque bond paper before, maybe it is time to change to trace and a sharpie. Or to go for a walk to see if I can find any interesting rocks. Or perhaps it is time to try building a model.

A quick model in any medium will force you to rethink assumptions you made on paper, from a fixed point of view. A model that you can pick up and manipulate is one that you have to think about in a wholly different way. The simple act of making it will force you to reframe your design problem, and in doing so, you may just find an unexpected new way forward with it.

Maybe you should make a model the first thing you do, even before picking up the pencil. Some architects just think more effectively about design in model form. They just think and work better with their hands. I think Frank Gehry may be one of these. His studio is full of models. Sketch models hide fewer sins than traditional sketches can. In a drawing, you can always stick a bit of shrubbery in front of that corner detail you haven't figured out before your client presentation. A physical model will show all such sins if you pay attention to what it tells you when you're building it.

Real sketch models are made by you for yourself, for your personal use. You might make one for any number of different reasons. Maybe you are having trouble conceptualizing the flow of terrain on a complex mountain site. Or perhaps you need to figure out some complicated formal intersection between big masses in your design. Or maybe you just need to figure out how a corner detail is going to get built for the first time. Sketching is a way for you to talk to yourself, for you to speak to your design. Never be precious about them prematurely.

Physical Modeling

As a student, the place I most loved working was the shop. Cooper Union maintained, in partnership between the Art and Architecture Schools, a shop filled with enough tools to build an entire house from scratch. In support of the tools, we also had access to shop technicians who were highly skilled craftspeople, experts in wood, metal, casting, and plastics. Through the course of my education there, I learned how to make many things with my hands, using a wide variety of tools and materials. I carry those skills with me everywhere I go now, and they are an innate part of the work I do every day. I'm not actually building houses very often anymore, but I maintain a garage full of tools and have worked my way in (and out) of all kinds of shared shop spaces over the last 20 years. I've learned a lifelong connection with making things.

It is truly a joy to make things with your hands. Drawing is a kind of gateway drug for the vast range of modeling materials that await the intrepid sketch modeler. Your new best friends are a hobby knife and a bottle of glue – with them and scrap cardboard rescued from your nearest recycling bin, you can create magical things that will take your ideas much farther than you might otherwise be able to take them on paper with a pencil. Beautiful little simulacra of your project that take on a life of their own as you're working. Like Gepetto, in his workshop, my designs come to life as I'm working on them.

I carefully cultivate collections of scrap material that have no significance to anyone but me. It is frustrating to the orderly minded people in my life that I need this, but I just can't feel creative without a scrap bin of stuff nearby. All the best shops have a scrap bin, though far too many of them call it the "trash" and try to throw it away all the time. They should quit doing that immediately. Or, call me, and I'll come to pick it up and take it off their hands. Because I need it to get anything worthwhile done.

Presentation-quality production model making operates under different rules from sketch modeling. Production models are carefully planned, made from materials purchased just for that one purpose. They are expensive, detailed, and take a long time to complete. Maybe they even have cute little LED lights in them to illuminate them at night. They are not intended to help find and resolve design problems. They are built to convince clients and their stakeholders that the design decisions you have already made were the right ones. I think these kinds of models are great. Mainly because after they have been built, there is a fresh boxful of scrap in the shop so I can start sketching on the next project.

Sketch models, unlike presentation models, are made in the same rough and tumble way that a paper and pencil sketch is made. They are made of worthless materials, meant to be built, considered, and thrown away. You need to be able to work fast. You need to be able to work cheaply. You may find you have some particular kinds of materials that you like to work with; museum board, maybe. Or little sticks of basswood. Whatever does it for you. So you might keep a selection in stock by your desk all the time. Or maybe, you just make models out of whatever stuff you have at hand. A reflection-in-action sketching process doesn't care what you have at hand. Whatever it is, when you need it, you'll put it in play.

Figure 5.1: Seymour Papert and Lego Mindstorms.

Seymour Papert (Papert 1993) identified two primary modalities of problem-solving. The first is analytical problem solving – where a designer might identify design requirements and then optimize known patterns and prior solutions until the requirements are met. The other is a process where the designer collects whatever materials are at hand and messes around with them until something interesting happens. Sometimes referred to as "bricolage" (from the French equivalent to "DIY") in academic circles, this is the strategy I suggest most closely matches the needs of a sketch modeler (Figure 5.1).

Papert is perhaps best known for his work on "constructionist"[1] learning theories, which led to the commercial development of the Lego Mindstorms construction kit. Prototyped in Papert's lab at MIT, Mindstorms kits provided a simple palette of Lego blocks that add basic physical computing capability to Lego constructions and a graphical programming language that allows students to snap together blocks of code to write simple programs to control them. Mindstorms were developed to encourage thinking by making. While designed for children as an educational toy, I believe Mindstorms represent strong evidence that all kinds of thinkers benefit from model making of some sort. As an architect, though, maybe you don't need your hand held quite as tightly, you are probably capable of sticking some bits of something together without the scaffolding of Lego's kit-of-parts approach.

The work of Papert's lab at MIT had wider-ranging results than just those that informed Mindstorms. He and his colleagues also opened up a new kind of digital model making that I think has relevance to architectural designers as well. Physical models need not necessarily be considered as dumb, static objects any more. An architectural model might also contain sensors or actuators, coupled with simple programmable microcontrollers so that the model might begin to acquire some simple "thought" capacity of its own.

The buildings of the future will unquestionably be imbued with considerable embedded computational capability. Even the simplest of buildings already have some kind of computer (if even just a basic analog feedback unit) attached to their

---

[1] Papert was following the work of Jean Piaget, who pioneered "constructivism" as a theory of early childhood education.

heating and cooling systems to control when and for how long the heat should run in the wintertime. My home today includes sensors for temperature, air quality, light and sound levels, and many other things. My lights, appliances, and home stereo system are all bound together in a "smart home" that (frankly) seems to have a mind of its own. A simple mind, I think. The last time I called out, "Hey Google, turn on the coffee machine," my robot vacuum came to life and started cleaning the hallway. The systems that bind all of this together are complex to design and work best when thought of in the overall context of the architecture in which they are doing duty.

Physical model makers may well benefit from sketching physically with these kinds of thinking things and materials in mind. They aren't science fiction anymore – you can buy them at IKEA. How will you consider these capabilities in the next building you design? And how well do you think you really understand them if you don't have any direct manual experience working with them or their control systems? You need to lay hands on them, learn by making with them what affordances they will genuinely offer, and what problems they will pose.

I find that I can't really understand any new material until I have had it in my hands. This holds true for the finished surface materials in a project like tile, stone counter surfacing, or the wood finishes on cabinets. Most designers will collage samples of important finishes onto a project finish board of some design, but there's considerable benefit to be found in having real material samples available to you while you're sketching. You might find that the wood finish your client loves is oddly heavy, or that it scratches easily. You might find that the factory finish on it has a particularly beautiful subsurface shimmer that is visible only when it reflects light at a particular angle. These kinds of things cannot easily be simulated in photorealistic rendering. You have to learn about them by manually working with the real material.

I have seen beautifully instructive sketch models built only by collaging together material samples acquired for free from a product manufacturer's sales rep hopeful that their product will survive through the sketch design phase and become considered a critical design feature to preserve through construction. These highly abstract models may not necessarily confirm many of the most specific design ideas about form, but they are unassailably precise when it comes to defining a material "feel" for the final project. There are few better ways to explore the material qualities of a particular stone next to a specific wood, contrasted against a paint color for the walls than to physically collage them together.

Sometimes, even when you really need to be sure about what you will get from a material, you may need to model a small segment of your project with original materials at full scale. This may seem expensive, but it is also an essential form of physical sketching. A full-scale mock-up of some unique detail condition will answer so many questions before construction begins. You can't avoid the information a mock-up will grant you and your team about the true constructibility of your project. If you can shape the materials yourself, even cutting or welding steel for the frame, you will also earn considerable credibility with your contractor.

By contrast, it may be that when you are ready to make a sketch model, you don't care at all about the materiality of your project. There is a beautifully abstract quality to non-materials like chipboard, foam core, and basswood. If you are mainly learning about the overall massing of your project, it may make sense to start from a set of wooden blocks, shaped only with a disk sander. Particularly early in your design work, the nonmateriality of a sketch model is essential. If you haven't chosen between

Figure 5.2: Froebel blocks.

brick or stucco for your project's exterior, why get hung up on picking the right color too early?

Frank Lloyd Wright is known to have played extensively with Froebel blocks as a child,[2] and many historians claim to see their geometric influence throughout Wright's extensive design career (Figure 5.2). A collection of blocks representing the primary programmatic elements of a design can be quickly fabricated and then combined and recombined in many different arrangements quickly by a designer iterating rapidly on 3D planning ideas. For program elements with spatial qualities beyond what can be thought through in a simple sketch, playing with wooden blocks may actually be the best way to learn about what the building wants to become.

I can never think about physical modeling without considering the tools at my disposal to shape whatever material I've chosen to work with. A hobby knife, bottle of glue, and recycled cardboard are readily available to anyone. They are perhaps the best sketch-modeling analog to a pencil and paper. With these three simple tools (possibly aided by a cutting mat, a metal ruler, and a pencil), models of great beauty and complexity can be built. Because the materials are cheap and common, it is easy to remember their sketch quality and easy to discard models that fail to carry a useful idea forward. They can be quickly broken down for scrap and recycled into the next sketch model.

I have almost always also had access to a few simple woodworking tools when modeling. A bandsaw and a benchtop sander are tools which I find are capable of most modeling tasks I might need to tackle. I find they allow me to sketch freely in wood in a way very similar to what can be done with cardboard (Figure 5.3). The two material languages, in fact, combine well together. They have a sort of cellulose-centered harmonic relationship that makes them useful companions. The mechanical cutting and shaping assistance of a decent set of power tools changes my working style, though. There's an element of danger in sharp tools, but that just helps me to focus on the work at hand without distraction.

---

[2] (Wright 2010) on pg. 17, "For several years I sat at the little kindergarten table-top ruled by lines about four inches apart each way making four-inch squares; and, among other things, played upon these 'unit-lines' with the square (cube), the circle (sphere) and the triangle (tetrahedron or tripod)—these were smooth maple-wood blocks. All are in my fingers to this day."

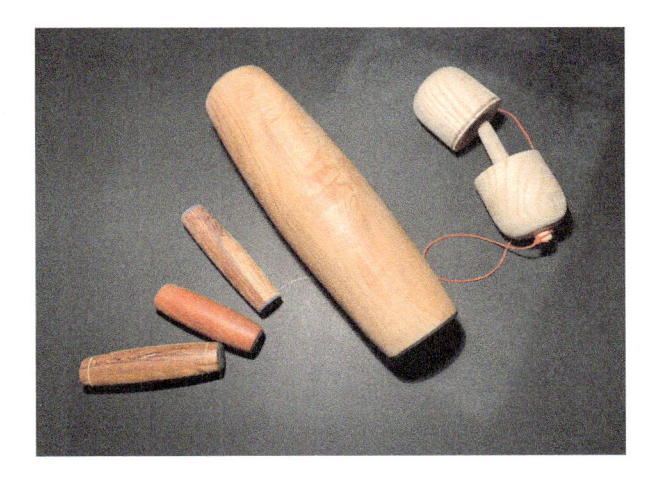

Figure 5.3: Lathed forms from my shop.

Recently, I acquired an old lathe in my shop, and that changed the kind of design work I was interested in for months. All of a sudden, I could make perfectly round things, not just prismatic forms with straight cuts and flat planar surfaces. Suddenly, all I could think about making were revolved forms. Ornate banisters, which I had until then dismissed as hopelessly romantic (I'm a child of the Bauhaus, after all), suddenly caught my attention. I was in love with a new kind of form because suddenly, I understood the magic of its making. With my hands, thanks to my new mechanical advantage, I sensed something new.

I had a similar experience the first time I played with a 3D printer. Suddenly, I had a quick way to physicalize the purely digital models I was building in SketchUp. I could print them out at scale and integrate them into a more substantial physical model I might have been working on. But the 3D printer also brought with it a new kind of physicality – models couldn't always be printed, because they had overhanging parts or parts that were just too small for the printer's extruder to resolve. And I could still see horizontal ridges, lasting evidence of the formal process of their making. But no matter – 3D printed parts in a physical model are pretty great for what they are, and they offer an elegant integration point between digital and physical modeling processes.

I think your choice of tools will always make some difference in the kind of design you're able to do. Maybe, even, in the sort of design you're able to imagine in the first place. Physical model making forces you to confront the physicality of the materials you choose for your project. You have to know how to shape those materials, how to connect them together, and how they behave under the physical realities of light and space of the real world. Models can't lie about these things; you can't fudge the details in quite the same way you can with a drawing. You have to take a little more time, you have to get your hands a little dirty.

And make no mistake about this. Sketch modeling is a dirty, dusty, stained, and sticky mess. You need piles of scrap material scattered around you. There will be sawdust, scraps of paper and sticky overshoot from your glue. You might cut yourself. You should wear safety glasses and ear protection. You're going to wear a hole in your jeans. Model making isn't something you do in a three-piece suit. Your client will know you've been at it because you'll earn some calluses and gunk under your fingernails. But don't worry, they'll thank you for it when they see how much better their design has become.

Sketching is always about finding not just the most optimal solution to a known problem, but ideally, about discovering something novel, something delightfully unique. A sketch doesn't know the answer to a question before you make it. You are using sketching to figure out what the thing wants to become through the process of its making. If you listen to it, a sketch model will tell you a story about what your building can become.

Figure 5.4: Ivan Sutherland's "Sketchpad" application, running at MIT Lincoln lab.

Beware of closed systems in which are encoded the full range of possible outcomes they can create. Closed systems – systems that presuppose solutions in the very nature of their system definition – are harder to use in unexpected and delightful ways. I think you always need some way to inject the unexpected in them, some kind of external source of randomness to instigate new thinking for surprising outcomes. Umberto Eco wrote on this subject in his 1962 book *The Open Work* (translated to English in 1989). He argued that the best and most rewarding books were those that invited interpretation. Works in which multiple interpretations might be found. An open work, to Eco, is one that invited the reader into it to build their own world (Eco 1988).

Great buildings are like this, too. They don't dictate, they invite. You are designing a building in which others will live their lives. You owe it to them to leave the design as open as it can be. They will bring their own meaning to it in time. They will shape and reshape the materials you choose, and they will arrange their lives in the carefully crafted plans you design for them as they see fit – not as you direct. That loading dock you tried to hide may become the place everyone wants to eat lunch on a beautiful spring day. Buildings learn what they will become every day. Your sketching should acknowledge that right from the beginning.

I find that sketch model making reminds me that I am not singularly in charge of anything about how the building will be when it is complete. I have a role in the beginning, and I should set things off on the right path if I can. But the material, the site, and a thousand other factors will come into play that I have little to no control over in the future. Just like what happens when I'm sketching by hand.

Digital Modeling

Perhaps more than any other single technological advancement in our industry, digital 3D modeling has changed the way architecture and design are done. 3D modeling is still at its infancy from a technical perspective – especially when compared to other drawing and modeling technologies that architects have employed for centuries. But it is hard to find an architect in practice today who doesn't at least acknowledge that they should be modeling their work in some digital 3D modeling application. The benefits of adoption are significant. If you're reading this book because you aren't so sure if you should be adding digital 3D modeling to your process – I think I'll convince you easily.

The first graphical 3D modeling applications were demonstrated only 50 years ago (Figure 5.4). It took another 20 years to get them into the hands of ordinary working professionals. It took another 10 years before the specialized graphics hardware that was needed to really put a model interactively on your computer's screen was really widely accessible to

DIGITAL MODELING

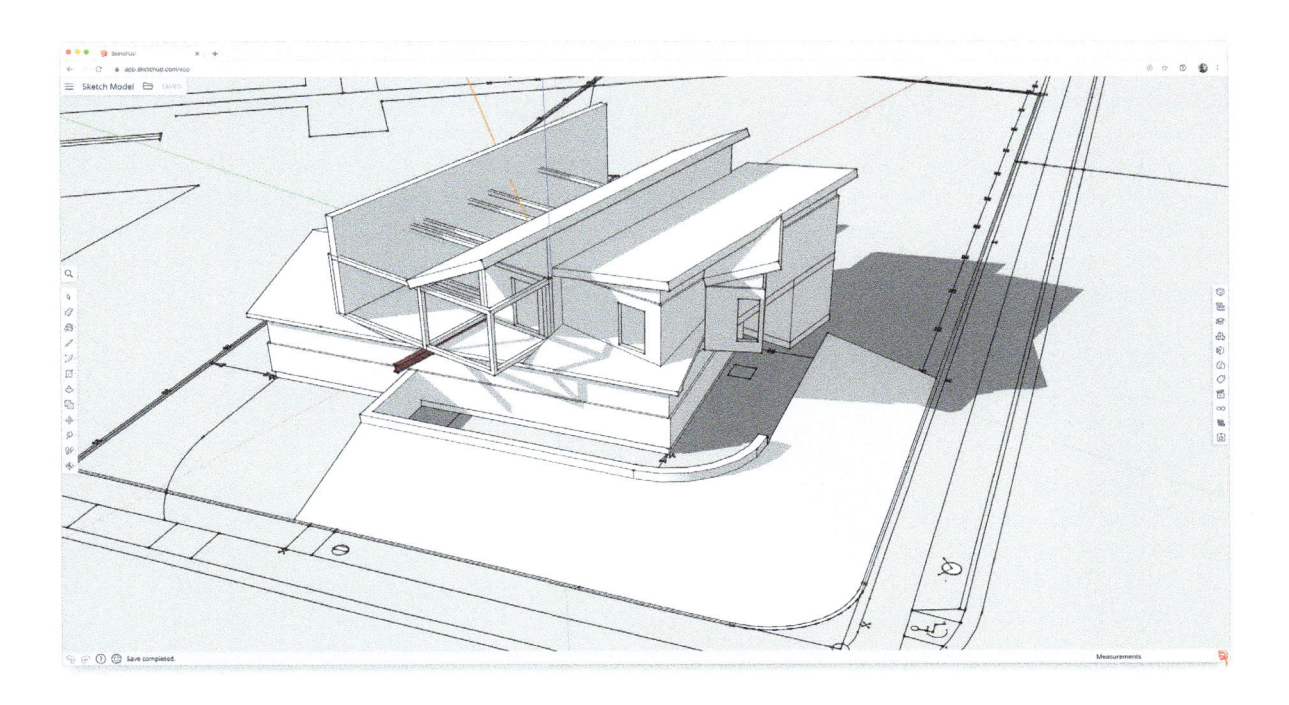

Figure 5.5: Modeling in 3D with SketchUp.

professionals. You're probably most familiar with 3D modeling that works inside an interactive perspective view composed of shaded, lit, and textured polygons that can be shaped with a palette of digital tools. Once inaccessible to any but the most advanced users, these applications can now run on a computer that costs only a couple hundred dollars (Figure 5.5).

SketchUp painted its first screenful of pixels in late 1999 and was launched to the public in May of the following year. When we launched it, we were leaning on decades of prior research and development from across our industry, like all software companies do. But we also had a unique advantage. SketchUp came to life just as cheap 3D graphics cards targeted at consumer video gamers came to the market. The computer graphics juggernaut that you know today as NVIDIA launched its first consumer graphics card in 1995 (NVIDIA Corporation 2020). In 1999, the same year that SketchUp began, NVIDIA unveiled its first graphics processing unit (GPU).

In your computer, the GPU is a special kind of computation unit that is dedicated to rendering 3D models to the screen at interactive frame rates. A computer with a GPU is typically capable of rendering scenes of high complexity to your computer's screen at 30 frames per second or faster. When you virtually pick up a model on your computer screen and orbit it smoothly to a new point of view without visual lag or stutter, thank the designers of your GPU.

Before the commercialization of GPUs, graphics on personal computers had been much slower and much more abstract. 3D models were typically edited in wireframe views, visible as solid forms with color and texture only after laboriously rendering them – a process that could (and sometimes did) take hours to complete. Real-time 3D graphics were reserved for the most advanced computers and software applications and were financially out of reach for any but the largest and most profitable of design firms.

The computing power available to the average architect today is so much higher than what was available to them even 10 years ago that close examination of prior capability is mostly irrelevant. You can model interactively today with exceptional fidelity, detail, and visual style. It is almost too easy to develop models to fully constructible levels of detail. Every single piece of rebar, every doorknob, and bathroom fixture, everything down to the anchor bolts in the foundation, can be modeled to full fidelity. Models at this constructible level of detail will undoubtedly revolutionize the construction industry. They will dramatically reduce surprises on the construction site to minor annoyances because the entire building and its construction can have been simulated and rehearsed digitally to perfection in the office before anyone even thinks about swinging a hammer on site.

You probably already know the names of several of the most popular 3D modeling applications, and maybe you even have a favorite. Or perhaps you're struggling with one that someone else, maybe a boss or a client, told you that you had to use. I'm unashamed to say SketchUp is my favorite, mainly I'm sure because it is the one I helped to design. But I think I'm not alone – there are profoundly large numbers of people modeling in SketchUp today. We must have gotten something right. I know what we were thinking was required for a digital sketching process when we designed SketchUp, and I know that those are things that every architect needs in their work.

Generally speaking, people who know SketchUp believe that it is easy to learn and easy to use. They may feel that it is missing some features compared with other modeling tools, but I don't care too much about that. We have deliberately kept SketchUp as simple as we could, and I believe that simplicity is at the core of SketchUp's power. It is simple, like a pencil, simple like chipboard and glue. Simple but powerful. SketchUp is perfect for making digital sketch models.

For many traditionally trained architects, the adoption of a 3D modeling tool is challenging. And they are predisposed to understand how it works. 3D modeling software is usually frustrating to most people because it is very unlike any other class of software that they commonly use. It doesn't really match any prior experience that most people have in the real world. There is nothing inherently "intuitive" about any computer software, though software designers like me often talk freely about how intuitive our stuff is. Most people find software tools frustrating to use ("un-intuitive") when they fail to match expectations set by some prior experience.

Word processors are "intuitive" to use because they are like typewriters (remember typewriters?). Photo editing tools are "intuitive" because they are like cameras and darkrooms. Painting programs are "intuitive" because they model paintbrushes and paint palettes. Spreadsheets are "intuitive" because they model ledgers and tables. And so on. 3D sculpting tools rely on clay modeling as scaffolding for "intuitiveness." In SketchUp's case, we relied on "drawing" for our scaffolding. Modeling in SketchUp is more like drawing with a pencil on paper than it is like anything else. Maybe even more than physical modeling.

Unfortunately, there are fewer people in the world with real-world experience, either clay modeling or drawing, than there are people who remember typewriters. Everyone has fiddled with a lump of

DIGITAL MODELING

clay at some point, but few people are capable of sculpting a car with it. SketchUp was a quick hit with architects because they are all taught how to draw – and (especially) how to "think by drawing." By matching a familiar way of thinking, we made the transition to something new easier.

When you're viewing 3D space on a traditional computer screen, it is as if you are viewing it through a window. This is similar to the experience of space in a drawing or a photograph. In both those cases as well, space is typically understood as being an abstraction projected onto a picture place, physically flat, but inhabitable by the viewer's mind. When you're modeling in 3D space, you've got to figure out how to reach past the picture plane and "touch" the elements of the model somehow. But you can't really reach behind the picture plane, of course – it is only a rasterized projection of 3D space painted electronically onto a pane of glass.

3D modeling on a traditional computer display is kind of like trying to model in the real world with a pane of glass between you and your model. You can only touch the surface of the glass, but everything you're working on is behind the glass. And to make matters more complicated, you can only use the set of tools that someone thought to leave behind the glass for you. Kind of like you're working in a glove box (Figure 5.6).

When you're drawing, you probably have an intuitive knowledge of know-how to work in the space implied by a perspective projection. You probably also know how to imagine the space of a building by chunking it up from multiple plans, sections, and details. This is one of the especially magical skills that architects learn to do, and it is one of the things that make the architect's brain identifiably different from others. You have doubtless seen your clients struggle to do the same things you do intuitively when working only from a set of drawings.

Architects learn how to design simultaneously from both the outside and the inside, at multiple scales,

Figure 5.6: Modeling in a glove box.

and varying levels of detail. From far away perspectives (at an appropriately low level of detail) to highly specific constructible interior details. They are also able to switch between all these different modalities freely and intuitively. In fact, switching between these modes rapidly is considered a good design habit. You should get so good at doing this that you don't even recognize you are doing it anymore.

At the beginning of every digital 3D sketch, at any scale and in any projection system, there are only lines. Lines that start and end at a point, lines that, in time and with more context, surround and imply surfaces and forms. Lines are the most versatile of sketching tools. When drawing by hand, lines can be made with a pencil on paper. When sketching in 3D, you need a tool that makes lines as versatle, as simply, and as powerfully as a pencil (Figure 5.7).

Most 3D modeling tools sold to architects and the construction teams they work with are too specific for the fast and loose practice of sketching I'm promoting. They ask the designer to answer too many questions about the design far too early in their design process. There is, of course, a time and place for detailed, constructible building simulations, and if your design isn't heading in that direction eventually, then you are missing out. But like any sketching tool, too much detail too early will block your creativity. Jumping immediately to the conclusion of the first idea you thought up will not give you the best design solution. You'll find you've given up your freedom to experiment prematurely and are now bound to the first hare-brained concept that came to mind.

Drawings made of points, lines surfaces, and forms leave room for interpretation, room for the design to speak back to you as opportunities are uncovered. A point might represent a point of view or a coordinate origin, or it might represent the beginning of a line. That line might be an axial vista with a reflecting pool eventually or a fence. A collection of lines that bound a surface might come to represent a wall, or maybe that fence I just mentioned

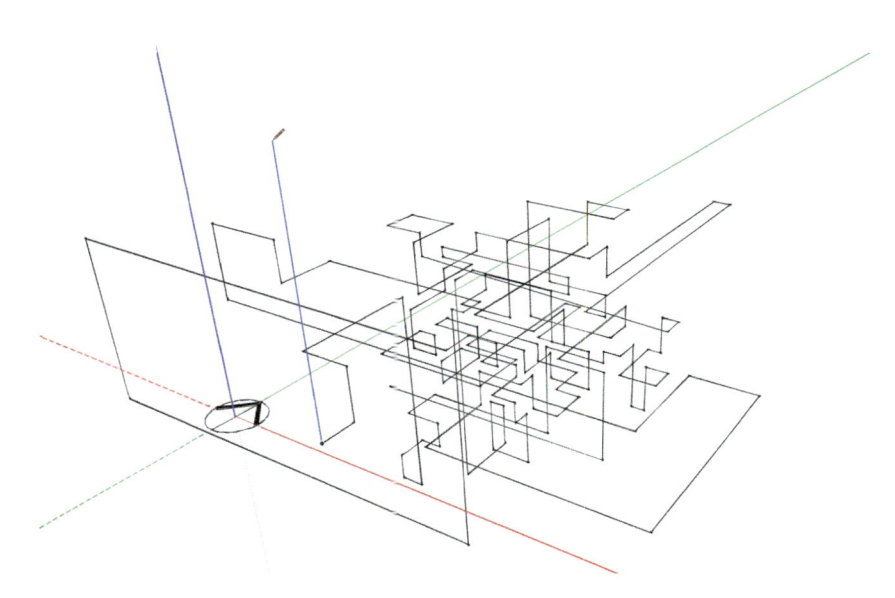

Figure 5.7: SketchUp's line tool.

DIGITAL MODELING

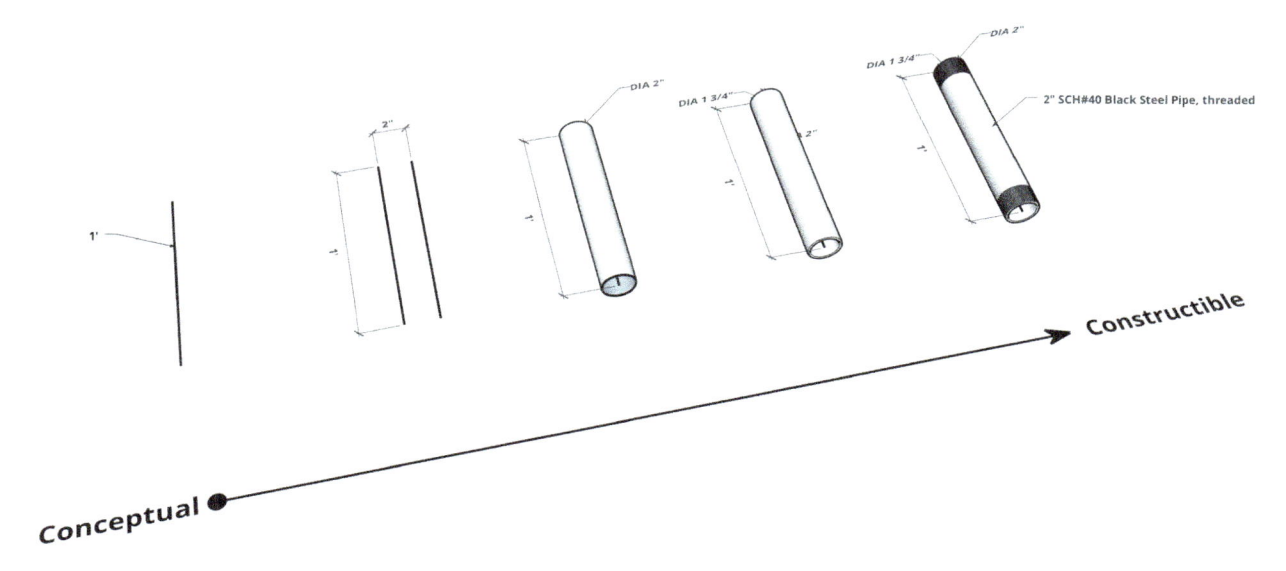

Figure 5.8: A model progression, from line to detailed AEC object.

seen from another point of view. Together, a collection of points, lines, and surfaces in time will come to represent a building, with all its depth and spatial character intact (Figure 5.8).

A line can become a "beam" when you're ready, and you need the additional constraint that will imply. Beams can be woven together in logical systems of truss work and then be detailed down to the individual gusset plate. A line can become a steel beam, a wood beam, or a post-tensioned concrete beam. Each is detailed in entirely different ways and demands totally different methods of construction. But you don't need to decide all of that right away. While you're sketching, let the beam be just a line for as long as you can.

Because once you've assigned the additional conceptual weight to that line, specifying details about its construction, it is difficult, if not impossible, to work it back into being just a line again. Decisions, even bad ones, are sticky. They are tough to make, and the people you work with will hold onto them tightly. Even if everyone knows they aren't wise. Nobody wants to throw work away, and decisions are a kind of work. Nobody wants to second guess, nobody wants to waste time rethinking things.

Nobody, except maybe you, because you know that sketching is a way of exploring ideas, of testing them for their fitness and throwing out the bad ones in favor of the ones that work best. You know that everything about the design is up for reconsideration while you're sketching. There is no penalty associated with discarding a bad idea, no sense of waste associated with replacing it for something better. You should never consider making a change just for the sake of changing something, nor would you willfully choose to make the design worse. But when you're sketching, changes come fast and furious. If you settle prematurely, you have to know that you are settling for a design that probably isn't as good as it could be.

Digital modeling tools like SketchUp are great environments for sketching in 3D. They are simple, like your pencil is simple. And powerful, too, in the way that your pencil is powerful. Once you accept the appropriate level of abstraction for the stage of design in which you find yourself, they are quick, nimble, and easy to use. You'll stop looking for more advanced tools until you actually need them in your practice. With experience, you'll learn to defend a level of abstraction in your 3D models previously reserved only for physical sketches with pencil on paper.

How 3D Modeling Actually Works

Digital space, in almost every case, is Cartesian in its measure. You almost always interact with space through a 2D projection of some kind. Usually, it is either an orthographic or a perspective view that is projected on the flat surface of your monitor. But behind that projection surface, the computer maintains a complete database of objects and their coordinate location. Points in 3D are addressed, by convention, with coordinate triads (x, y, and z). A standalone point may exist in the database, but more often, they are associated with some other class of objects. Maybe they are the insertion point for a component, or perhaps the endpoint of a line segment. Or they might be a special kind of grip or control vertex for a complex free-form surface. Points alone aren't perhaps graphically interesting until they have a higher purpose to serve of some sort. More than anything else, points are a way of encoding spatial location for some other object.

To be the most effective 3D sketch modeler that you can be, you should have some understanding of the mechanics that govern the 3D environment in which you will be sketching. And to understand how 3D spatial interaction works on your computer, you first have to know how points are picked in the virtual space behind the projection you see on your screen. Picking is how you tell the computer which of the set of available objects in the model is the one you want to work on next (Figure 5.9). For example, how would you pick just the table from this model to work on next?

It may be intuitively obvious how to pick the table in this scene. Just move the mouse over the table, and click to pick it. Easy, right? Think of it as the cursor location in your text editor. The computer needs you to tell it where you're going to work next. Maybe you use a mouse to point at the right spot, or perhaps you use the arrow keys on your keyboard to move it to the right place. Of course, you're doing that in the context of a screenful of paragraphs, words, and letters. But this is more complicated in 3D.

Let's skip the bit where the computer renders a 2D image of the scene for now. To pick the table, you look at a pre-rendered image of the scene presented to you on your computer's 2D screen. As you move your mouse across the 2D space of that rendered image, the computer registers a screen-space coordinate, a pair of integer numbers in the form (x, y). Discrete math at its purest. To you, looking at the projection of the scene, it may be intuitively obvious when you should stop moving the mouse – you move it until it is over the image of the table in the scene, but the computer has only (x, y) to work with. There is no simple way for the system to understand how deep you intend to pick in the scene. As you hover over the array of pixels that (to you) look most like the thing you want to select, the computer infers a line that passes from the point you indicated with your mouse on the 2D plane of your screen through to the vanishing point in the scene's perspective projection. It runs an algorithm to determine a list of objects through which the line passes. And then it sorts the results from nearest to farthest from the camera's point of view. The first object in that list, the one closest to the picture plane, is assumed to be the one you meant, and when you click the mouse, it is picked.

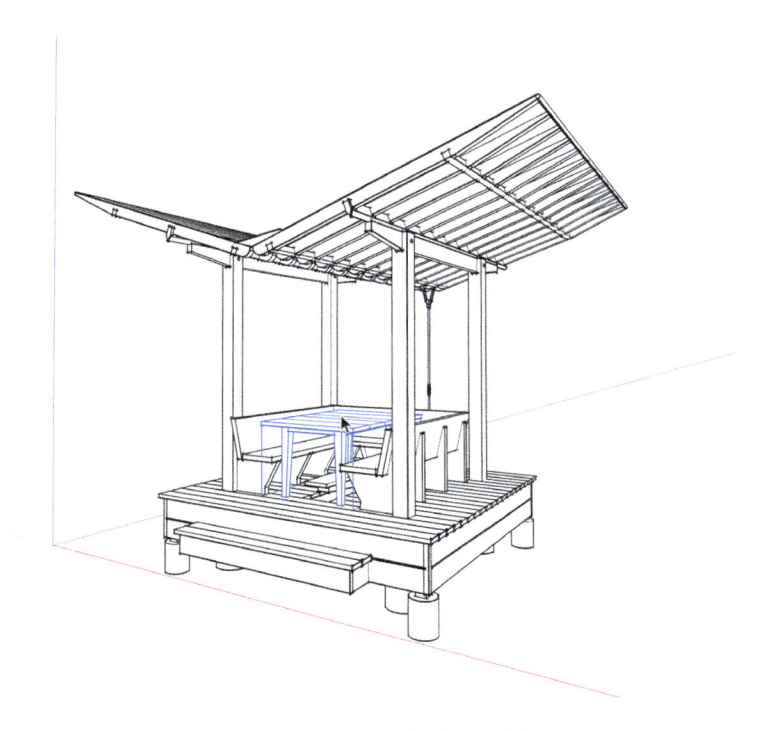

Figure 5.9: How would you pick the table out of this model?

This is a little more complicated than you thought, right? What if you actually wanted to pick the one in the back? Or maybe just the "right" one (from your mental model) from somewhere in the middle of the scene. Even though the computer has, in its memory, a nicely complete spatial representation of the scene and all the objects in it, it can only receive input from you through the 2D coordinates of your picture plane.

There are many other ways you can pick an object in the scene, of course. You might, for example, have some piece of UI that has a list of all the named objects in the scene. Here's how that looks in SketchUp (Figure 5.10).

Or your system may have something different that serves the same purpose. A layer system, for example, or maybe some kind of whole-model search field. Visual interaction in the model view clearly isn't the only way you can pick in your model. But when you're sketching, it is undoubtedly the most direct, and therefore probably the most straightforward and most convenient for you as a designer. Picking an object in the scene is really a way of drawing a point that represents either a simple declaration of ". . .this one," as input to the next operation you'll initiate, or as the starting or ending point of a process. For example, if you intend to draw a line, you'll need to pick a point where it starts (Figure 5.11).

A completely infinite empty space is the hardest place imaginable to pick a point. Without any geometric context, how does the computer know which of the infinite possible points along that line between the mouse click on your screen and the vanishing point at infinity is the one you want? It

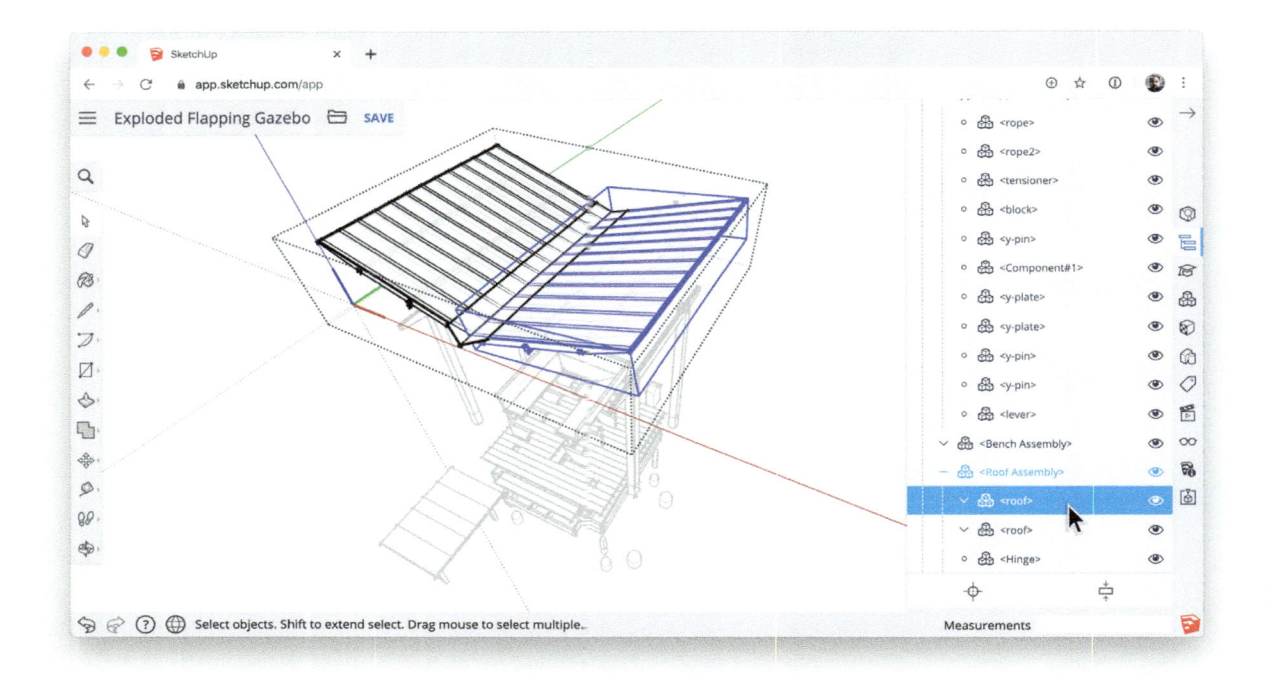

Figure 5.10: SketchUp's outliner.

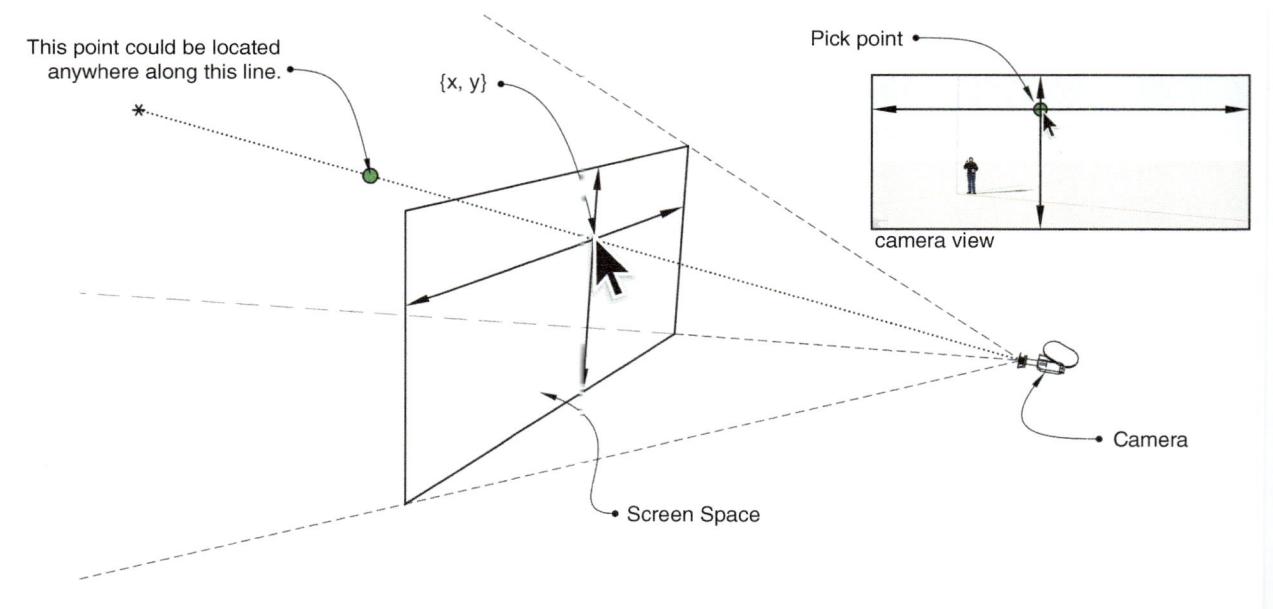

Figure 5.11: Picking a point in free space.

really could be anywhere along that line, mathematically. But it has to be somewhere. The computer has to make some decisions for you, based on an assumption of your intent at the particular moment of picking. We call these kinds of decisions, *heuristics,* and getting them right all the time is tough. It takes a designer to teach the computer how designers think.

In the case of a completely empty scene with no geometric context available to hint the decision, SketchUp will assume you mean to place a point on the ground plane. Given that as an assumption, there can be only one point that is both on the picking vector (between your mouse and the vanishing point) and also lies on the ground plane. So that is it unless a better choice can be made evident. Sometimes you might be aiming for a vertical plane, or maybe just trying to hit a point in free space somewhere that makes sense visually to you at that moment.

As you can imagine, this is a wickedly complex problem, and it is one that has no perfect solution for every designer and their unique mental model of how space works in their project. It gets easier, however, as you add more geometry into the scene. Context provides valuable hints to the picking system, making it more and more likely that it will be able to return you the point that you were, in your head, trying to pick. Once there is some geometry available, the system can infer location in 3D space relative to points, lines, and surfaces around your picking location.

Once you have picked an object, there are several things you might choose to do to it. You might move it or rotate it. You might decide to change its dimensions by scaling it. You might delete it, or make a copy. Or you might choose to create an array of copies. Sketching is a process of definition and refinement – of framing a problem, then reframing it over and over again until something new emerges. Something unexpected, something delightful. When sketching in 3D, you'll likely find yourself spending more time modifying things you've already modeled (or that someone else has modeled) than you might spend modeling entirely new things from scratch.

If your intent is to rearrange the components in a scene, how do you tell the computer where you want the one you've picked to move? It is going to be at least a two-step process – you have to tell the computer where you're starting the move and where you're ending it. You'll have to have some way to tell the computer which direction and how far you want to move your selection.

The start point is easy enough to define. It is the place where the object you picked up was before you decided to move it. But the second point can be tricky to place. Especially given you have only a 2D coordinate to offer. You're missing a dimension in the system. How do you tell the computer how near or far you want the object to be placed in the depth of the scene? Remember, when you click your mouse, the computer casts a ray through the scene to translate the 2D screen coordinate to a 3D model coordinate. Your object could end up anywhere along that line without additional information from you. It turns out, moving an object is a lot like drawing a line. You have to move a point through space.

Sketching Lines in 3D

At the beginning of every sketch, there are only lines. It is the same in 3D modeling as it is sketching by hand. Models are made of lines, too – lines that start and end at points, lines that surround and imply surfaces and forms. Lines are the most versatile of all sketching tools in any medium. When drawing by hand, lines can be made with a pencil on paper. When modeling digitally, lines are usually made with some kind of Line Tool (Figure 5.12).

SketchUp's line tool is unique in its ability to sketch in 3D space with minimal setup and prior construction. In a tool like Solidworks, modeling operations begin with the creation of a 2D "sketch" in a separate drawing environment (Solidworks 2020). This is done because it removes all the picking ambiguity from the tools, making it apparent to the system precisely what every mouse click from the user means. SketchUp isn't like that, though, because we figured that every time you mode-switch from one kind of view of the model to another, you run the risk of losing visual continuity and getting lost in the space of the model. The more continuity we could preserve, the better for our users. Continuous experiences are less burdensome cognitively, and they make it easier to focus on your design.

So SketchUp's line tool works in context, right in the main model view. To make it work, we implemented a system of geometric hinting we call an "inference system." This system helps you tell the computer where in space you intend to draw, based on a contextual understanding of the geometry adjacent to wherever you are working in space. SketchUp may not know what you mean specifically when you're working, but it is pretty good at "inferring" your intention from nearby spatial context. We have tuned and retuned this system endlessly over the years, and I'm sure we'll continue to optimize it for as long as SketchUp exists.

SketchUp's inference system is not unique in the world of 3D modeling, but I think it has been better tuned than most other similar examples. It is deceptively easy to build a basic geometric inference system, but wickedly complex to get it to work well enough that designers consider it "intuitive." If

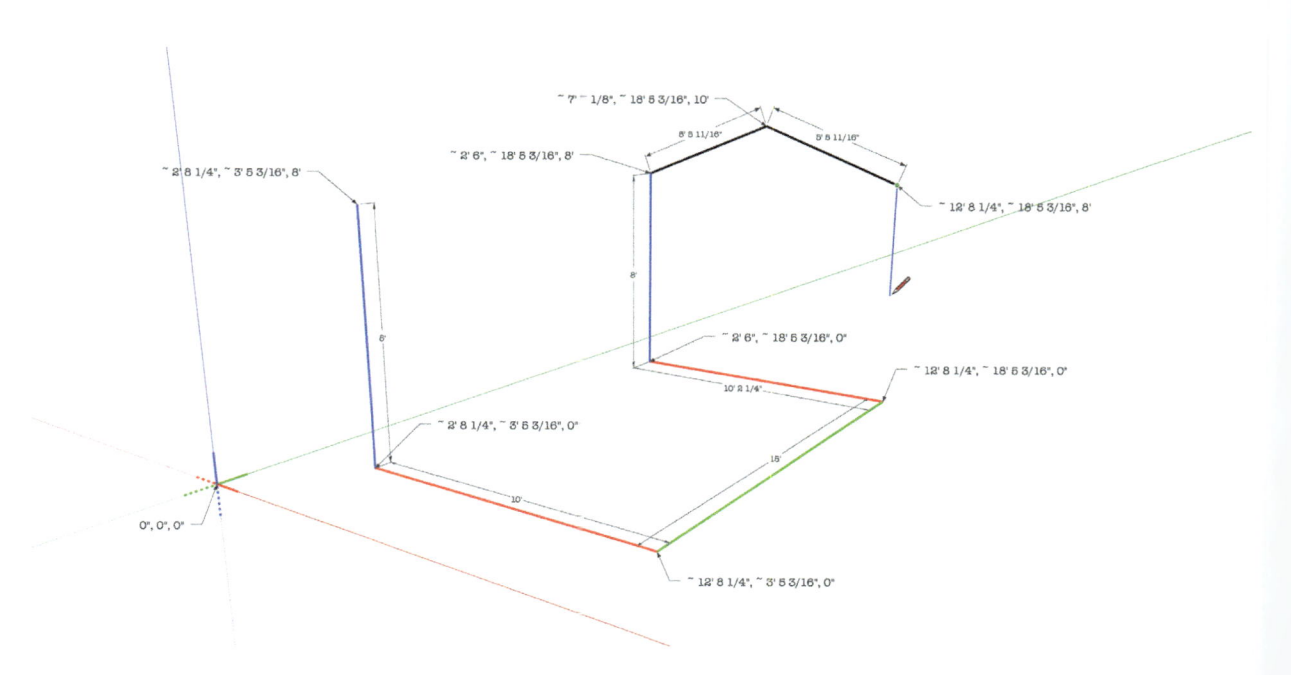

Figure 5.12: Drawing precise lines in free 3D space.

SKETCHING LINES IN 3D

we weren't careful, SketchUp might feel as though it was under- or overconstrained. Either extreme is unworkable, though for different reasons. Even after careful tuning, SketchUp may still sometimes feel as though it is wrestling with you over control of the cursor.

If a line is understood as the residue of a point moving through space, drawing a line with a line tool is closely related to moving an object through space. It begins with a point, indicated by a cursor hovering in screen space over some point of significance, and sealed with a mouse click. Now unique among the set of all possible starting points in the model, SketchUp begins tentatively drawing all possible lines from that point of beginning. As you move the cursor across your screen, SketchUp makes guesses about where you might want to place the second point, and therefore where you intend to draw a line (Figure 5.13).

But now there's a bit more context to inform SketchUp's guess. For example, you may intend to draw a line that is aligned with one of the cardinal directions in SketchUp's infinite Cartesian space. You might mean to draw in the red direction or green. Or maybe you intend to draw straight up, in the blue direction. Alternately, if your line started on a plane, you might mean the line to stay in that plane. Or maybe you want it to be perpendicular to that plane. Or perhaps parallel to one of the plane's edges. Or any of a myriad of other possibilities. For every screen space coordinate, there may be a half dozen possibilities to the inference engine, and hopefully one of them matches with your mental model, with your design intent.

With every line you place in the scene, SketchUp gets more hints to work with and gets more likely to pick the "right" one for you at the moment you're picking. Correct line placements lead naturally to more accurate decisions in the future. Gradually,

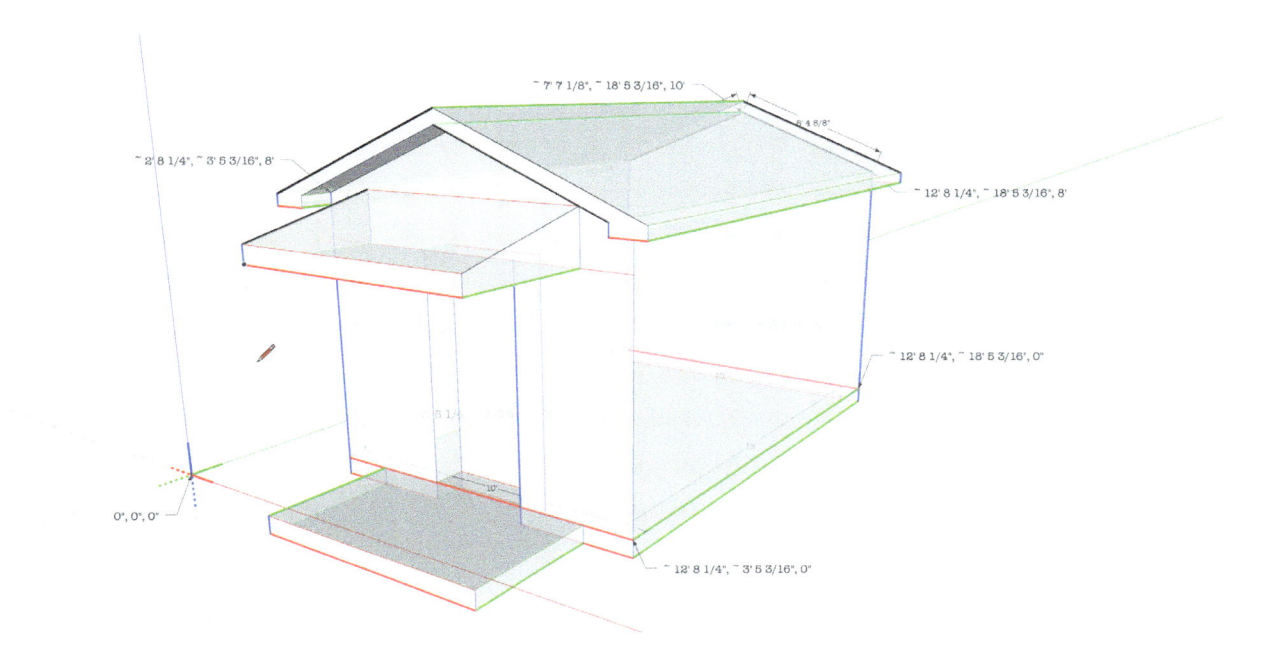

Figure 5.13: Using SketchUp's inference system.

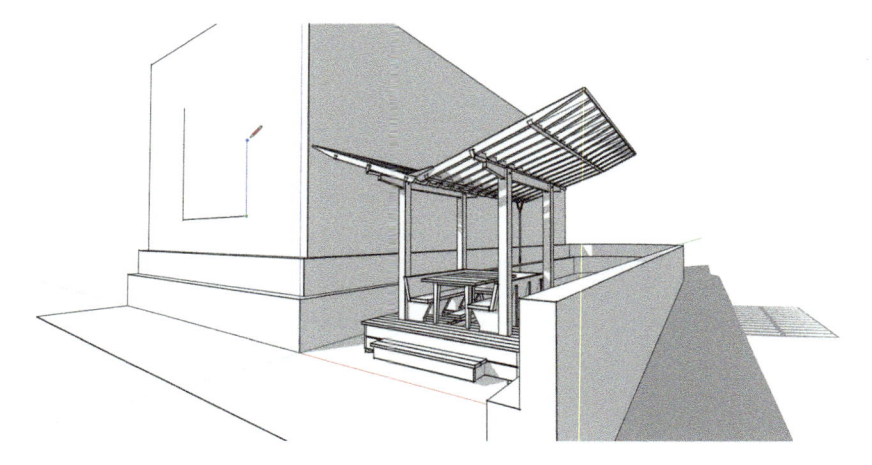

Figure 5.14: Modeling in SketchUp, one line at a time.

you can build up information in the model, line after line. What the lines represent to you and to your design may become increasingly clear over time as well, though, in the beginning, every line is just a line (Figure 5.14).

For all the varying types of lines that may be drawn in a 3D modeling system, the same rules apply. Your input to the computer model is restricted to the points you can pick. You have to have a way to tell the computer where you mean to be working, and (maybe later) what the lines you have drawn may come to represent in the context of your design. Arcs may be drawn with arc tools, curves with curve tools. Lines might be infinitely long as intended to show some unique alignment and nothing else. Or they may be the spine of some higher-order model object like a beam or a wall. Underneath all of these lines are some system of picking points in 3D space.

Sketching Surfaces in 3D

Geometrically, planes are an artifact of the lines that bound them at their edges. In SketchUp, this is literally true. Any three lines drawn in SketchUp, connected together at their endpoints, will automatically generate a triangular surface. More complex surface shapes are also allowed – any number of connected lines that form a closed loop together in a single plane will create a surface.

Delete one of the lines that bound a surface, and the surface will disappear. Add the line back, and the surface will come back, too. Surfaces in SketchUp, as they are in a hand-drawn sketch, are an implication of the edges that surround them, not an entity on their own to be manipulated with other specialized surface-manipulating tools. Surfaces can have holes in them, as many holes as you like. A hole in a surface is just another set of enclosing edges that can be manipulated like any others. And, at least in SketchUp, surfaces can be bounded by as many coplanar lines as needed (Figure 5.15).

By tradition in computer graphics, surfaces are visible only from one side. A surface is defined by its bounding edges. The "winding order" of their connections defines a front (and back) of the face. To save on computation and maximize scene rendering speed, back faces are typically not rendered

SKETCHING SURFACES IN 3D

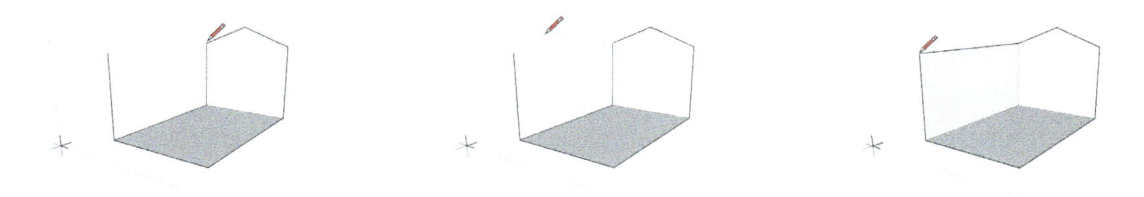

Figure 5.15: Surfaces in SketchUp are a product of their bounding lines.

at all. This can be confusing for people who are not computer graphics experts, and so in SketchUp, we render both the front and back faces of all surfaces all the time (Figure 5.16).

Unlike surfaces you might construct from paperboard in a physical model, SketchUp's surfaces are mathematically without thickness – they have no more thickness than the lines that define them. But on screen, when rendered for you to examine, they have a visual presence. At a minimum, they are always at least one pixel thick. All SketchUp surfaces are planar, even those that are rendered to look like they are smoothly curved. SketchUp is great for modeling "boxy" things. Most buildings, even those with undulating freeform facades, are ultimately composed of flat surfaces (Figure 5.17).

When modeling with straight lines, any three connected lines can form a triangle. This much we know from Euclid. If there are more than three lines, then an additional requirement must be met – all the connected lines must be coplanar with one another. If they are not, then other boundary lines must be added until planarity is achieved. Any free collection of points may be interconnected in at least one way to create a triangulated mesh. Usually, there are multiple possible solutions to this.

Freeform surfaces, like those found in organic objects, cannot as easily be modeled in SketchUp, but you will find that the basic principles of their modeling are quite similar in any freeform surface modeling environment. Surfaces are still defined by their boundary edges. In some cases, and with some caveats, those edges may be Bézier curves rather than straight line segments (Figure 5.18).

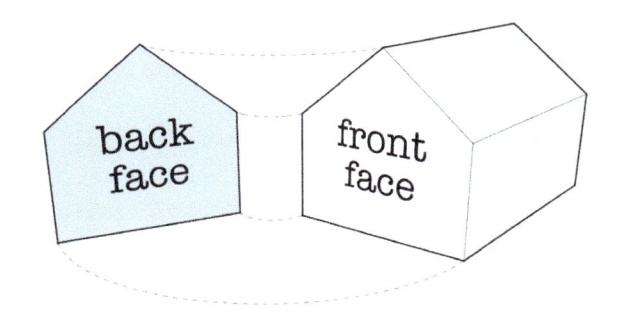

Figure 5.16: Front face, back face.

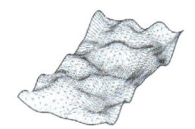

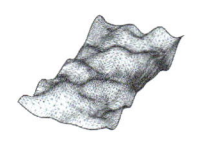

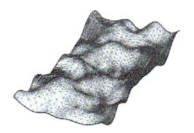

Figure 5.17: A free-form surface, faceted for construction, at five different levels of detail.

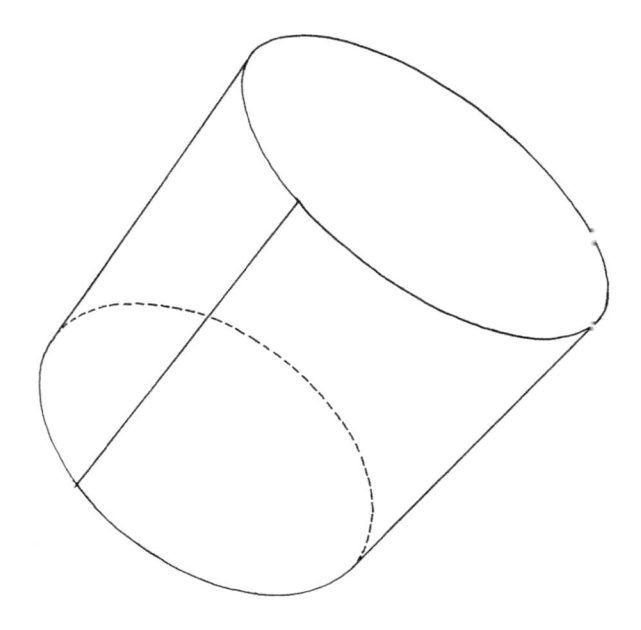

Figure 5.18:   A simple NURBS patch; simple extrusion.

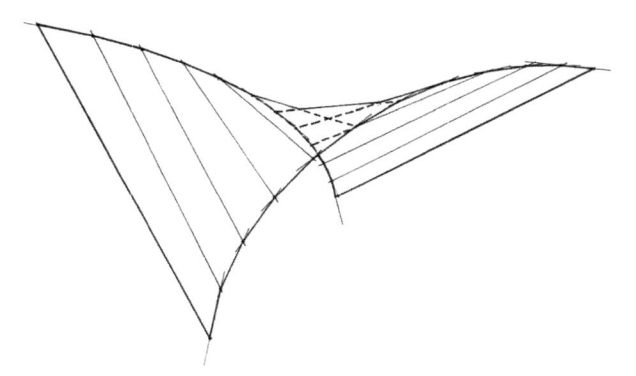

Figure 5.19:   A ruled surface, between two different curves.

A Bézier curve swept through space can define a four-sided Bézier patch. More generally defined as a NURBS patch, any of the four bounding curves may be either a straight line segment or a curve. If a single curve is swept through space, a simple extrusion is created. These forms are simple to build on a construction site and are also easy to build in SketchUp.

If two opposing sides unique curves, with the other two edges straight, the resulting surface will be a "ruled" surface. Ruled surfaces are convenient in construction because they can be built easily from simply manufactured linear structural members. For similar reasons, they can be efficiently constructed in SketchUp, even though SketchUp doesn't model with an underlying NURBS geometry model (Figure 5.19).

When three or more edges are uniquely curved, the resulting surface is of much greater complexity and requires special consideration when both modeling and building physically. When modeling such a surface physically, you would need to use some plastic modeling material like clay, or physically carve it from a block of something like foam. Freeform surfaces of this kind are mostly useful for modeling organic forms, or forms destined for machine tool fabrication on a 3D printer or CNC router. One day these kinds of tools will be more common on construction sites, but for now, they are pretty tough (and expensive) to build at construction scale (Figure 5.20).

Any of these freeform surfaces may be efficiently approximated with a polygon mesh – a connected network of triangular faces. Converting from NURBS to a polygon mesh needs to be done with some final purpose in mind so that an appropriate level of detail can be chosen. The conversion will inevitably result in an irreversible loss of geometric fidelity and must be done with a responsible understanding of the appropriate level of detail. That said, converting from NURBS to a polygon mesh is a necessary step on the way to field fabrication. In the final analysis, buildings are mostly made from flat surfaces, not freely curved ones.

SKETCHING FORMS IN 3D   147

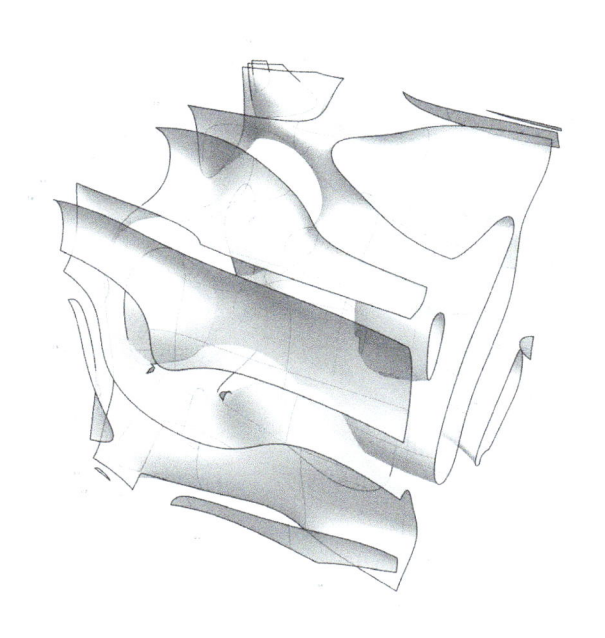

Figure 5.20: A complex freeform surface, using Mathematica's "BSplineFunction."

Sketching Forms in 3D

When a collection of four or more surfaces are drawn connected together in such a way as to create a completely enclosed volume, a form is created. Called variously *manifold, solid,* or *watertight,* these forms can be used in more complex geometric ways than any single surface on its own. A solid can, for example, have a calculated volume. With no holes out of which material may leak, the computer can be told the form is made of concrete, or steel, or wood. Or the frozen tears of unicorns. Whatever makes sense in your design at the moment.

As lines are defined by points and surfaces are defined by their bounding lines, forms are defined by their bounding surfaces. In computer graphics, there are many different conventions for the representation of 3D forms, each with their own advantages and disadvantages. NURBS, T-splines, boundary representations (BREPs), and winged-edges. Solids and Surfaces. As a designer, you shouldn't really have to care how this all works, so long as you can build the forms, you imagine in your mind with them. What differentiates these kinds of models only really matters to you, technically in a few basic ways.

The purest forms are those formed by a flat plane swept through space on a perpendicular vector. SketchUp's PushPull tool makes these primary prismatic extrusion forms quickly from any single surface. The originating surface is duplicated along the vector, and each of its bounding lines is replaced with new surfaces. Coupled with its simple Line Tool, you can likely see what PushPull is SketchUp's second most useful modeling tool. Buildings are naturally modeled using not more than these two tools alone (Figure 5.21).

To model something more complicated than a prismatic extrusion, different techniques must be employed. This is especially true for models that include freeform surfaces. The data required to maintain a freeform surface definition, particularly when you care (like an automotive stylist cares) about just how continuously smooth the surface really is, requires considerable modeling care and attention. If you are going to paint your curvy building with glossy automotive finishes, you might care how smoothly reflections play across the surface when you walk by. Think of the "pillowing" in reflections on the vast expanses of curtain wall glass in the reflective 1980s skyscrapers of Phillip Johnson, for example. Or the reflections on a gently moving pool of water (Figure 5.22).

Freeform surface modeling is particularly relevant, as well, if you intend to use machine tools of some kind to manufacture your design. It was

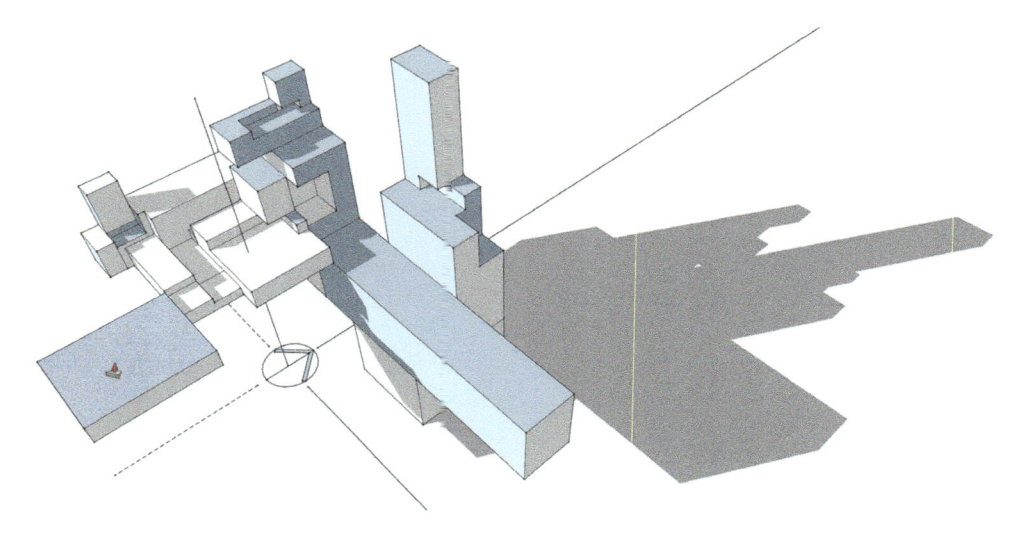

Figure 5.21: SketchUp's PushPull tool.

this problem, again in the automotive industry, that lead Casteljau and Bézier to innovate their freeform surface-defining math in the first place. If you don't plan on doing this, your freeform surface designs will have to be rationalized for construction in some other way, likely to a faceted planar definition like you might build in SketchUp.

Figure 5.22: Pillowed reflections on water.

You can be pretty effective modeling faceted form using only the simplest of tools – a Line tool to add geometry and a Move tool to move individual vertices around. I have found that I can model quite organic-looking things in SketchUp using only the primary Line, PushPull, Move, Rotate, and Scale tools. For example, I sketched the ant in Figure 5.23 in that way. No special tools, no additional plugins or extensions, are required.

If you are finding yourself needing to do a lot of organic modeling, you may find one of the big digital content creations tools from the visual effects industry a better fit. Tools like Maya, Blender, and 3DSmax have NURBS and Sub-D modeling tools built-in. To use them effectively, you will have to spend some time learning their limitations. NURBS, for example, must have four and only four bounding edges for every surface patch. You cannot make a triangular patch, nor a five-sided patch. Consequently, you need to be somewhat strategic about how you lay your model out. Also, it is essential to understand that these modelers cannot guarantee a "solid" status for anything you model either.

SKETCHING FORMS IN 3D

Figure 5.23: An ant modeled in SketchUp, using only basic modeling tools.

To be considered a "solid" form, a 3D model needs to be able to unambiguously determine if any arbitrary point in space is either inside the form or outside of it. Put another way, the model needs to be able to ensure that there is no question of what the inside of the form is. Topologically speaking, a solid model is also one that is "closed." For example, a collection of six square polygons connected together seamlessly by their edges might form a closed cube. Remove one of the faces, and the remaining five faces are no longer closed. The insides and outsides of the cube have become ambiguous to the computer (Figure 5.24).

A human designer, looking at the same condition, might easily imagine something like a cardboard box with an open top. Surely this is a form that can provide some sense of "enclosure" to an occupant. But mathematically, to the computer, ambiguous conditions are anathema. The form is either closed or it is not. Luckily, tools like SketchUp include a quick system for determining if any given form has been successfully closed. And, when a form is determined to be closed, additional analysis becomes possible. Where lines have length and surfaces have an area, solids have measurable volume (Figure 5.25).

In practical modeling, the topologically closed nature of geometry is challenging to preserve. If you need to ensure that your model is at all times made of only mathematically solid components, you are going to need to find a "Solid" modeler. These are most common in products designed for the manufacturing industry, for example, Catia, Creo, or NX. These products are stratospherically expensive, but also incredibly versatile in the creation of machined parts and their assemblies. They are uncommon in the AEC industry, with notable exceptions in Frank Gehry's practice and a small number of others in his technological sphere of influence. If it matters to you how much these tools cost, they are not for you. Nor for that matter, in most practical circumstances, are the tools you'll find you really need.

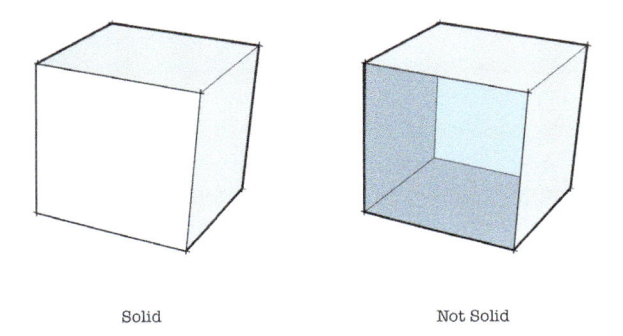

Solid   Not Solid

Figure 5.24: Closed cube, open cube.

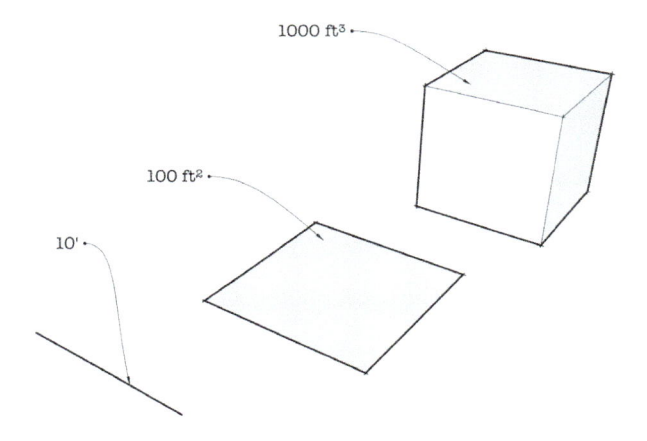

Figure 5.25: Length, area, and volume calculations.

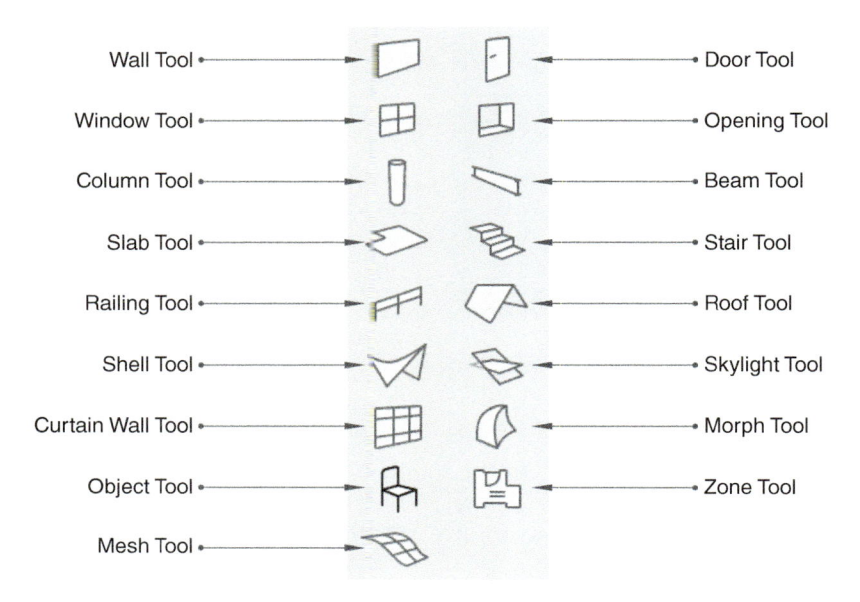

Wall Tool ·―――――――― ⬜ 🚪 ――――――――· Door Tool
Window Tool ·―――――― ⊞ ▭ ――――――――· Opening Tool
Column Tool ·―――――― ◯ ╱ ――――――――· Beam Tool
Slab Tool ·――――――― ◇ ▤ ――――――――· Stair Tool
Railing Tool ·―――――― ⊟ △ ――――――――· Roof Tool
Shell Tool ·―――――― ▽ ◇ ――――――――· Skylight Tool
Curtain Wall Tool ·――― ⊞ ◗ ――――――――· Morph Tool
Object Tool ·――――――― 🪑 🏠 ―――――――· Zone Tool
Mesh Tool ·――――――― ▦

Figure 5.26: Some typical AEC-specific modeling objects, from ArchiCAD 23 for macOS.

There are cheaper alternatives to these, which are capable enough and affordable enough that I wonder why more architects don't use them. Solidworks, for example, or Fusion. Even Rhinoceros can work as a solid modeler if that's what you need. However, the overhead required to learn these tools is still very significant. And the constraints under which you must use them may feel restrictive to you when you're "just" sketching. The assurance that the forms you're modeling can be machined comes at the cost of a more stringent editing workflow that prevents you from making "mistakes" that would cause the geometry to become "degenerate."

Assembling 3D Models from Parts

Buildings, like any complex manufactured object, are made by assembling smaller parts into larger systems. These systems might be structural systems (walls, floors, roofs), mechanical, electrical, and plumbing systems (HVAC, lighting, water) or something else like that. Sketches of buildings often include such systems in diagrammatic form, or may just add a space in which they may be designed later. But there's a simple logic of inclusion, classification, and enclosure that can make sense while you're sketching on any aspect of your project (Figure 5.26).

Many 3D modeling tools designed for architects include palettes of specialized tools that automate the creation of conventional building systems. For example, they may include a Wall tool that makes various kinds of walls quickly and efficiently. These objects, by design, represent every kind of wall the designer of the modeling tool imagined you might want to be able to make. And nothing more. But every wall created with this kind of tool is complete in its representation of a wall. If that's what you know you want to add to your model, this sort of tool makes it quick (Figure 5.27).

ASSEMBLING 3D MODELS FROM PARTS

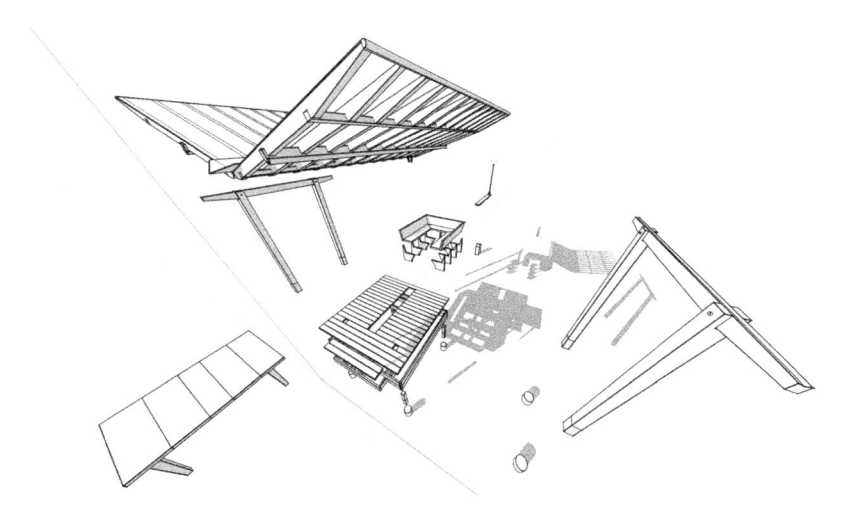

Figure 5.27: Components in SketchUp.

In SketchUp, we included a simple method of grouping individual geometric elements together into logical units that could be manipulated together as one unit. Called *components,* they serve an important conceptual function while you're sketching. If you want, you can use components to represent a wall. Or anything you want. Entirely up to you. It only matters how you choose to frame the model. Which pieces relate to which other pieces?

When you're sketching, you may be laying lines and other geometry into your model without much thought about what relates to what at first. Before you have a formally defined wall, you may have only a single line drawn on the ground, or maybe just a single vertical surface defining a division between two programmatic elements in a diagram. When you're ready, however, to commit to a collection of lines and surfaces as representing some higher-detail object in the building, it is as easy to put a box around them and call them out as something unique in the model. A wall, but maybe a special kind of wall that really makes sense in your project. You should be free to reframe the definition of "wall" as you want.

Once identified as a component, you can begin to treat the encapsulated geometry differently. You can give it a name. You can determine what kind of thing it represents. You can copy instances of it around your model, all of which retain reference back to their standard definition. Components let you begin to think of your design as being composed of parts, but without breaking your flow and without requiring too much data before you're ready to supply it.

Components can also contain other components, making it possible to dynamically build up a hierarchy of sets, subsets, assemblies, and subassemblies to begin to bring systems thinking into your design as early as you are ready to do so. Components remain quick to edit; just double-click to edit them in place without ever losing your sense of context. You can have a component that represents a stud, another that includes multiple stud component instances and represents the structural framing for a wall, which combined with sheathing and cladding for the wall, as well as maybe a window and the trim around it which all together make up a component for the wall. Which, in turn, might be a part of the "second floor" component, all wrapped up together with hundreds of other components that make up a "house" component.

But there's no reason why your model has to be organized in that way. For you and your design process, you may have an entirely other conceptual organization in mind. For every conceptual framing, there should be an opportunity to reframe. It might be that you want to group things together based on who is responsible for building them. Or for when they might be built. Or maybe you just want to group items together based on what color they are. Or by how they impact the building's program. Or by the material from which they are made, or even how they should smell. No judgment on my part, you should be free to organize in whatever way helps you to understand the project best (Figure 5.28).

Learning to frame the problem, and to reframe it over and over again, is among the most essential

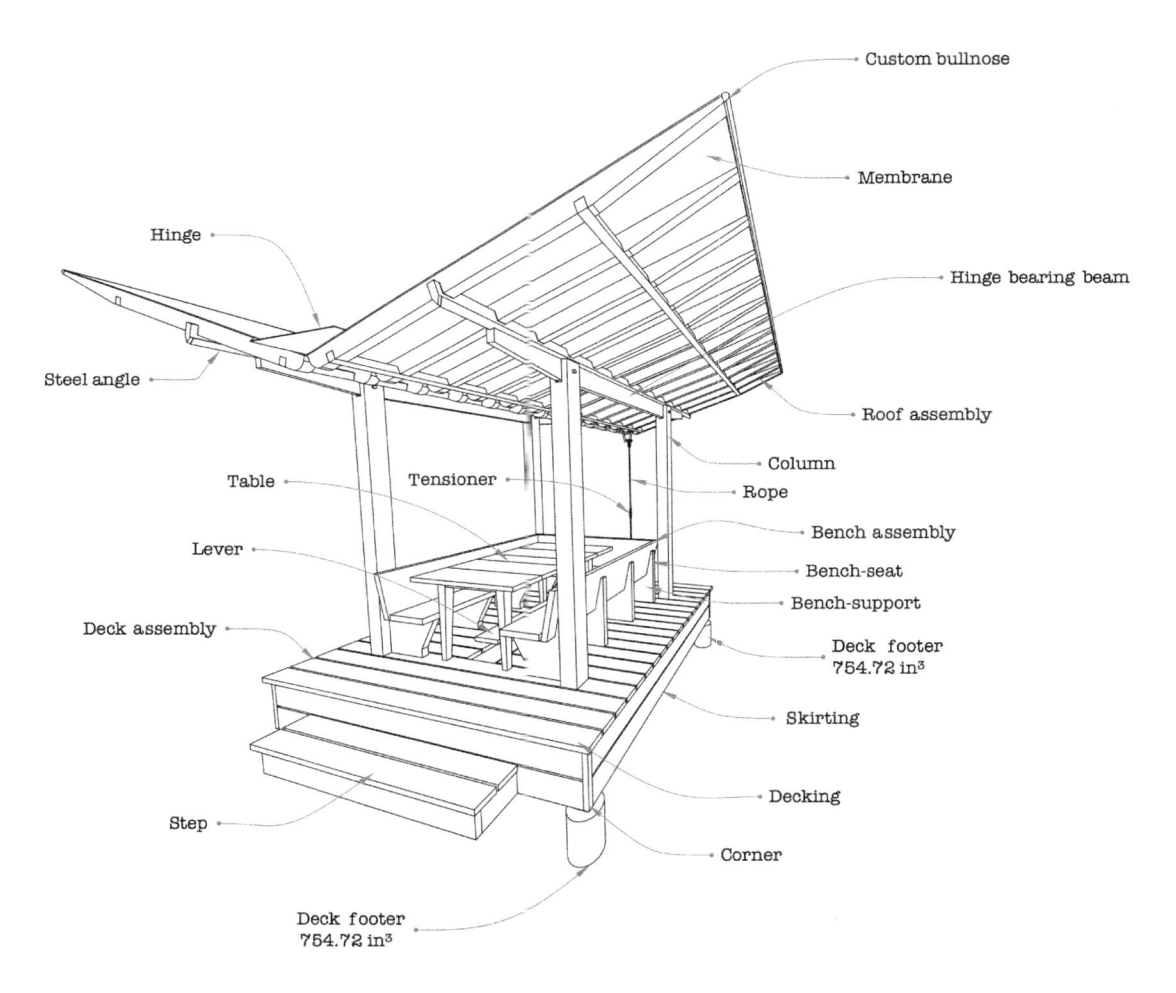

Figure 5.28: A well-organized sketch model.

elements of a functional sketching process. You do not know what the building will be like until you figure that out. If you think you know what the building will be before you have set the first line into the model, how much design work are you really planning on doing? Are you really thinking, really creating, and really being as critical and inventive as you could be? You should never be satisfied accepting someone else's framing of the problem without the ability to reframe it in your own terms.

This represents a process inversion from most profession-specific 3D modeling tools. It is far more common to find tools that ask you to define what kind of thing you want to model before you commit to modeling it. When you want to model a wall, you choose the Wall Tool to do it. And that tool is excellent at making wall entities that have all the right kinds of wall-appropriate attributes attached to them by default. But what if the wall your design needs can't be defined in terms of that predefined wall entity? What do you do then? A better system is one in which the characteristics of the object are allowed to emerge, bottom-up, over time. In a system like that, you can build walls that are like every other wall, or, if appropriate, you can design a completely unique new kind of wall just perfect for your project.

The naming and classification of those named things into sets according to some taxonomy is a powerful design tool all by itself. When you imagine something that the world has never seen before, you limit your thinking about it by wedging it into someone else's a priori classification. But by giving it your own name and by identifying it with other similar or related things in your model, you can begin to say quite rigorous things about what it will become as you add further detail to its definition.

A true sketching process frees you to add detail over time, as you explore alternatives and follow opportunities as they are revealed. The method of grouping, collecting, and naming helps you to do that gradually and with great flexibility. You want your sketching work to be open to interpretation, open to reframing.

Of course, less and less of new building construction is created from scratch from raw materials on the construction site. Building product manufacturers are increasingly creating building components in factories offsite, to be delivered and installed rather than site-built from raw materials. Working with these manufactured components can provide an opportunity to move faster and get to the final on-site assembly quicker. And there are considerable advantages to working in this way. However, this comes with different opportunities to consider while sketching in 3D.

In the Hollywood visual effects industry, model makers have long known that you could quickly reach a high level of visual detail in a model by bashing together parts from different plastic models to create something new. Called *kit-bashing* in the industry, this 3D bricolage technique is done without particular regard for the intended depiction of the originating model parts. Parts from automobiles, airplanes, and trains might be combined to create a spaceship for the latest sci-fi epic (Figure 5.29).

While still practiced by some effects houses, physical kit-bashing has been reconstructed by a digital kit-bashing analog. With resources like the 3D Warehouse available, literally, millions of freely-available models can be scoured for cool-looking bits and bobs, to be assembled together in recombinant ways to create something entirely new. As a designer, you should always be open to the creative misuse of someone else model. And, you should be similarly open to someone else using a piece of your work to make something else entirely new on their part.

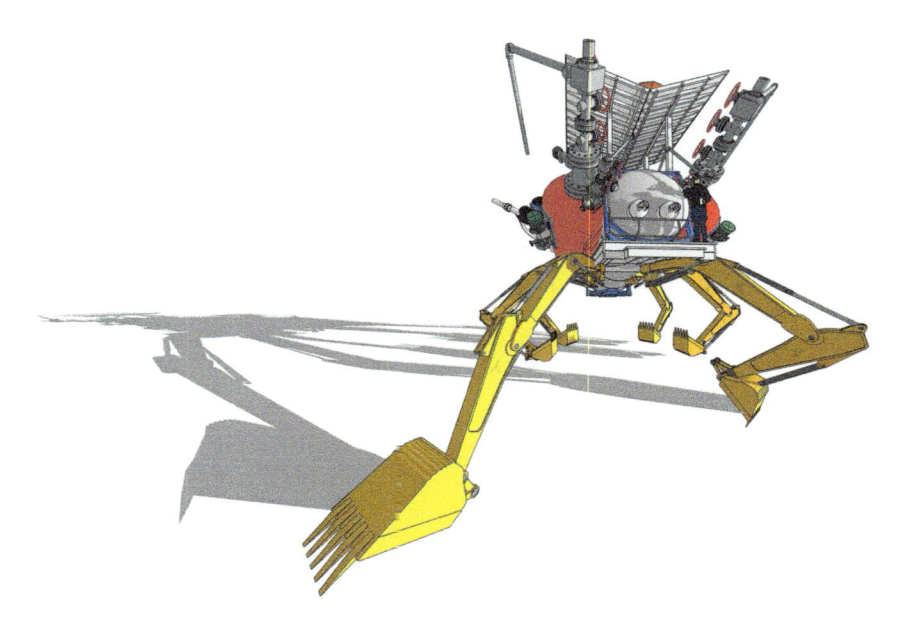

Figure 5.29: Kit-bashing in SketchUp.

There are several repositories of model geometry on the internet, but 3D Warehouse is by far the largest and richest. Because it lacks a rigidly browsable system of object classification, searching it for something specific can feel frustratingly like searching for a needle in a haystack. Still, it is full of all kinds of wonderful things that might not be exactly what you thought you were looking for . . . but that turn out to be far more useful to you than you may have expected.

Claude Lévi-Strauss (1974) wrote about "bricolage" as a methodology for cultural reinvention, where ideas from one society might be appropriated into another. In the process of appropriation, they would be reshaped, re-contextualized, and sometimes would come out looking nothing like they went in. From a designer's perspective, this can be magically productive. Nothing is more exciting than misappropriating someone else's concept for a new purpose. Kit-bashing is incredibly fun when you're looking to quickly build up detail in a model.

Like any other sketching technique, however, it is easy to jump to a premature conclusion when you are sketching with other people's models. This is particularly apparent when staging an interior scene for a design presentation early in a project's development. It is all too common to find beautiful pieces of furniture – for example, a historically significant chair from the Eames Office, or maybe Le Corbusier's signature "LC4" chaise lounge, complete with pony hide upholstery, situated prominently in the scene. You think you're just setting the right tone for the rendering, and maybe just sharing what you hope might eventually be chosen in the furniture package. But this just isn't realistic. Too much detail, too soon. You are building in more questions than you need to.

A similar problem exists when you prematurely choose a large appliance or perhaps a fancy sliding patio door system. Your client may fall in love with these things, but find out later they are simply unaffordable in their budget. By adding something definite and highly detailed to your model, you run the risk of writing a check your client can't cash. When sketching, you are much better served with generic objects at low levels of detail.

Cautions aside, however, it may be that your client is, in fact, well prepared to add a specific very particular piece of furniture to their budget. And they love it so much that they want you to design the project specifically around it. In this case, having the most detailed representation possible of that object will help you when you're sketching. A prebuilt model of that chair can become a fixed point around which your other decisions must orbit. The object defines a design context, and your design must respond to it.

chapter 6 Sketching in Code

Throughout the history of design, there have always been architects who idealized their design work through systems of rules. Consider Vitruvius' *"De Architectura"* or the works of Andreas Palladio. In each case, they worked to describe all ideal architectures in a carefully designed system of rules and standards. By applying the rules of the system rigorously, an ideal design solution could be discovered for any site. In this plan for Palladio's Villa Rotunda (Figure 6.1), a highly systematized design algorithm was applied to create an ideal design solution according to Palladio's vision of perfection.

Rules-Based Design

With the availability of cheap computers and accessible programming environments, entirely new kinds of rule-based design are possible. Where in previous systems, the level of complexity was gated by the designer's ability to manually compute its permutations. Today, we can use computers to automate, enabling the management of many more interdependent parameters in a model than might ever be realistic to compute by hand. The computer's ability to rapidly iterate on whatever design system you devise makes it possible for you to explore such systems deeper and more comprehensively. Systems that might have been studied to a dozen iterations manually can now be run for thousands, for hundreds of thousands or even millions of iterations.

I think designers should always provide some kind of a formal rationale for their work, although some feel that they need that more strongly than others. There is a pang of residual guilt that comes from the belief that only that which can be rationalized can have a quantifiable value. Digital form-finding techniques popularized by the "Parametricists"[1] are likely in some small part a result of this thinking. But there is also a delirious, joyfully playful quality to the work that deserves closer inspection. Even those who most rationalize their work are still spending important quantities of time reflecting-in-action in highly intuitive ways.

Greg Lynn, in his essay, "Animate Form," reminds us that the computer is not a brain, that it is not capable of independent critical thought. That, *"...the failures of artificial intelligence suggest a need to develop a systematic human intuition about the connective medium, rather than attempting to build criticality into the machine."* Designers need, in other words, to learn to use computational techniques intuitively, just as they do other tools for design. Computation can and should be deployed into an architect's practice, he continues, like a pet that

---

[1] Of course, they have a Manifesto (Schumacher 2008).

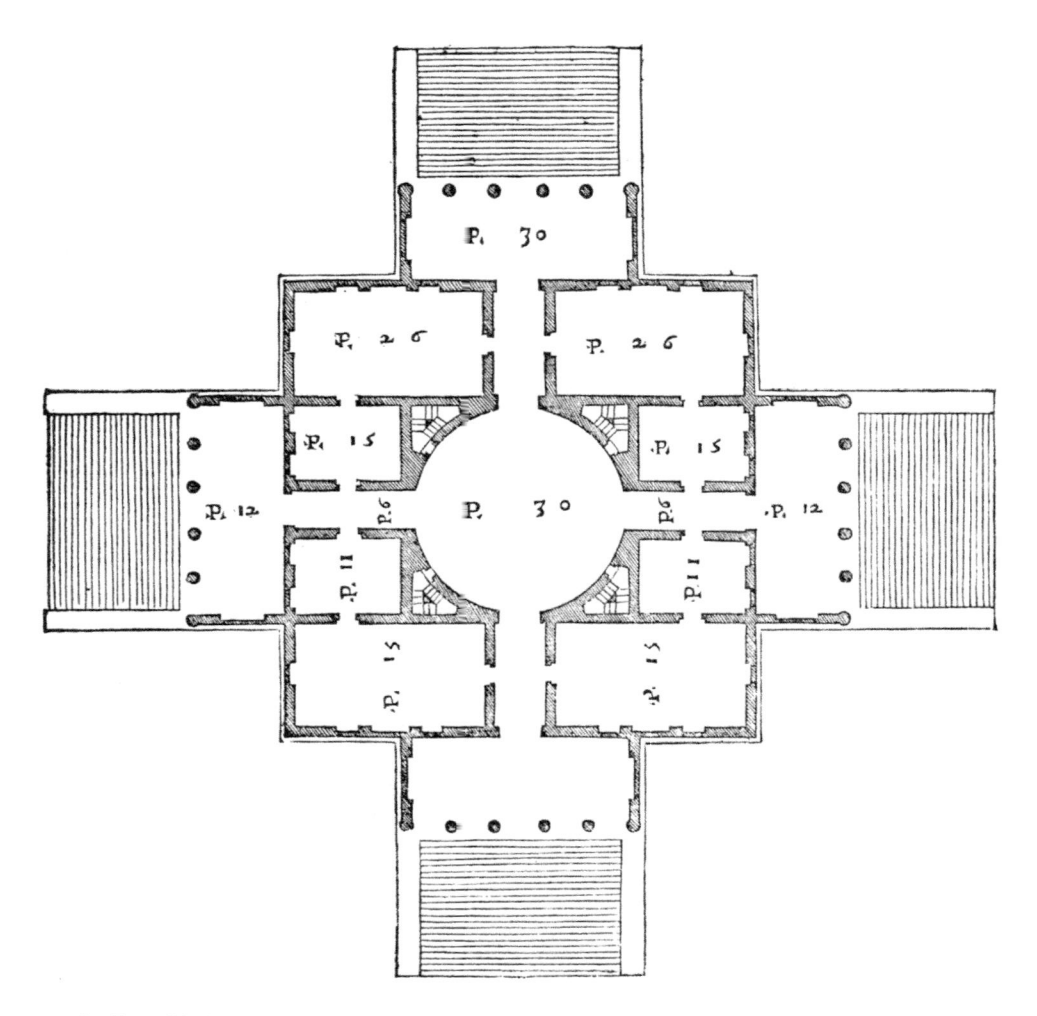

Figure 6.1: A Palladian Villa, in plan.

"...introduces an element of wildness to our domestic habits that must be controlled and disciplined, (it) brings both a degree of discipline and unanticipated behavior to the design process" (Lynn 1999, pp. 19–20).

But if you have an idea to explore that is best represented as a system, the best way to explore it might be to write up a couple quick lines of code that implement the idea in an algorithm that can test every variation in the system at once. Similar to drawing and model making, a designer who learns to sketch in code is one who can think through many variations and opportunities in a design quickly and disposably. If the practical cost to explore every possible variation in a scheme is the same as exploring only one variation, why not try them all? If the design idea you're trying to study can be described in a set of rules, you can write an algorithm to express it.

Materials for Digital Sketching

As with any new sketching practice, you have to start by getting yourself set up to work. Where designing with paper and pencil is relatively easy (you just need a pencil and some paper), designing with algorithms requires you to set up a programming environment. Depending on the configuration you choose, this can be either moderately or incredibly complex to start up. You are going to have to learn some new things, and probably you are going to have to spend some money.

First, you need a computer. If you are already drawing and modeling digitally in your practice, you already have this. Next, you need a programming environment. There are many of these, some more capable than others and some more straightforward to set up than others. Some of them are wired into modeling applications you might already be using for your design work. For example, Greg Lynn got started in his digital sketching practice with the "Mel" scripting language in the Hollywood visual effects package Maya. The Grasshopper IDE, running on top of Rhino, is another popular contemporary choice. SketchUp has Ruby, and AutoCAD has Lisp. These are all very powerful and capable programming environments.

Environments like these focus on automation of the core modeling features of the application into which they are wired. Mel helps to automate the generation of NURBS geometry, for which Maya is widely appreciated. Grasshopper automates the creation of NURBS geometry as well, but with a Rhino flavor that is a bit more suited to machine fabrication. SketchUp's Ruby interface makes it easier to build utility functions that add new modeling tools to SketchUp's toolbar, allowing for NURBS, Sub-D, and other modeling paradigms to have a primitive implementation. But in general, think of this class of programming environments as essentially automating the geometry that your modeling application can already make.

As an alternative, you could settle down with a more full-featured professional IDE ("Integrated Development Environment") of the sort that a professional software engineer might use. While you will have to learn quite a bit more to be effective, you will also be able to implement your own stand-alone tools that are capable of taking you much farther creatively. Without question, this is the most general and powerful choice. Some of the most innovative design tools got their start with a frustrated designer who thought, "How hard could it really be to learn how to program?" Realistically, especially if you have never really programmed before, it is going to take a long time to come up to speed.

Learning to program is hard, but not impossibly so. There are countless examples of programming languages and accompanying IDEs that are designed to make it easier for beginners to learn how to write code. Since Ada Lovelace wrote the first identifiable computer program in the 1800s, it has been clear that people needed help learning how to give instructions to computers effectively. In general, people are more intuitive than the machines we have designed to help us work. Computers are bloody-mindedly rational, and they can't infer much of anything by themselves. As the ultimate children of the intellectual rationalization of the Enlightenment, computers need carefully stated instructions before they will do anything. The challenge in learning to program effectively is not about the syntax or the basic structure of any particular language. It is, instead, in learning how to structure the underlying logic of the instructions.

When choosing a programming environment to commit to learning – and make no mistake, you are going to have to put some work in to get good at programming – be sure to pick one that will not just be easy to learn, but also will be rewarding to grow into. Like any tool, you want to pick one in which it is easy to get started, but which will provide more headroom than you'll ever need in the future as your skills and interests grow. Look for an IDE that has a low threshold but a high ceiling.[2] You're going to need proper documentation and lots of easily accessible examples.

Many simple languages make it easy for you to learn by reducing the degrees of freedom in the system. Tools like the graphical programming language Scratch, designed to accompany Lego Mindstorms, are examples of this. Scratch has only a small catalog of pre-built functions available, and they can only snap together in predictable ways. For the basic kinds of tasks Mindstorms designers expected students to want to accomplish, a more straightforward system was a great choice. But it is easy for students to come up with goals that can't be easily achieved using only Scratch for programming.

Resist the urge to choose something you think of as the language everybody else is using. To be sure, learning something new is easier when you do it together with other people. And almost every programmer I have ever known has learned to program by hacking on code someone else has written. Having access to the work of others to follow and learn from is wildly useful when you're learning. But if you pick a tool of any kind (IDEs are just a tool, after all) just because it is what everybody else is using. . . you're going to end up missing out on some important stuff.

For myself, I like Mathematica and the Wolfram Language that ties it to the richly curated data sets in Wolfram Alpha. For my way of thinking, it resonates well. Mathematica is profoundly deep, but (as computer programming goes) it also has a shallow barrier to entry. I first used Mathematica almost 30 years ago – making it almost eternal in the continually shifting landscape of computer software development. Mathematica runs on just about every computer you can imagine, from the largest of supercomputing clusters to the lowly but fantastic Raspberry Pi. It is comfortably at home on my Macintosh, and I know that if for whatever reason, I needed to move up to something more powerful, the transition would be smooth.

Mathematica has its roots in a particularly unique form of computation that made it possible for computers, which are inherently discrete in their operation, to perform calculus computations. Stephen Wolfram, the creator of Mathematica and the Wolfram Language, is by training a mathematician and a physicist. So you will find that Mathematica is most at home when chewing away at mathematical problems. Which is fine – actually, this is precisely what we really need when sketching in code. Think of Mathematica as a palette of thousands of basic computational functions that you can quickly collage together to try out an idea.

Programming in Mathematica is done in a notebook interface that reminds me of an architect's sketchbook, as shown in Figure 6.2. Lines of code can be interspersed with rich text, and graphics, including fully interactive 3D models. You can quickly write a line of code and execute it to immediately see the results output into the same document. The next iteration of your program is only a quick copy-and-paste away, allowing you to sketch ideas rapidly while keeping some history of past work behind you as you proceed. For an additional layer of management, Notebooks are entirely self-contained, making them easy to check into your version control system without any extra configuration work. If sketching in code is anything at all like sketching

---

[2] This concept comes from the research of Gerhard Fischer and his "Center for Life-long Learning and Design" at the University of Colorado.

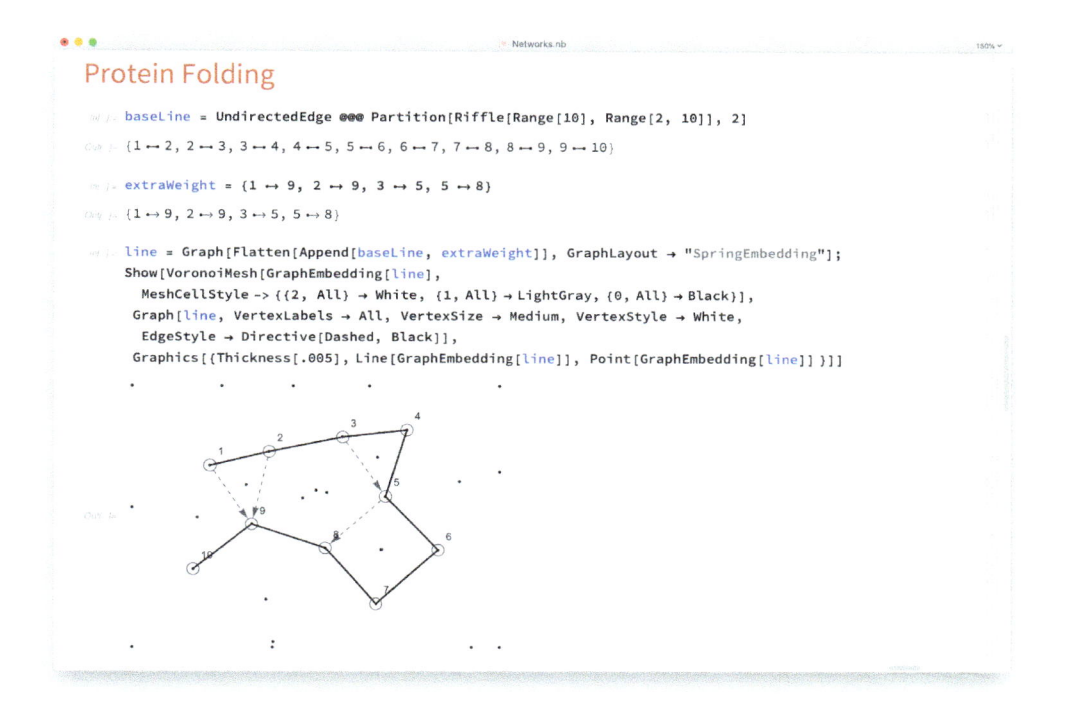

Figure 6.2: Mathematica's Notebook user interface.

on paper, Mathematica's notebook interface is the right way to work.

There are surprisingly few alternative examples of a notebook interface for programming that is quite like Mathematica's, though the open-source Juypter Notebook front end for Python is worth a look. I'll offer more analysis of alternative programming environments, especially some of the graphical "boxes and wires" IDEs like Grasshopper in Appendix B. For now, I think it is most comfortable to get started with a compact language that is well supported.

There's little to no configuration required to get Mathematica set up, and (as a language), it includes thousands of prebuilt functions that let me go from hypothesis to test quickly. You don't have to spend a great deal of time tracking down special third-party libraries to explore even the most esoteric of functions, though you can, of course, also extend the language with your own functions quickly as well. Writing a program is about as simple as it can be. The Wolfram Language is rich and complex, but you can express ideas with it in compact, easy-to-write lines of code. Most importantly, it has a very rich online help system that goes much farther than merely documenting each of those thousands of functions – providing context, theory, and clear examples for each as well. But really, there's no reason why you have to choose one particular environment and stick with it forever. But I'll be using Mathematica for the rest of the examples in this chapter.

Getting Started with Programming

Once you have your programming environment set up, you are ready to prove to yourself that everything is working. By tradition, the first program written in any computational system is called "Hello, world." Here's what it looks like in Mathematica (Figure 6.3).

The Wolfram Language is considered a "functional" language. This means that it expects you to tell it what you want it to do by first specifying the function you want it to perform, then passing that function the data you on which you want it to operate. Functional languages expect you to direct them to do complex things by building them up from simple atomic functions. A function, abstractly defined, takes some defined input (data and arguments) and passes them through a predefined operation to produce an output.

In this statement, the function `Print [ ]` (which prints text to the screen) is told to print the text that follows in brackets. The actual text to be printed is enclosed in quotes to make it amply clear what the computer should print. Computers need to be told what you want them to do in very precise terms.

Here is the output when the program is run (Figure 6.3).

Exciting, right?! Probably not. Think of "Hello, world" as being the first line you have drawn on a piece

Figure 6.3: Hello World.

GETTING STARTED WITH PROGRAMMING

of paper. Everything builds from here. What would "Hello, world" look like in a 3D model? Let's start by moving the output of our simple program into a 3D spatial context. We'll use the Graphics3D[ ] function to do that. To most computer programming environments, graphics of any kind are special contexts that have to be explicitly entered. Most computers are expecting to interact with you by way of text, at a command prompt of some kind. So the first thing we have to do is to tell the computer we're entering a Graphics3D[ ] context. Once there, we can tell the computer we want to draw a Text[ ]. Both of these are built-in Mathematica functions that I don't have to define by myself. All I have to do is collage them together in with the right syntax (Figure 6.4).

Since we're drawing in Cartesian coordinate space, coordinates have three values: *x*, *y*, and *z*. In this case, since I didn't specify anything different, we're drawing at the origin. Most functions in Mathematica have reasonable default settings. If you don't understand all the options you can configure for a function, just the simple version of it will likely give you a visible result of some kind.

Let's try drawing a line next. Logically, the line will require specifying two points; one at the beginning, another at the end. To draw a line, I invoke another of Mathematica's built-in function, called Line[ ] (Figure 6.5).

This Line[ ] is a little too thin, visually, and I'd like to style it differently. Drawing instructions are processed serially, one instruction after another. To draw this line as a thick, black line, I give a series of commands as in Figure 6.6.

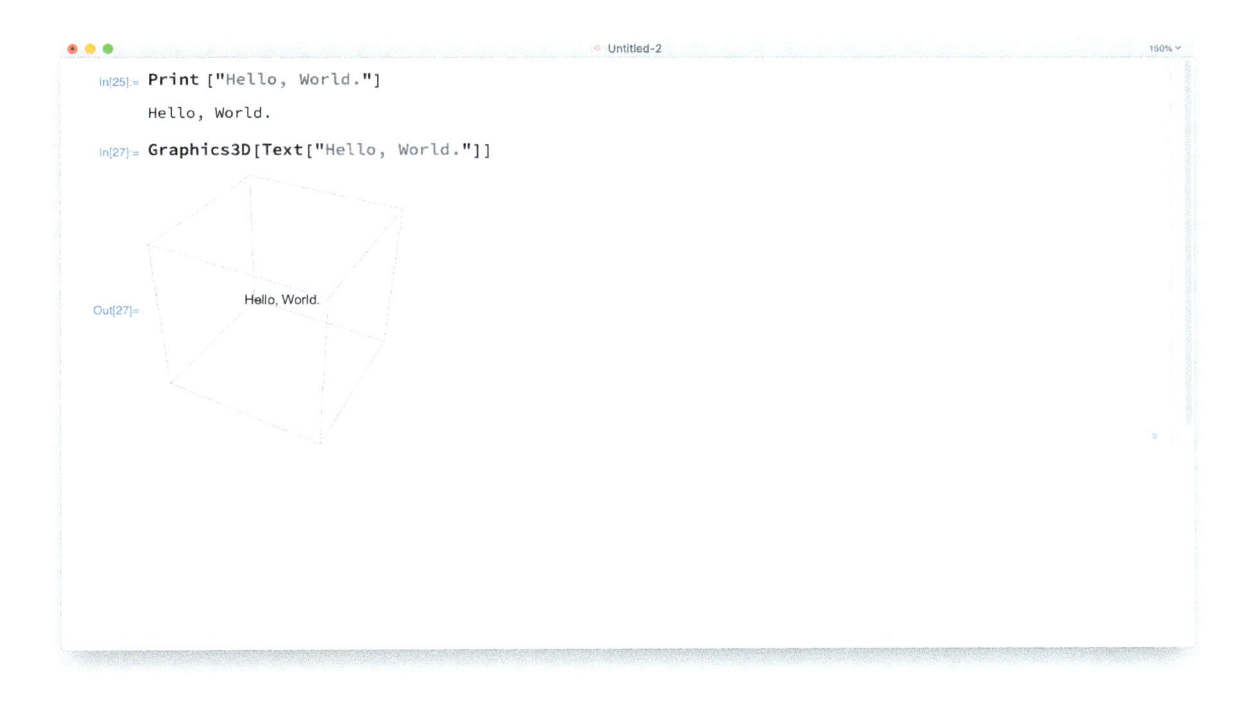

Figure 6.4: Hello World, in 3D.

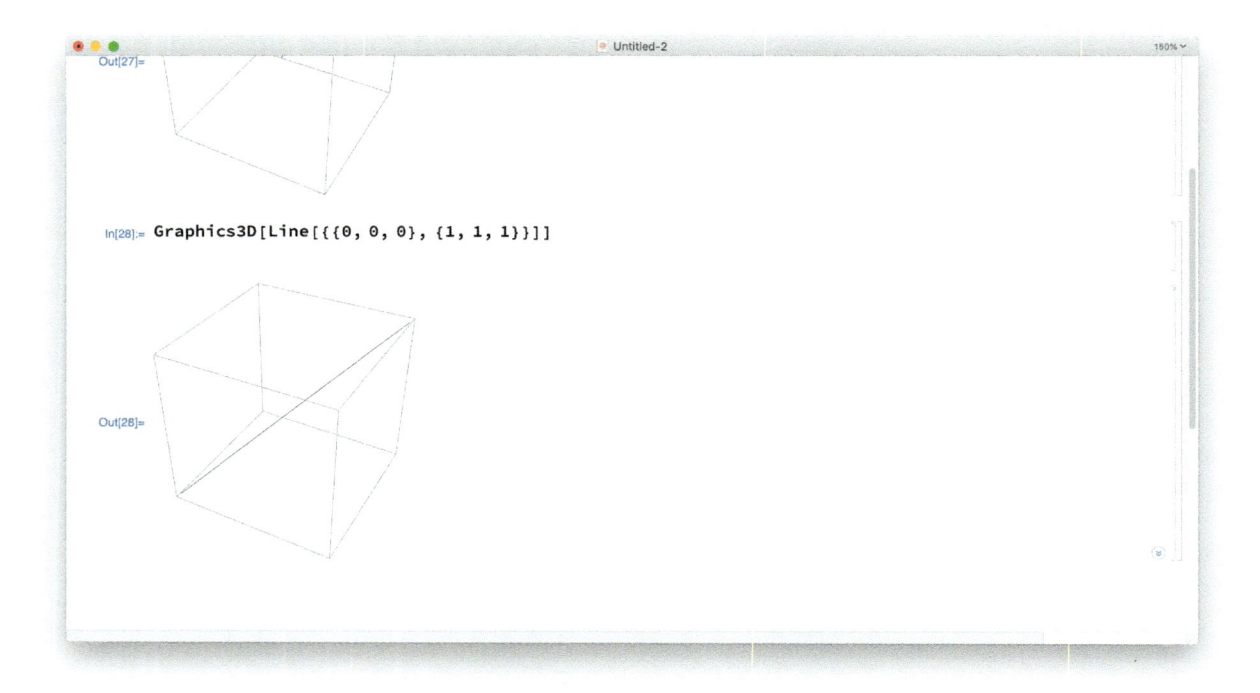

Figure 6.5: A simple line, in 3D.

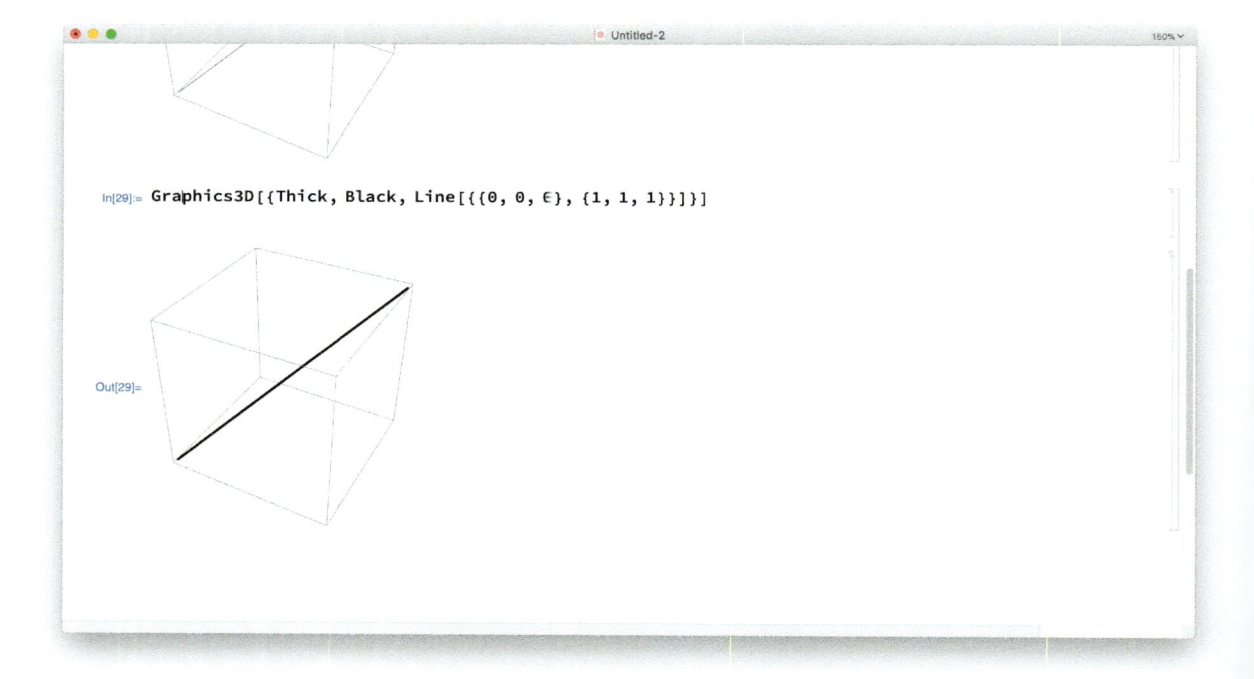

Figure 6.6: A more carefully styled line in 3D.

All the fancy curly brackets are just there to enclose logical sub-components of the data. Without them, the computer wouldn't know that we meant to give it two separate coordinates, one for the start of the line and one for the end. And of course, in 3D space, each coordinate has an *x*, *y*, and *z* value. The outermost brackets are there to warn Mathematica that there is a list of instructions coming, not just a single one.

If I can pass a list of instructions, then I should be able to pass a list of lines to draw, too, right? Let's try drawing four lines, equally spaced. To do this, I need a way to tell the computer that I'm going to step through an instruction a certain number of times. I need to iterate on my instructions. Coordinate points in 3D space are a list of three numbers; an *X*, *Y*, and *Z* value. And I need to have two of them, one for each endpoint of the line. To build a list of four lines, I need three nested lists of numbers – three numbers for each coordinate, two coordinates for each line, and four lines. I think of lists of lists as being like tables.

Conveniently, there is a function called Table[ ] that does just what I need. I can just wrap it around my previous function and add a bit more information about how many times I want it to iterate. In this case, I want four lines, each separated by one unit from each other. I add a new variable (named i to remind me it is for "iteration") and substitute it into the *y* position in each of my lines' endpoints. To combine the lines together in a single Graphics3D[ ] context, I then wrap the whole mess in a function called Show[ ] (Figure 6.7).

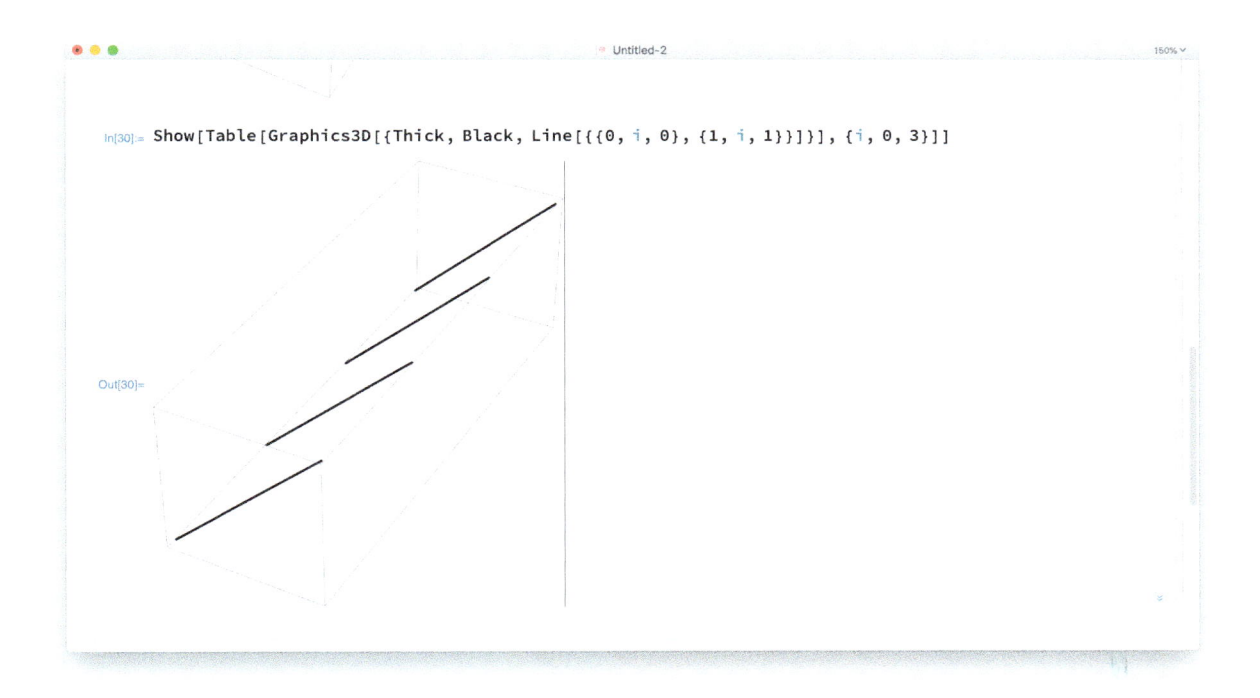

Figure 6.7: A table containing four copies of a line, equally spaced.

This still isn't very interesting, though. Maybe I should just draw a bunch of lines and see what happens. Since I don't know what I'm modeling yet, maybe just something random would be the best thing to try. I'll replace the hardcoded coordinate values with Mathematica's `Random[ ]` function. Random just returns a random numeric value in a range and of a type you specify. For now, the defaults for that are fine because I don't have anything better in mind. I'll just make everything random and see what happens next. Let's make six random lines (Figure 6.8).

And now let's combine them all together in the same space (Figure 6.9).

I wonder how this looks if I make a lot of random lines? What if I tried a thousand lines, instead of just six? (Figure 6.10)

That is kind of interesting, but I don't know what I've learned from seeing a bunch of random lines in a box. I need to have a bit more rationalization. Rather than something purely random, maybe I can try tracing something. Mathematica has an interesting function called `AnglePath[ ]` that takes as input a list of instructions in the form of distances and angles and gives as output a list of coordinate points that I can use to draw some lines. How about a square to start things off? That is, four lines, each the same length, rotated 90° from each other (Figure 6.11).

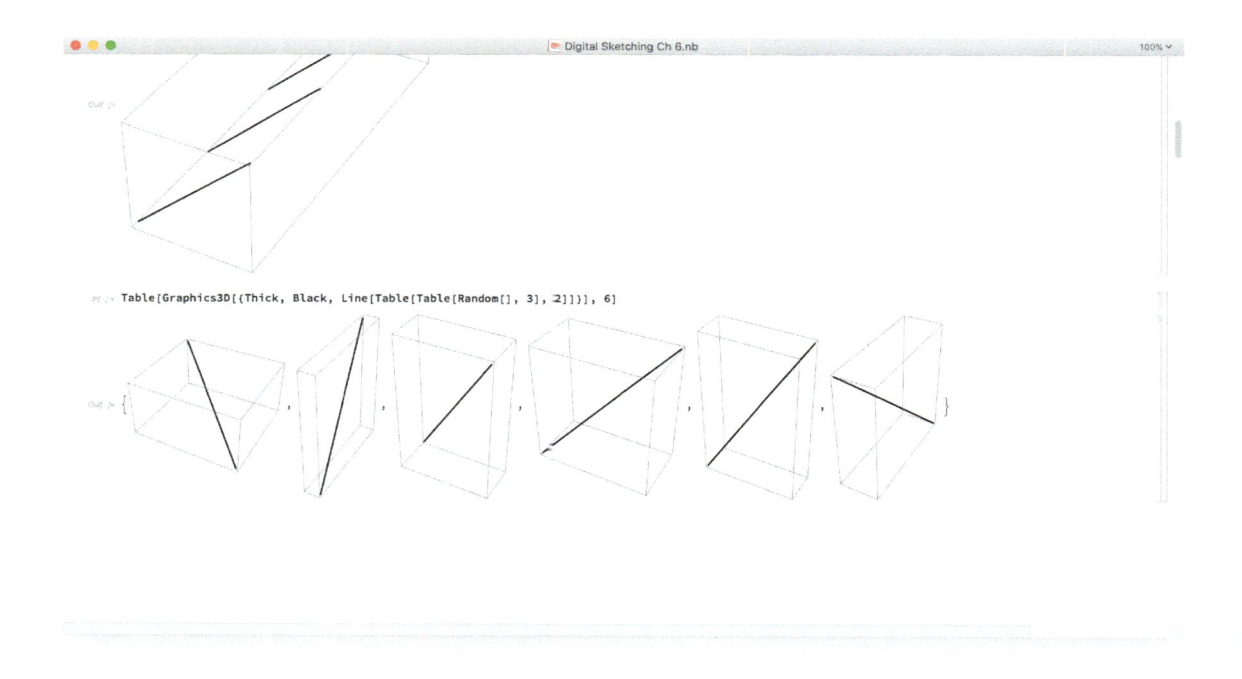

Figure 6.8: Six random lines.

GETTING STARTED WITH PROGRAMMING

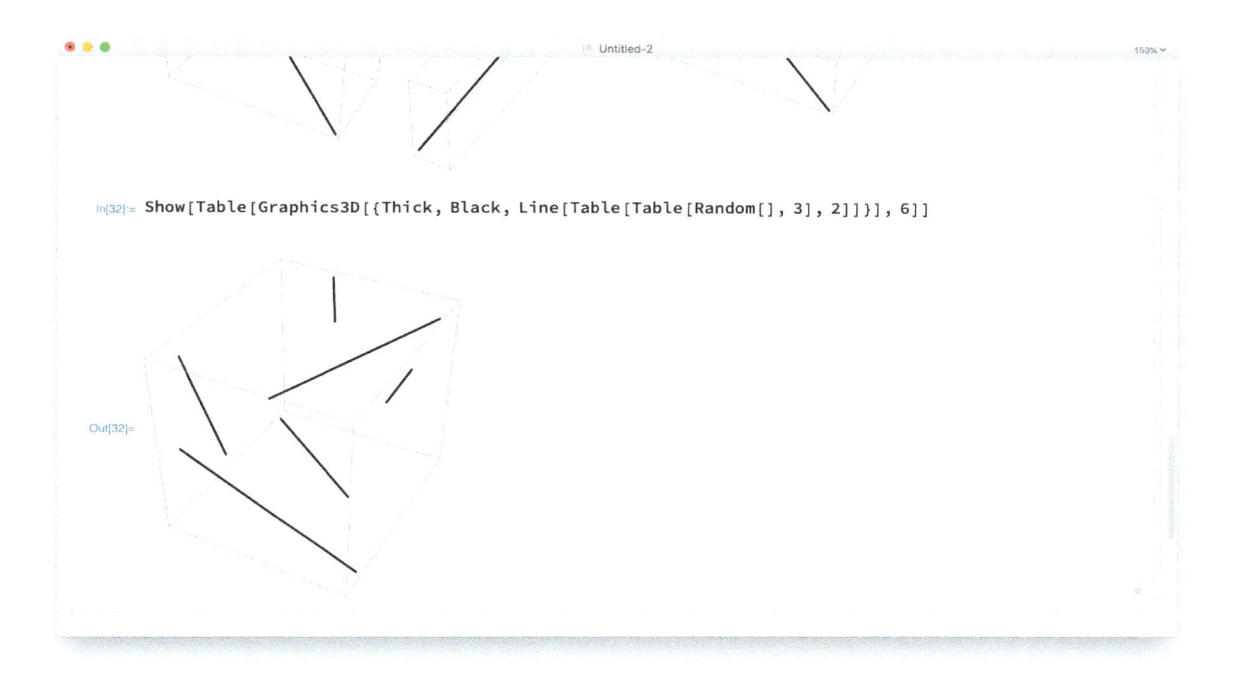

Figure 6.9: Six random lines, combined together in the same space.

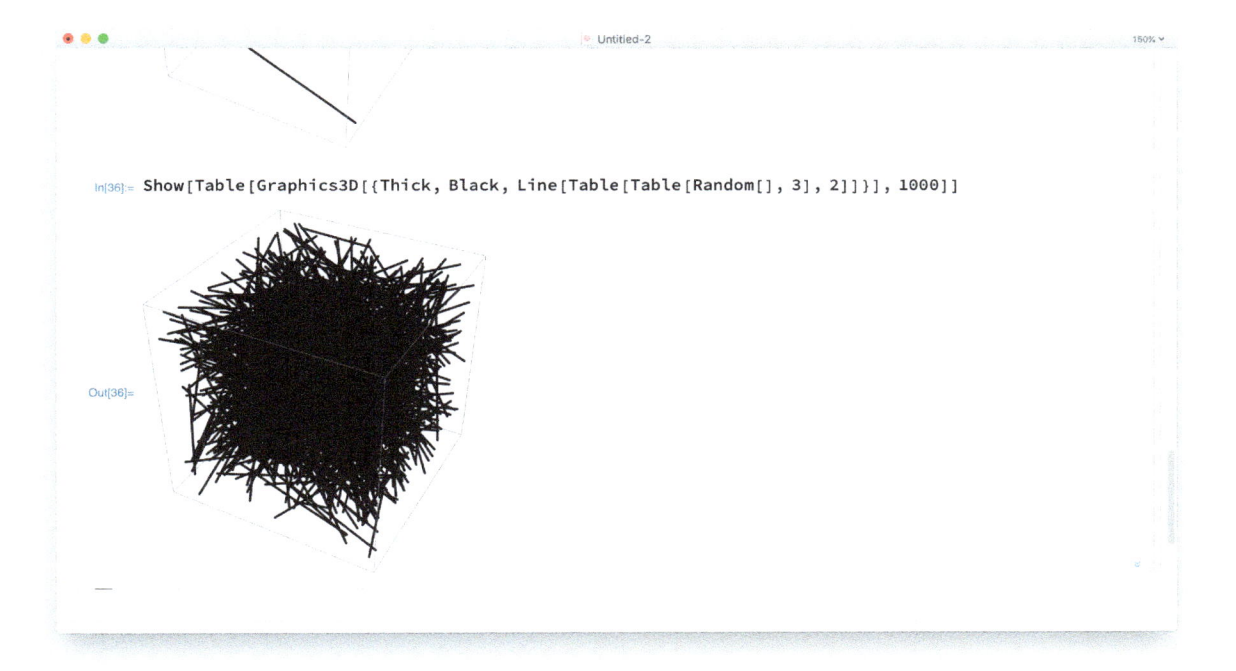

Figure 6.10: A thousand random lines.

```
In[37]:= Graphics[Line[AnglePath[{90°, 90°, 90°, 90°}]]]
```

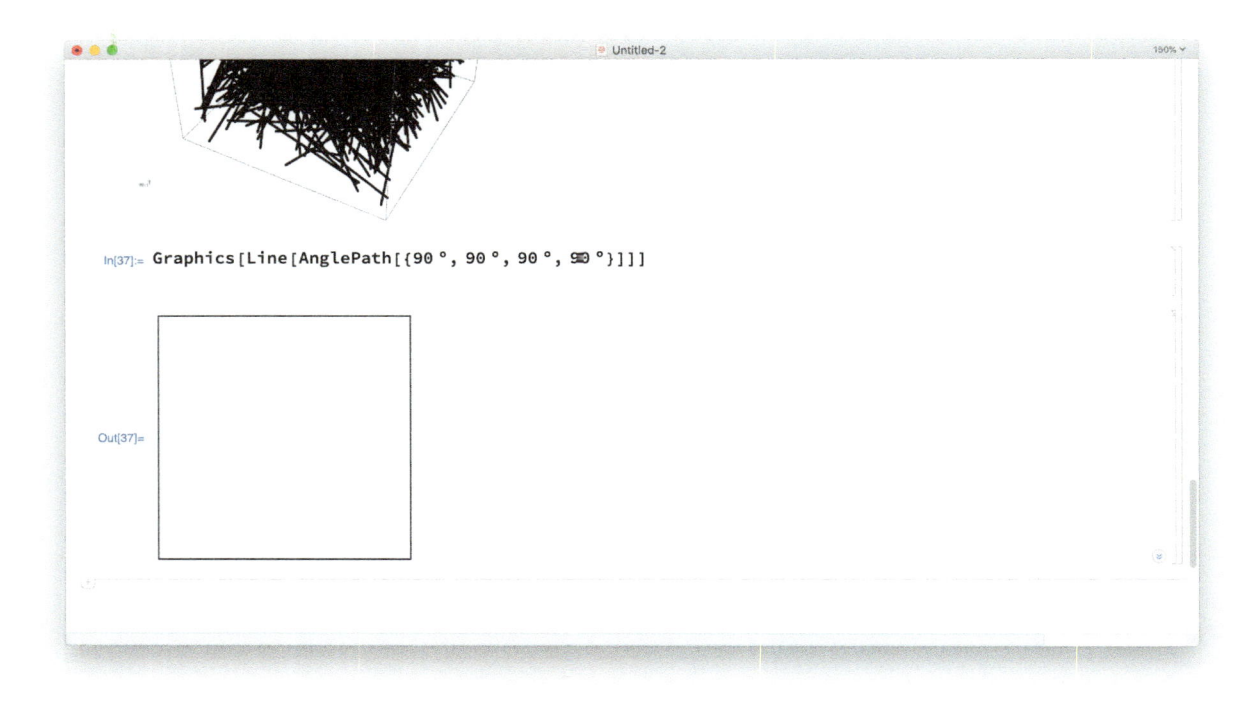

Figure 6.11: Experimenting with the `AnglePath[ ]` function.

I can easily draw a square by hand, digitally, or in a 3D modeling application. But if I had a long list of drawing instructions, maybe hundreds or even hundreds of thousands long, what might emerge? The `AnglePath[ ]` function takes as input a list of angles and lengths. Any long list of numbers should suffice to give it something interesting to draw. Clearly, any list of the same number over and over again will just close back in on itself eventually. Any alternating list of numbers will end up drawing something like a straight line. I need something more complex than that. The function `Range[ ]` gives me a list of integer numbers, from zero to however high I want to go. Let's try drawing a line with "`Range10`" and see what happens (Figure 6.12).

Something interesting here. Maybe a repeating pattern? Perhaps running it for a longer list of steps will reveal a hidden structure. And because I'm tired of looking at straight line segments, I'll try drawing curves instead of lines, replacing the `Line[ ]` function with `BSplineCurve[ ]`. I know from the Help that both functions take the same kind of list of points as an input (Figure 6.13).

GETTING STARTED WITH PROGRAMMING

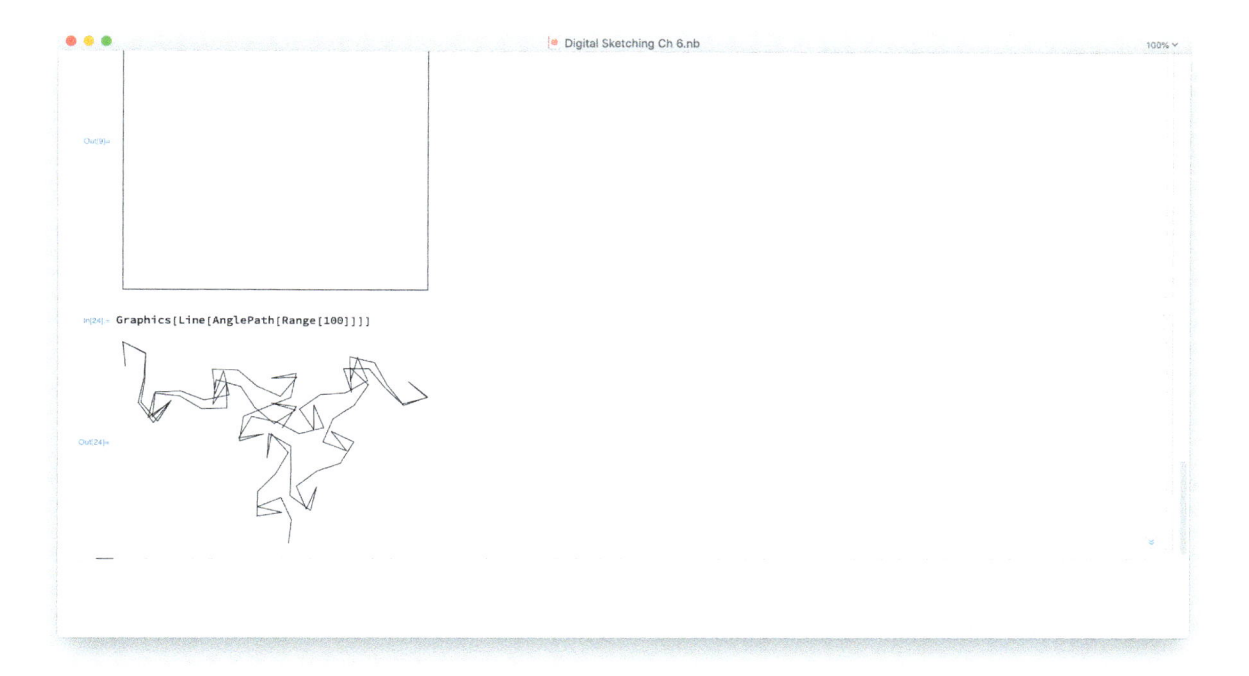

Figure 6.12: A line drawn with AnglePath[ ] from the range of integer numbers from 0 to 10.

Figure 6.13: 100k steps, drawn with curves instead of straight lines.

Clearly, there's something emerging here, which I didn't really expect. I wonder what happens if I let it run even longer. Maybe 10 million steps long, and I'll see something new come out. There is no way I would ever try to draw that by hand (Figure 6.14).

But enough of these lines. How about a 3D surface? Easy enough to draw by hand, but much more complex to draw with a computer. To draw a curved line in space with a computer, we have to actually define a mathematical equation to represent the line. Let's try a conic section first; a parabola. You probably remember the equation for those without much prompting. I hope you were paying attention in Trigonometry class. For this one, I'll use a `Plot[ ]` function to draw the curve, rather than a `Graphics3D[ ]` (Figure 6.15).

Without remembering much about the rest of the functions you learned in Trigonometry, what simple experiments can we try with this tiny program? What if I changed the value of the exponent in the function? (Figure 6.16).

What if I instead swept this curve through space to make some kind of 3D surface? I'll have to tell the computer I want to draw in a 3D context, not 2D. And I'll have to account for the extra dimension in the function as well. Something like Figure 6.17.

It could be a nice parabolic vault, only it is upside down. Let's keep experimenting. To flip the function, I need to give it a negative somewhere that was positive before. This is an elementary function, so there's only one place to try (Figure 6.18).

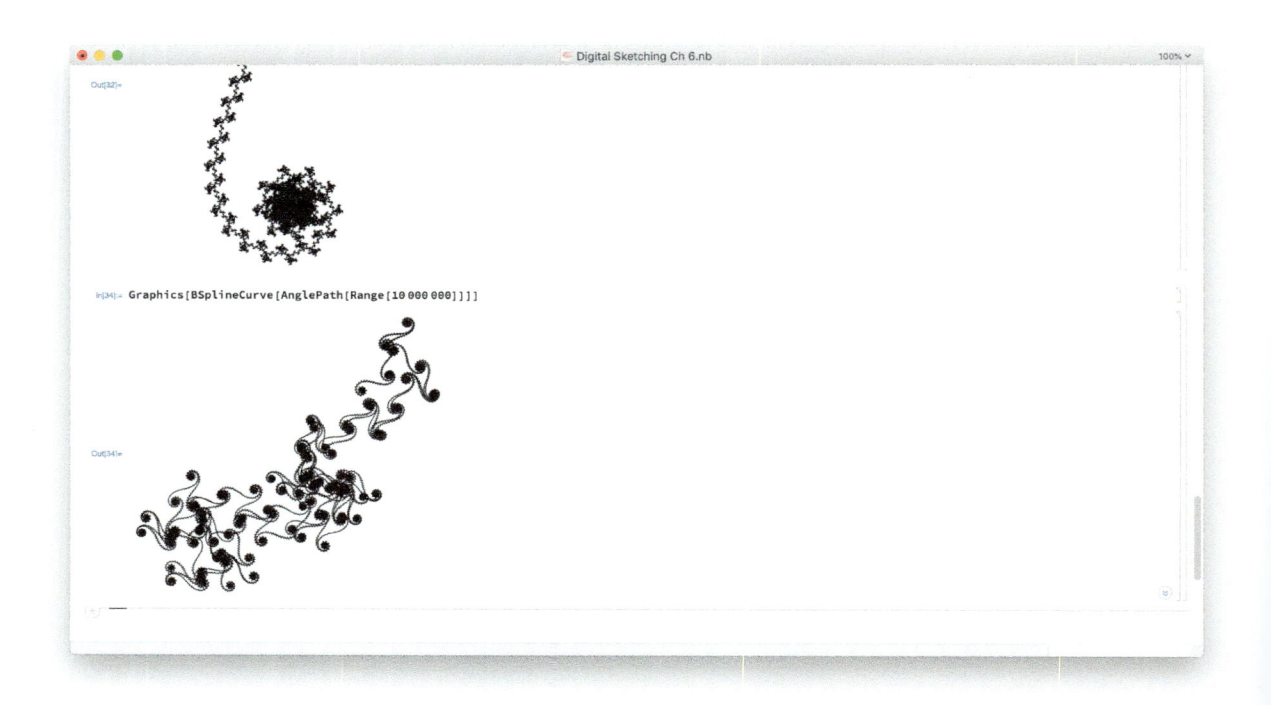

Figure 6.14:   Ten million steps.

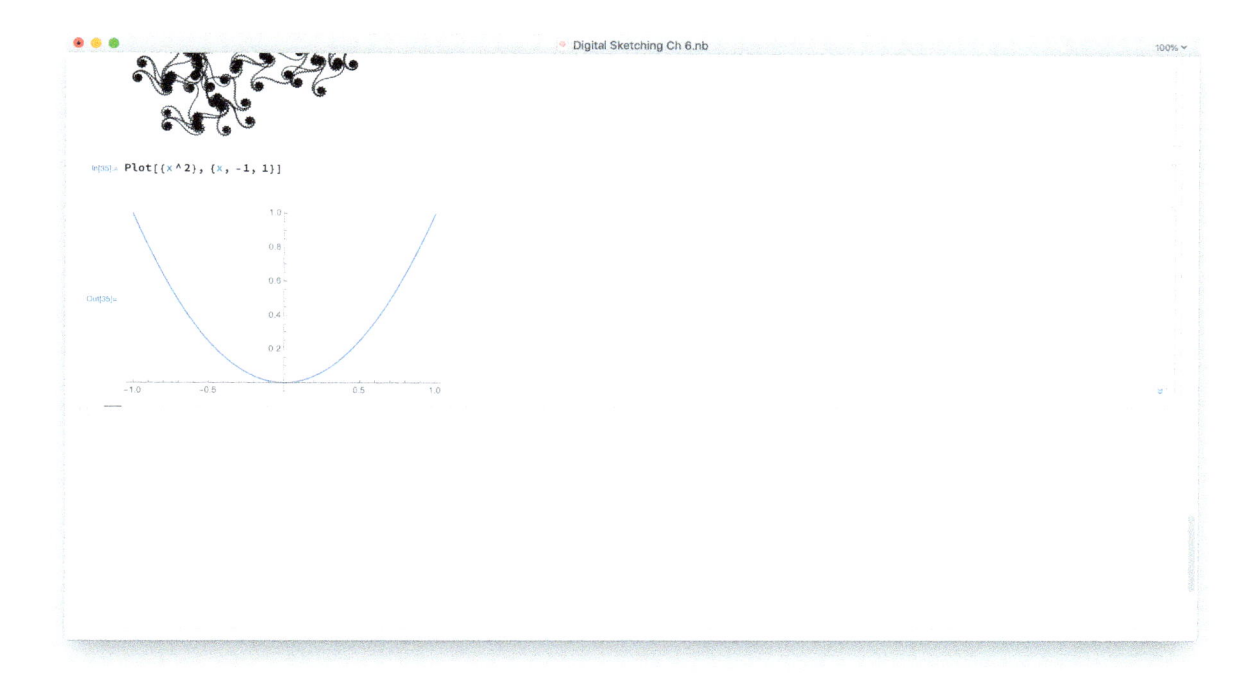

Figure 6.15: Plotting a parabola.

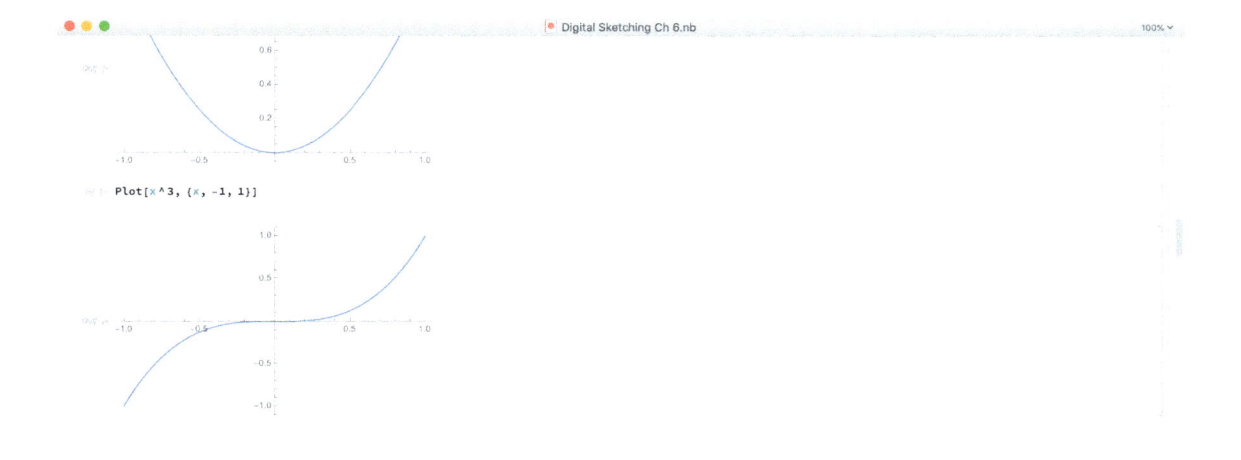

Figure 6.16: Replacing ×2 with ×3.

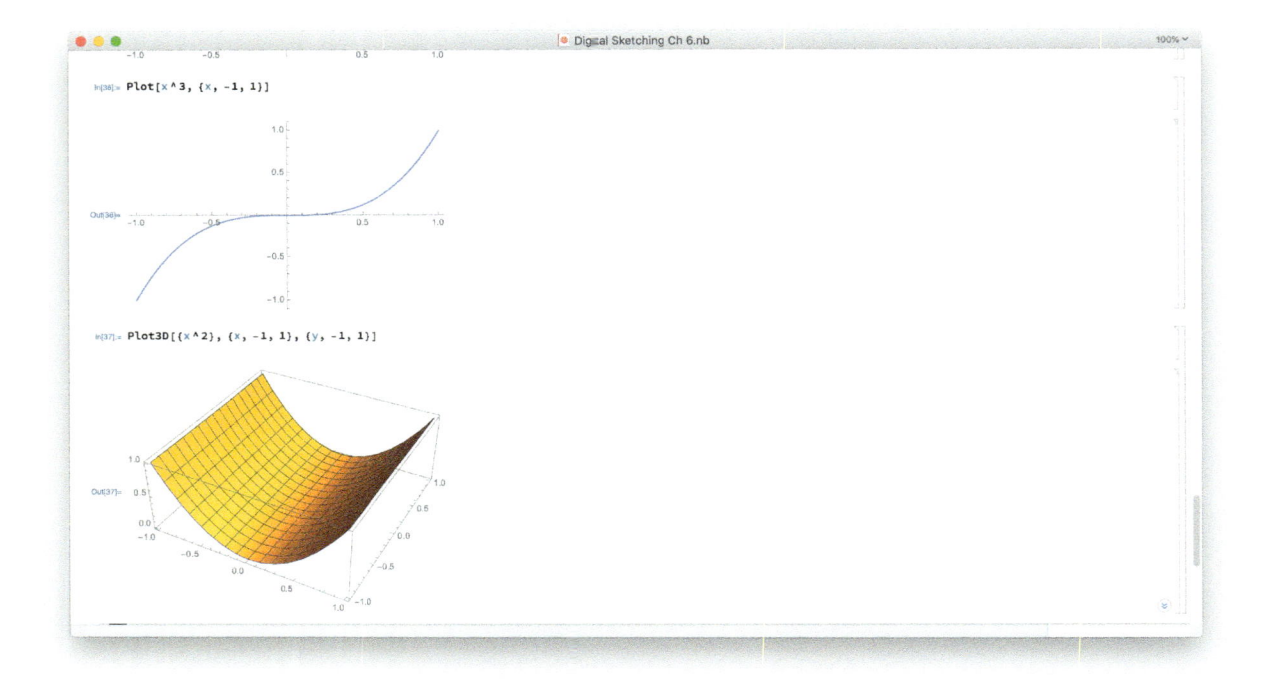

Figure 6.17: A parabolic surface function.

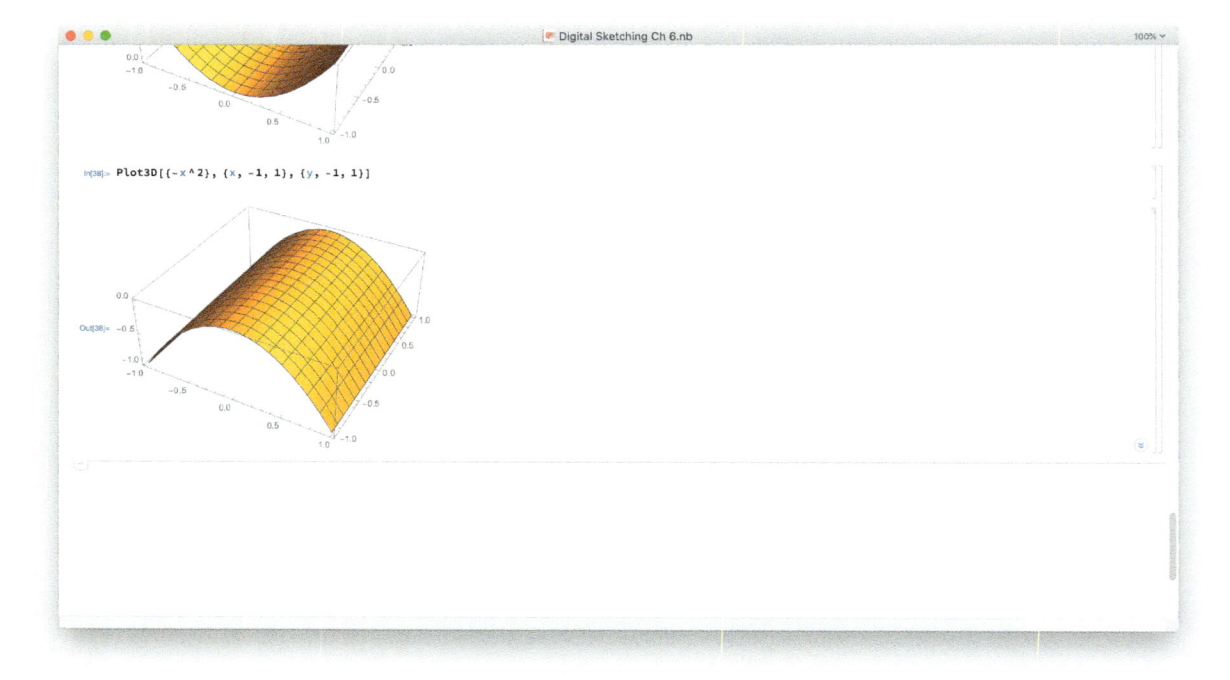

Figure 6.18: Flipping the vault right-side-up.

I wonder what else I could try? Maybe I should see if I can make a dome instead of a vault. In my function so far, I only have variation in one axis, the *x* direction. The other direction is constant, which is what makes this a simple swept form. Let's try varying in the *y* axis as well, as shown in Figure 6.19.

Neat! I didn't get a dome (what did I do wrong?) but instead, a saddle of some kind. A happy accident; maybe there's someplace for that later on. I think what I wanted to do was a negate both the *x* and *y* values, though. Let's try this again (Figure 6.20).

OK, there's a nice parabolic dome, spanning equally from −1 to 1 in both *x* and *y* axes. Now let's mess it up a little. It feels too mathematically perfect. Let's just try some stuff. How about I take the sine of that whole function. I don't know for sure what we'll get,

but I think it will be a little curvier. I like curvy, so let's just try it and see what happens (Figure 6.21).

And it is a little curvier at the edges. Also, less impressive than I hoped. Let's mess around a little more. What if I change the range some, make it less equal on both sides? (Figure 6.22).

Whoa. I think that was too much. I can't really tell what's happening anymore. Let me try moving in the other direction. Instead of 10 times more, let me try 10 times less, as shown in Figure 6.23. Interesting. Now my dome is high on one end and low on the other. This has possibilities, I think. Kind of reminds me of the saddle curve I didn't expect earlier. Let's fiddle with something else. Maybe try a higher exponent value and see what happens there.

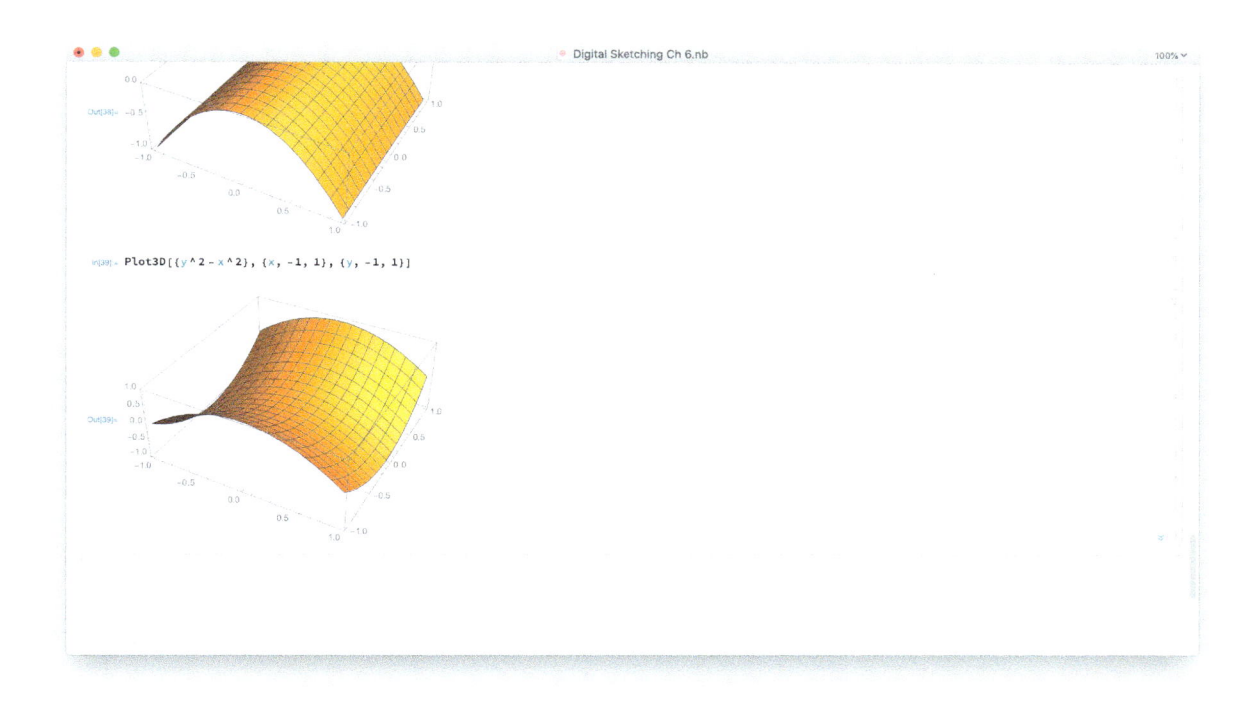

Figure 6.19: Varying both *x* and *y*.

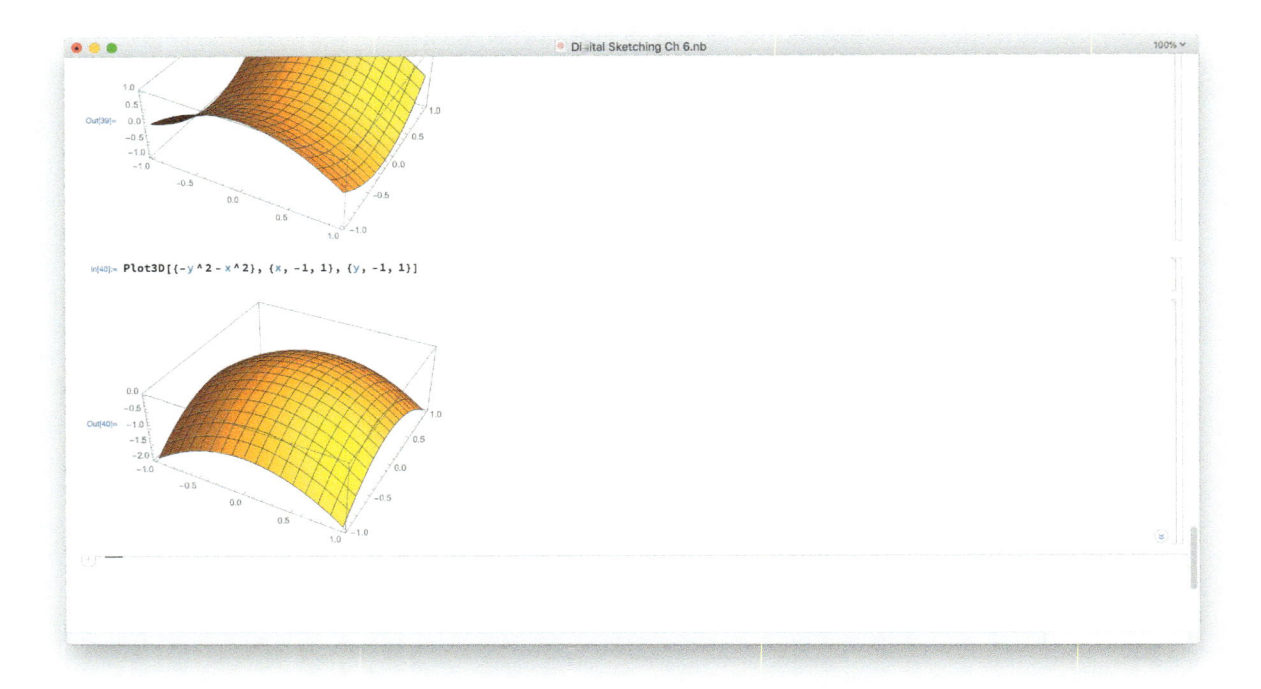

Figure 6.20: Trying for a dome...

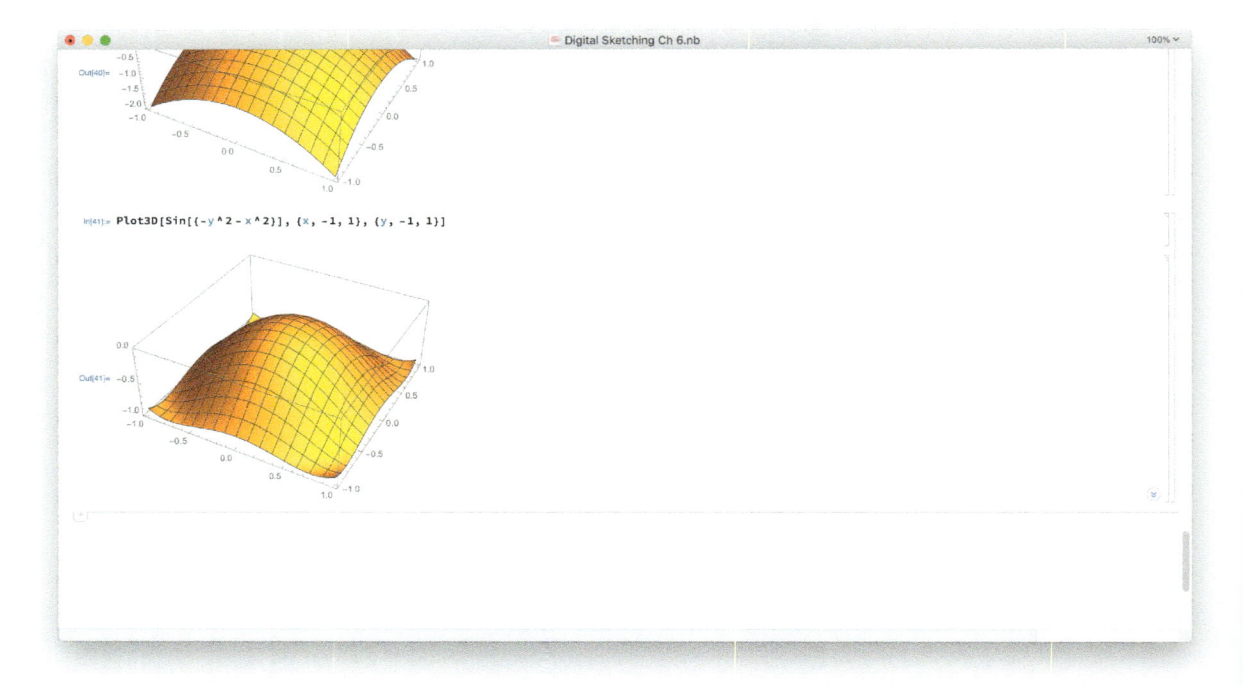

Figure 6.21: Add a Sine function to make it a little curvier?

GETTING STARTED WITH PROGRAMMING

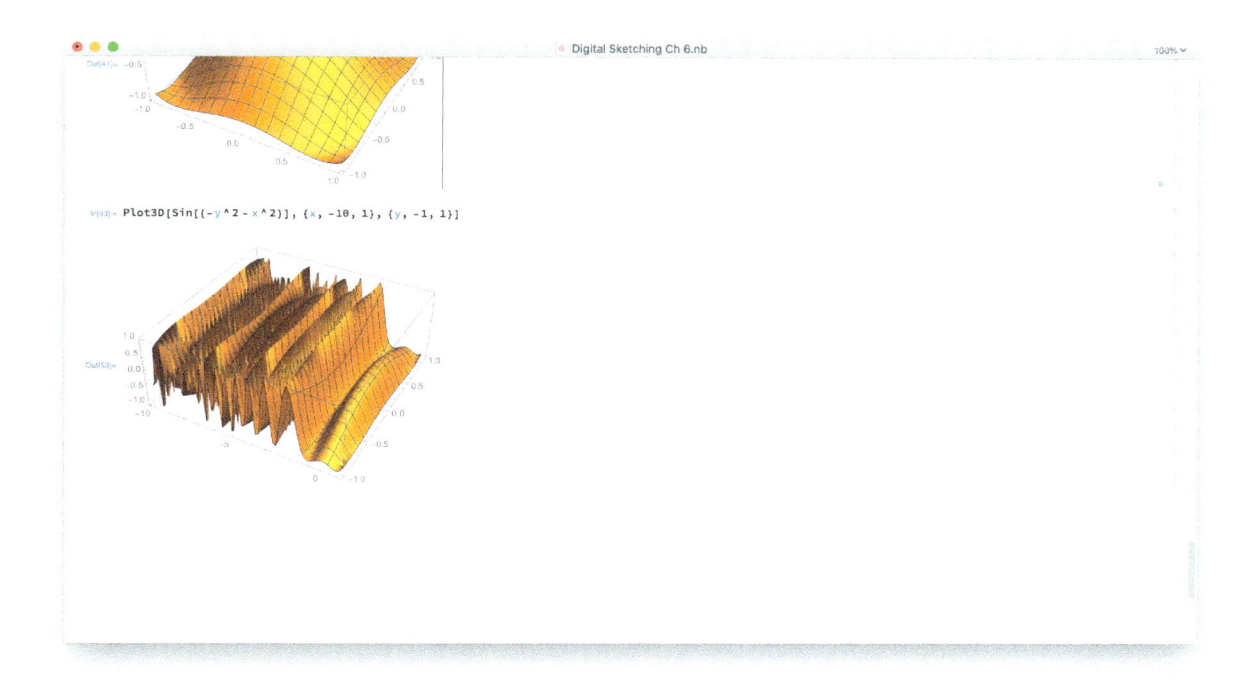

Figure 6.22: Trying to make it even curvier.

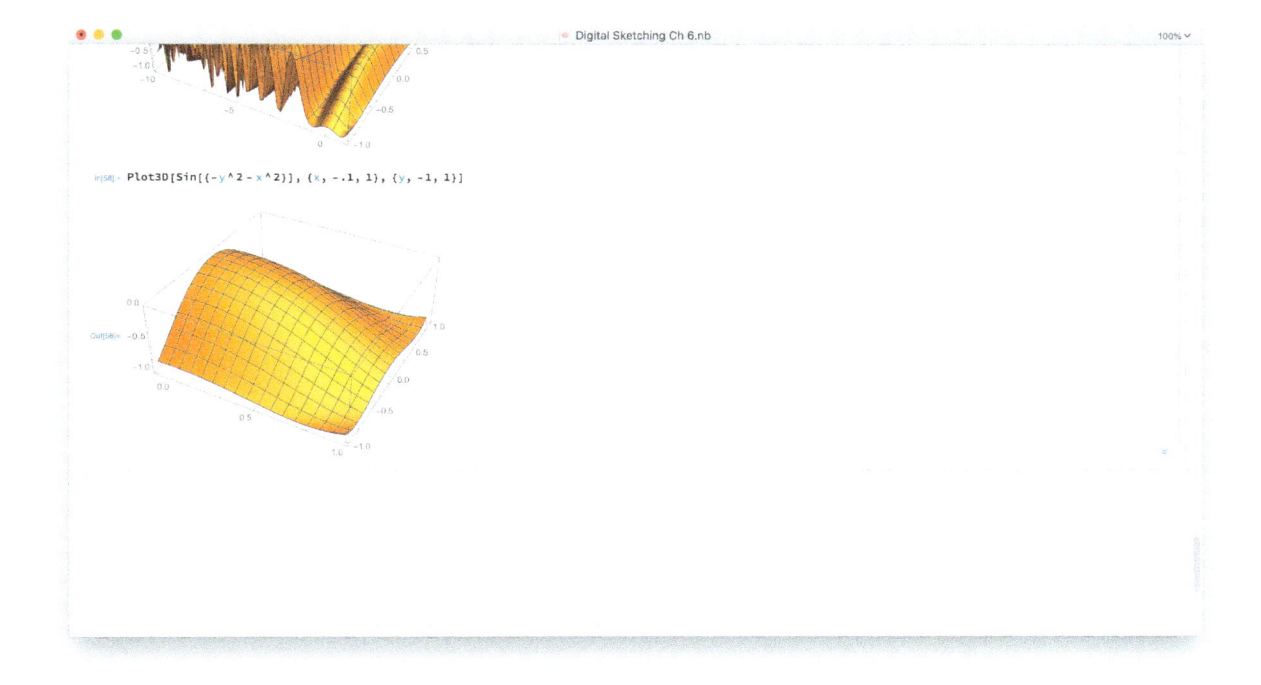

Figure 6.23: Trying a lower value.

Not sure what happened here, but that is OK. Maybe I just need to widen the view a little bit. There's not much going on in this part of the curve, so maybe I should snoop around a little. I'll increase the range just a little bit. I remember what happened when I went all the way to 10 before, and that got really noisy for my taste. I'm going to experiment a bit and see what kinds of forms are in there. Because I don't know for sure what I'll like, I'm going to ask the computer to try a range of values and show them all to me on the screen at once (Figure 6.24).

The Table[ ] function gives me a list of variations. I've replaced a number in the original function with a variable and then asked the program to iterate through a range of variations of that variable's value and return all the results to the screen so I can have a look.

I like the fourth one in from the left the best, and I'm ready to try it in context somewhere else. Time to switch modes back to my 3D modeling environment so I can collage this form into some other work I have from before. Export from Mathematica to another modeling format is also just a function. There are a dozen choices for export format that can be shared between SketchUp and Mathematica. I'm going to choose the DXF format because I know from past experience that it works fine (Figure 6.25).

This leaves me with a file I can import into SketchUp, where I can rotate it and scale it like any other piece of normal SketchUp geometry (Figure 6.26).

Since it was easy to get one surface to try, why not try a range. There might be something I am not

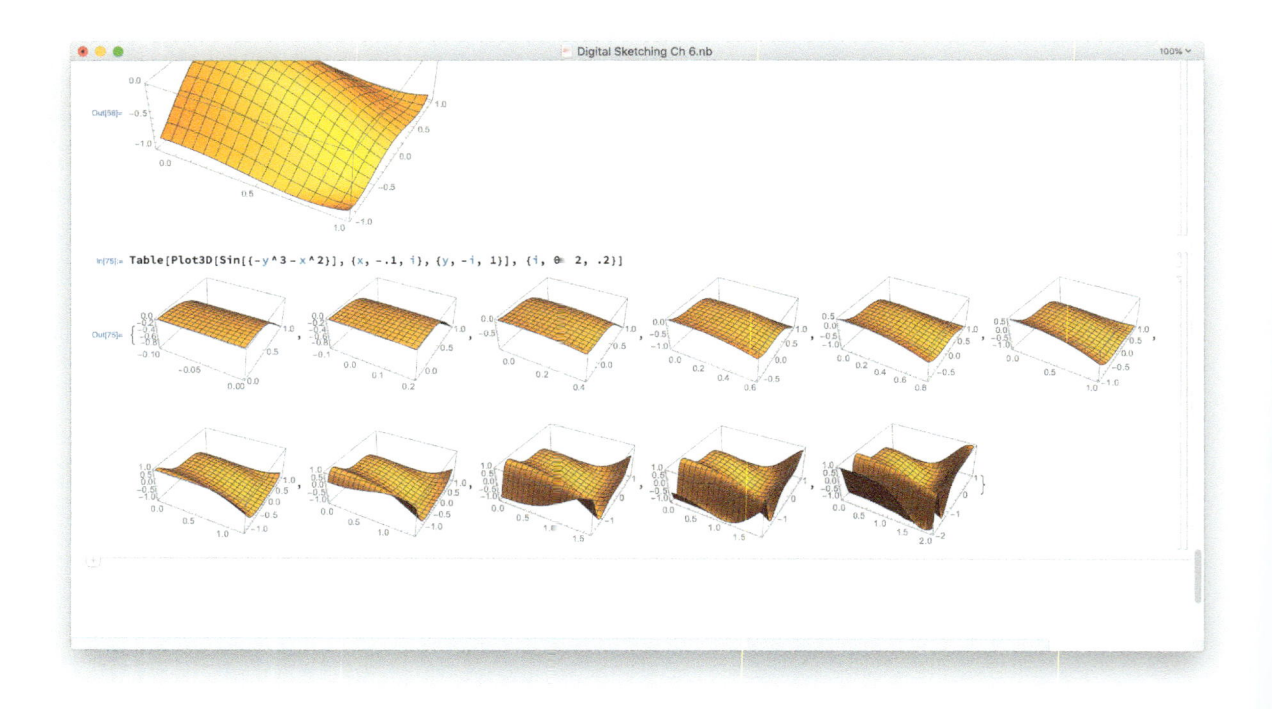

Figure 6.24:   Trying a range of variations.

GETTING STARTED WITH PROGRAMMING

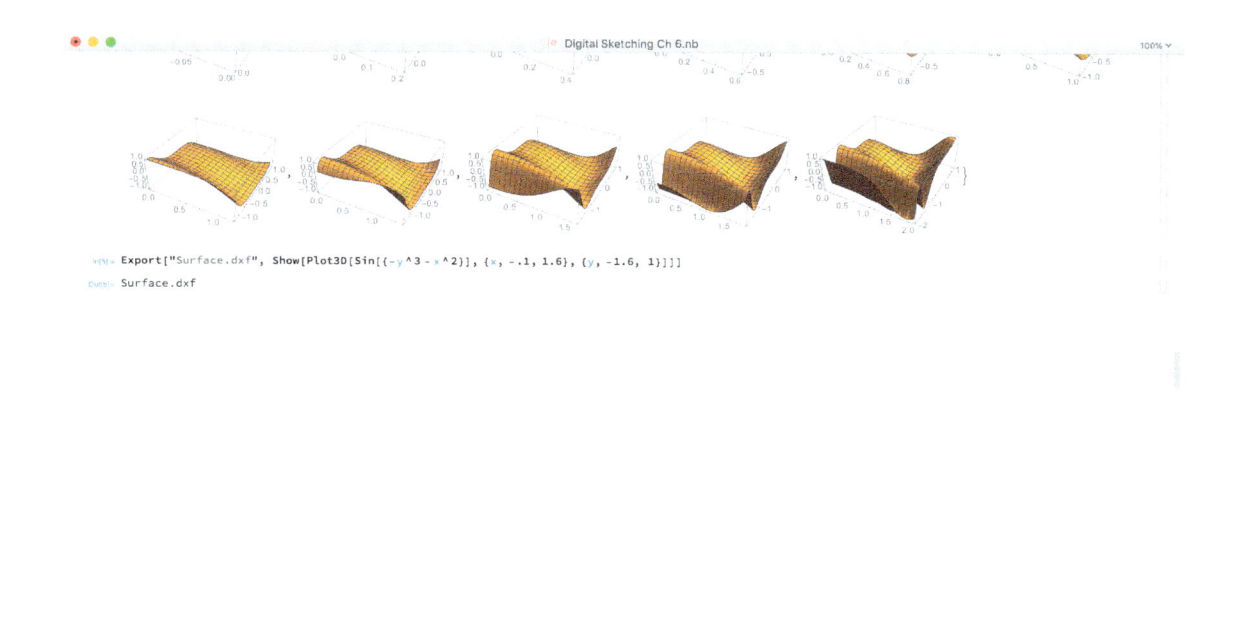

Figure 6.25: Exporting a surface from Mathematica using the Export[ ] function.

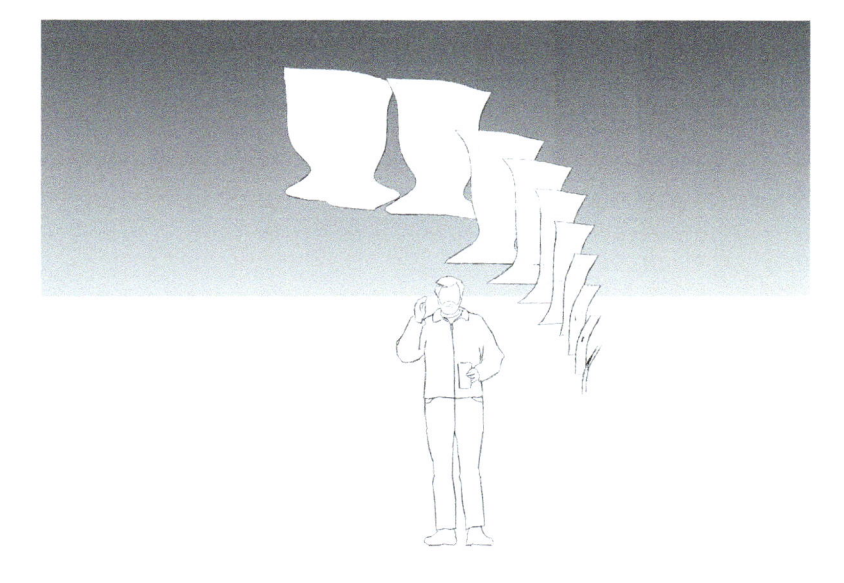

Figure 6.26: Mathematica surfaces, imported into SketchUp.

seeing about one of them in isolation, I need to be able to collage them in context to really know what will work. I can modify my program just a little to let it do the exporting work for me. Then I import all in SketchUp, and I'm ready to go to try whatever I decide comes next.

Not bad for a quick experiment with a trigonometric function that you learned in High School. But if I wanted to have more direct control over the surface, for example, if I knew there were some specific points that I needed it to pass through in another model, I need a different kind of math. Free-form curves like Béziers and NURBS are a bit more complicated to define, but they might be more useful when I'm modeling like this. I could write a quick function of my own to wrap up that math neatly so I can use it over again in a larger program, just like building a custom SketchUp component rather than searching through 3D Warehouse to see what I can find.

freeformCurve[t_, p0_, p1_, p2_, p3_] := (1 − t)^3 p0 +

3 (1 − t)^2 t*p1 + 3 (1 − t) t^2 * p2 + t^3 p3

If you look at this with two halves divided by the equal sign, the left-hand side defines the function freeformCurve, followed by the three variables that need definition to run the function. The right-hand side contains the math that will be run with those variables. Conveniently for us, because this math is really too complicated to mess with when I'm sketching, Mathematica also has a predefined function that draws Bézier curves. In fact, we've used it once already in this chapter. No need to go all the way back to first principles every single time. Mathematica's built-in function is called `BSplineCurve[ ]` (Figure 6.27).

There is also a built-in function called `BSplineSurface[ ]`. Can you guess what it will make for you?

This is going to look a little more scary, I bet. But really, this is just functions built one on another in a hierarchical way (Figure 6.28).

Computers aren't great at coming up with solutions on their own, but they are great at running through all the possible variations on a theme with great speed. If you can learn how to encode your design problems in a computer-readable form, there are a number of different ways for a designer to lean on the single-minded computation power of code to explore variations. We did that once before to explore a range of values for a single variable. But what if you have no idea what range is going to give you the best result?

If you don't know where to begin, any value is the right place to start. In this case, something random could be appropriate. Let's try the simple Bézier surface example again, but this time we'll replace the predefined geometric points with something random, and then we'll tell the computer to try some variations on the theme.

This looks superficially simpler, but it is actually more powerful. Rather than specifying exactly which points I wanted to be drawn, I gave the computer some freedom to choose numbers randomly. Here I let it run 10 different times, with varying results (Figure 6.29).

I had a chance to study with Stephen Wolfram and his team during his inaugural "New Kind of Science Summer School" many years ago. In that course, I experimented widely with different kinds of simple programs to generate patterns and forms, trying always to keep a designer's perspective. There's quite a bit of deep math in NKS, but you can easily put that to work for you as a designer in Mathematica. I can spend hours, for example, playing with complex tile patterns based on 1D Cellular Automata (Figure 6.30).

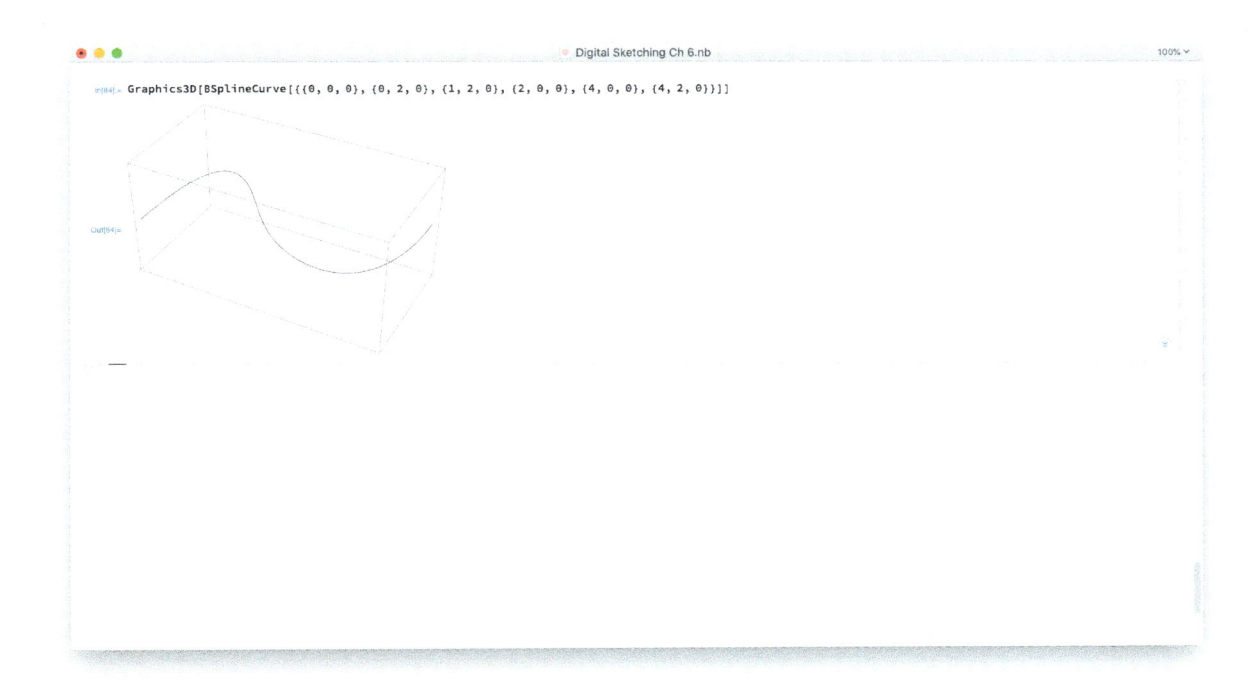

Figure 6.27: A basic BSplineCurve[ ].

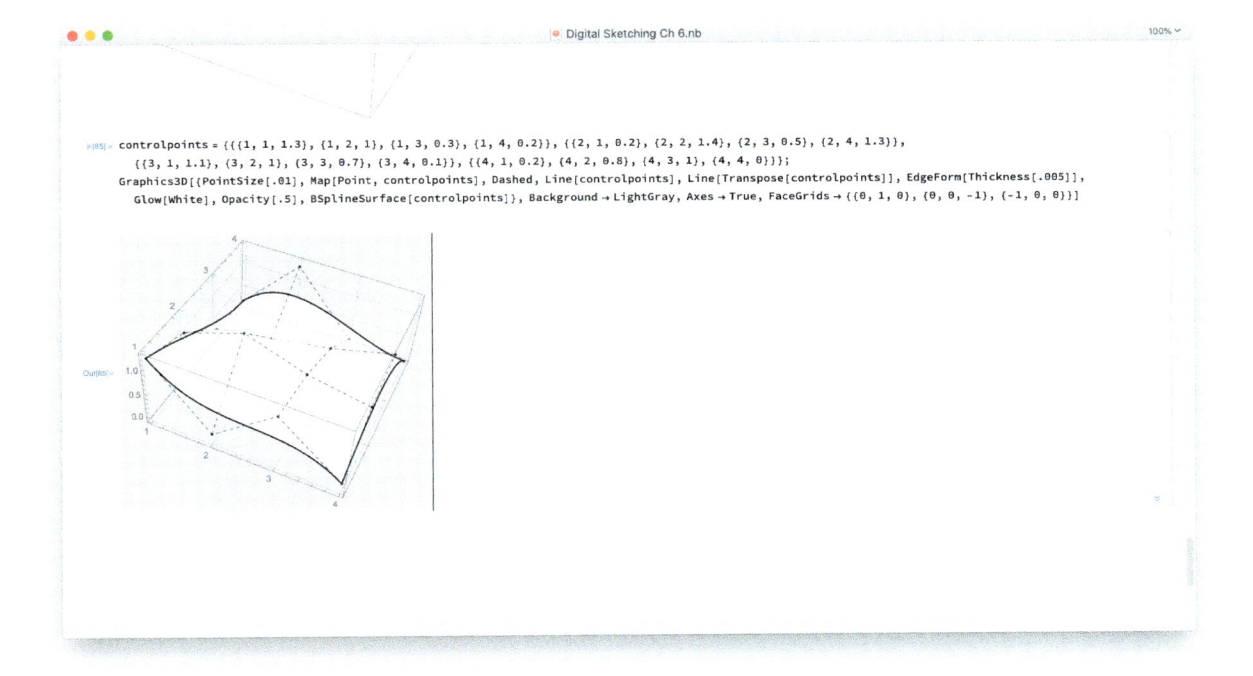

Figure 6.28: A BSplineSurface[ ].

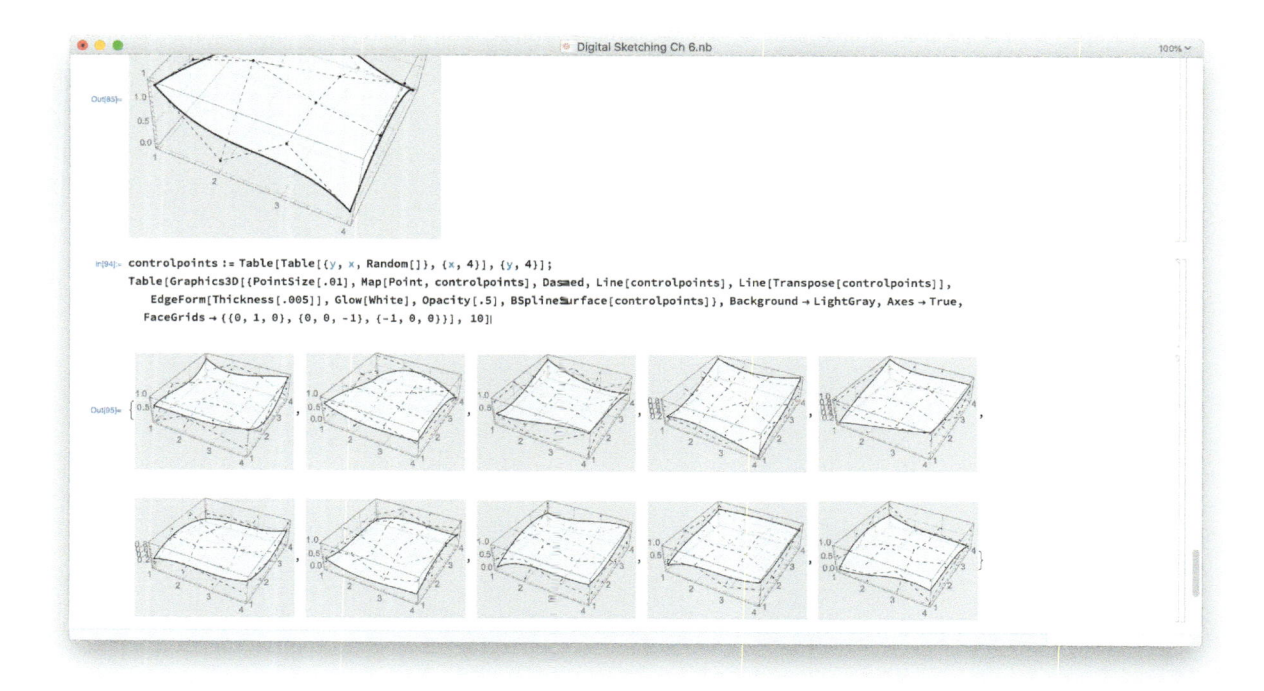

Figure 6.29: Exploring some random surface variations.

Figure 6.30: Ten variations on Rule 110 from random initial conditions.

From this work, it is easy to pipe the output of a Cellular Automaton into a line-drawing algorithm like we used at the beginning of this chapter. What would it look like if the progressive evolution of one of these raster patterns was turned into line drawing instructions? Maybe something interesting? (Figure 6.31).

There's something interesting happening here. I wonder what happens if we let the computation run even longer? Here's a set of variations, each running progressively longer. I don't know what to make of this, exactly, but there's a thrill in discovering something that the world may well have never seen before here. No combination of parameters have ever before quite drawn this set of lines. Not exactly, not this way (Figure 6.32).

Learn to Program, But Creatively

My old design computation mentor, George Chaikin, left a memorable impression with his design computation class at Cooper Union. In it, we were taught to think about computers as we thought about any other tool for use in the design studio. We didn't really have any "applications" installed on them. The engineering computer lab had copies of AutoCAD installed, but they were installed in the PC lab across the hall, and we didn't use them. Instead, we had an aging pair of Silicon Graphics workstations that provided access to IrisGL (the precursor to OpenGL) and a C compiler. Our first class homework was to come up with an algorithm for drawing straight lines on a raster display. On graph paper, by hand. With a pencil.

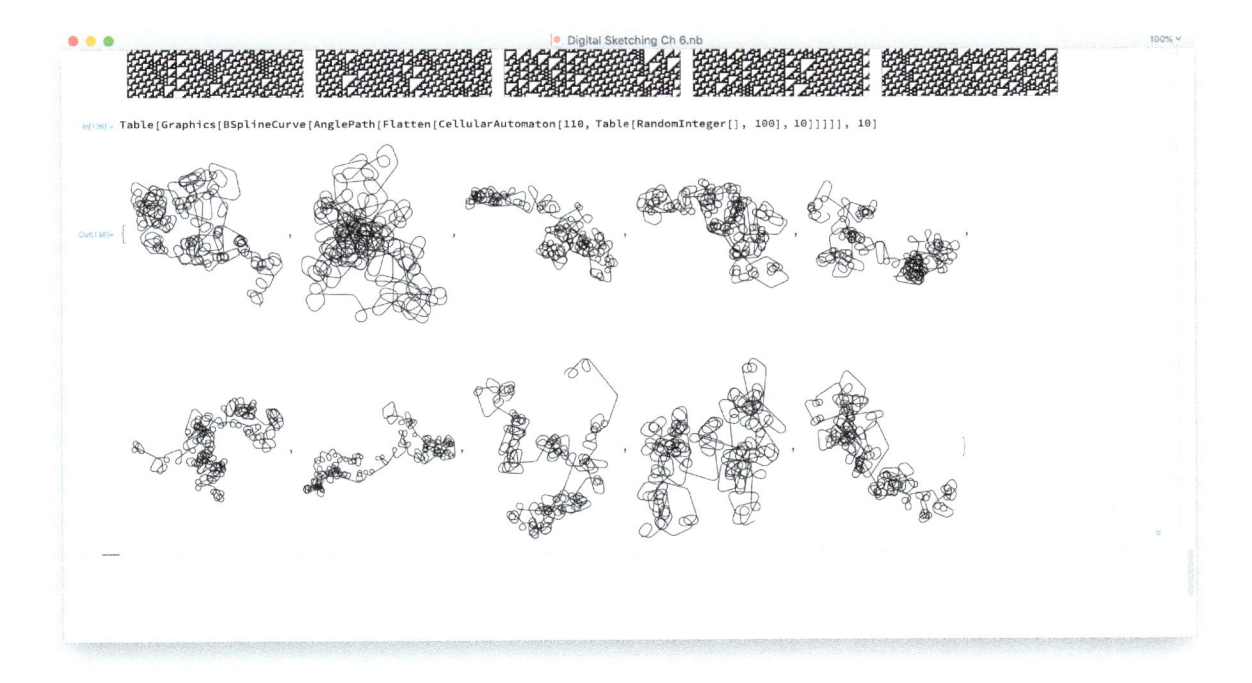

Figure 6.31: Variations on a Rule 110, drawn as lines.

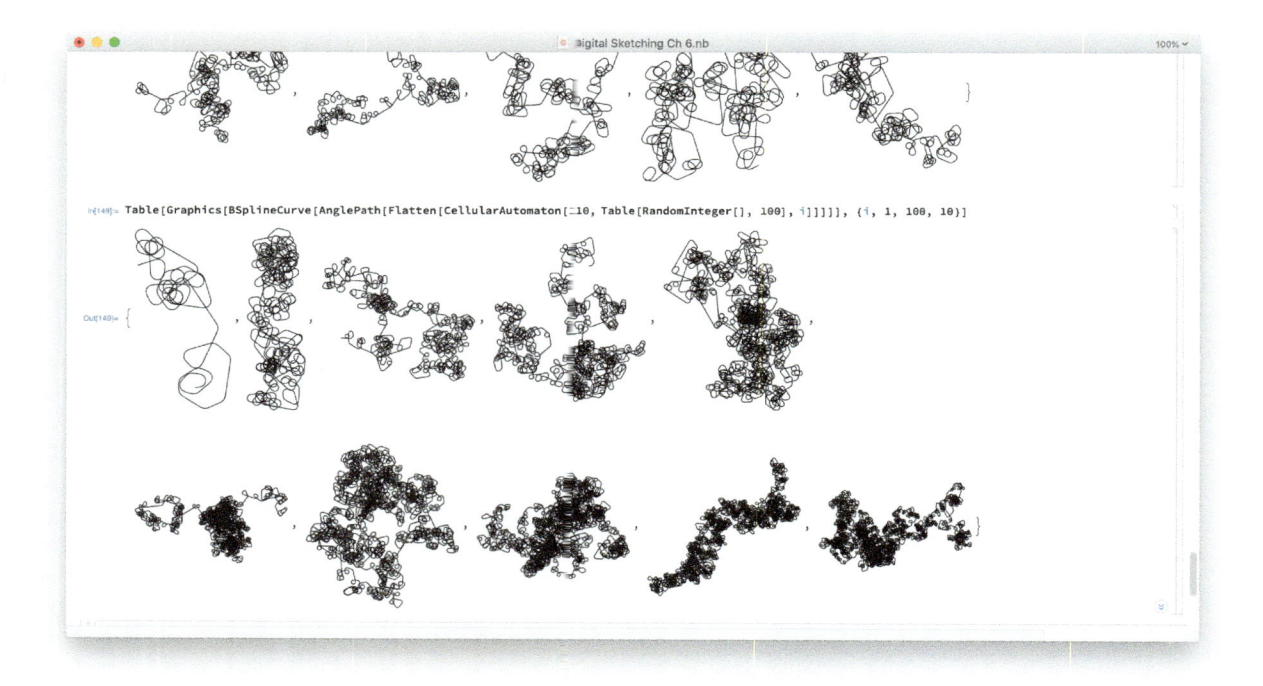

Figure 6.32: Progressively longer line drawing computations.

What I learned from George's class was that computers are useful tools for design, but not only because they can run high-level applications. They are also valuable because they can compute things in algorithms far faster than I can. Learning to program was, I learned, very much like learning how to draw. And also, like learning to draw, programming is mostly something that anyone can learn. There's no guarantee that you could get yourself a software engineering job at Google. Still, most folks who can structure a rational argument are capable of putting together a simple program.

I had the pleasure of meeting John Maeda many years ago at the Aspen Design Conference. He had just published his book *Design By Numbers* (Maeda 2001) at about the same time.. "DBN" was a simple programming environment targeted explicitly at graphic designers. In it, Maeda started students with. . . drawing a straight line on a raster display. Just like George taught in his class.

In the years since then, computer languages have grown increasingly humanlike in their syntax, and as a consequence, have become more natural for a broader audience to learn. DBN gave way to "Proce55ing", explicitly developed for multimedia artists by two of John Maeda's students. Processing lead to Arduino and the "Wiring" language. . . furthering the mission of bringing simple programming within reach of anyone with a computer.

Personally, I'm still fond of Mathematica, which is among the broadest and richest IDEs for scientific computing. As any creative will tell you – sometimes the most delicious of tools are those built by someone else for some other purpose. Mathematica has been around for a long time. I used the first build of Mathematica in 1988 – the same year I graduated from high school. I can't claim to be an expert, but I can drive Mathematica around the block without getting in too much trouble. I appreciate that notebooks I made with Mathematica years ago are still computable today, though the computer I'm running them on now is just spectacularly more capable.

Most designers who think about design computation today are probably thinking first about Grasshopper, a visual programming language wired into the modeling core of Rhino. With it, designers can build parametric design machines that can be puppeted by data to explore a wide range of parameter spaces. I'm reminded, however, of a lesson John Maeda taught me in Aspen. Specifically, that the advantage of designing with algorithms comes from being able to simultaneously solve for all possible design solutions in a parameterized space at once. Once a design problem can be turned into an algorithm, the designer can simultaneously explore every variation, not just the ones that can be manually drawn in the time available.

Programming feels intellectually quite distant from the simple, visceral act of drawing with a pencil on a piece of paper. How do you keep your code sketchy? At the root of a sketching practice in code are a few simple principles. First, the computer will only do what you tell it to do. References to AI and "machine learning" aside, algorithms are designed by people with some intent. To work with algorithmic design, you are going to have to learn how to talk to machines in the languages they understand – and that means learning how to program. But if you're clever about which IDE you use, the setup and get started time can be minimal. A snippet of sketch code should be cheap and expendable, just like a physical sketch on a throwaway piece of trace. If you had to spend hours getting set up before you could start to explore, you will never be able to get into the reflection-in-action state. Don't be afraid to cut some corners and fiddle with the code, even if you don't really know what it does. Surprising outcomes may occur.

I like Mathematica because I can readily experiment in its notebook interface. I can pull in a line of code that I don't understand, then just start poking at it to see what happens. What happens if I change this variable from a 1 to a 2? Or maybe if I just try piping the output into some new function. I have found that some wonderfully delightful things happen when I do this. Of course, often I just break the code, and it gives me an error without producing any output at all. Sometimes I get stuck when sketching in other media as well. Time to reframe the problem and move on.

Mathematica does expose its mathematical roots pretty quickly, and if you're threatened by that, you might want a different starting point. Since programming languages are a particularly fast-moving part of the world of technology, it makes little sense to go into too much detail about any of them in a book that may last longer than any given technology it describes. I'll confine the more detailed discussion to an appendix. Luckily, computer languages are growing increasingly similar to human languages. In many cases (for example, think about the voice interfaces to devices like Apple's Siri and Amazon's Alexa), it is even possible to hold something like a conversation with a computer today. This will only improve in time.

It is increasingly common for new students to graduate into our design workforce to come with a high degree of computational literacy. This needs to be understood as more than just knowledge of the latest high-level design applications. Some degree of programming competence needs to become a part of our society's basic package of literacy. Everyone should know how to read, to write, to speak well in public, perform basic math, and write a couple lines of code. As generations go by, we are growing together with our digital counterparts. I, for one, welcome our cybernetic future – especially if it helps me to sketch, experiment, and to iterate toward new kinds of design practice.

chapter 7 From Sketch to Production

. . .the house pre-existed in the mind of the architect: and this can be designated as the Idea of the house, because the artist intends to assimilate the real house to the same form that he has conceived in his mind.

St. Thomas Aquinas

Sketching for Presentation

Every successful construction project begins with a sketch, but it ends in a building. The intuitive freedom of a reflection-in-action sketching process must eventually resolve into an actionable, rational, and credible plan for construction, or there wasn't much reason to do it in the first place. It is by now, I hope, obviously important to preserve space in your process for free thinking at the beginning of a project, and not to jump with too much haste to the fully constructible level of detail. But you must, in time, reach that point if you wish to actually make a building.

The precise medium of informational exchange used to tell a construction team what the building should vary from job to job, depending mainly on your contractual relationship to the construction work. Traditional methods of construction documentation, while still highly relevant and practical today, will undoubtedly evolve into more efficient digital replacements in time. But for now, you probably will need to have some combination of 2D drawings, 3D models, and other representations of your project delivered before your work is done.

But whatever the medium of collaboration you choose, there are some basic principles to consider as you progress your sketches into a production process that includes a broader team of collaborators with increasing levels of accountability for the final construction project.

Every time you show a sketch you have made to someone else, you are making a presentation. A formal presentation closes the loop on your reflection-in-action and brings an external perspective from the world back into your work. With every presentation will come, hopefully, some validation that you are on the right track with your work. It may also expose your weaknesses both personally and in the completeness of your design. But more importantly, presentation opens your work up for critique, and with that, you will learn things about your work that you didn't know before. Nothing, in my experience, is worse than a presentation where the only response is positive. Platitudes don't improve the work. Honest, open, and carefully considered critique, particularly in ways that expose weaknesses you didn't see in your design, helps you to reframe the problem you are working to solve.

Some designers fear presentations more than any other aspect of their work, and for understandable reasons. At no other time are you as exposed and vulnerable as you are when you are presenting your design work. You have spent countless hours working, pouring your heart into piles of sketches. It is unlikely that anyone to whom you show your work will be even partially as invested in it as you are. In the 5–10 minutes of attention you will get from them, they will form an impression of your work, which may have little to nothing to do with the intention you had. It is your responsibility to steer them in the most useful direction.

If you believe in your work, you can't be passive about it. You owe it to yourself to make a good pitch. You owe it to your project to represent your best thinking at the moment when you are called on to talk about it. You owe it to your audience to clearly and concisely tell what it is that is interesting about your work. Your audience wants to like it and to have a real conversation about it. You have to help them find their way into that conversation. Your presentation is a performance, like a musical performance, like a dance performance. Like any great performance on any stage.

The key to an excellent presentation is how you tell the story of the project. People love a good story, and the best stories are those that provoke in your audience thought about something more significant and more important than themselves. A great story is a gift given by you to your client. If you do it well, the story will live on to be retold by them over and over again. No single thing you will do as a designer will be more useful to the project than the story you craft about what the project will be.

The best stories, the most universal and compelling stories, are those that follow the journey of a hero as he encounters challenges and moves through experiences that develop him into something new, something different than he was. Joseph Campbell, in his exhaustive study of comparative mythology in "The Hero with a Thousand Faces," put names and structure around a universal notion of mythical human storytelling. It may seem like overkill, but if you understand the formation of a story, you are a better and more convincing storyteller. It really doesn't matter if your project is large or small, an essential building, or just a minor renovation. Every project has a hero who goes on a journey. You can use that to tell a story that holds the whole project together.

I caution that the hero in your presentation must not be you. To be sure, you fought against the odds, did battle with dragons and saved maidens to come up with your design, but that is not the story your client wants to hear. In the story you will tell, the hero whose life is changed will be the one who lives in the building you are proposing. Tell that story, tell it well, and you will have the room's attention every time you speak. Save the story of personal struggle for another time.

At Cooper Union, we thought endlessly on the "program" for the project. We learned to tell the story of the project through the eyes of its users. Who was the project for? Who was, in a sense, the hero of our story? John Hejduk taught us to think is archetypes, what might now be referred to as a "user persona" or a "target market." But fundamentally, the question is the same, no matter what syntax you choose to use. Your project matters to someone. If you do your work well, you will change the life of that person positively. This person, not you, is the hero of the story you will tell in your presentation.

At Communication Arts, we told the Hero's Journey differently, but the same story was behind every one of our best presentations. The most successful projects are those that are built for people, to help them come together in community, or to enrich their lives through some other personal adventure.

At the beginning of your presentation, before you have shown a single sketch, your project exists only as a part of the everyday world. You should start your presentation by talking about what that world is like and how it operates before you have proposed anything. Before your project, where did your hero go for coffee in the morning? How did they take their kids to school? It might make sense to talk a bit about the existing site conditions. What dangers did they face crossing that busy road on the north edge of the site? In what state is the site today? How is it being used, by whom, and for what purpose? If you are working on a renovation project, what observations can you make about the history of the building or its prior uses? If you have a blank green field to begin with, it has a story to tell as well. What is its condition today? The ordinary world at the beginning of your project sets the context into which you will intervene with your new design. Normal world conditions are invisible to your audience until you spend the time to call them out and expose them as essential to consider. They are just the way the world is.

If you have presented on your project before, the normal everyday condition is whatever state your audience remembers from the last time you presented. Remind them if you need to, but know that you may not be entirely in control of what they remembered from last time. What part of the story resonated with them before? How might it have changed in their memory since the last time you spoke? What is their normal every day, and how sure are you that you understand it? Like the foundation that every building must have to stand up properly, every presentation needs a beginning. Spend some time setting it up well.

Once your context is set, you can begin to propose the state of change you want to see. People understand good ideas in the context of unfavorable ideas that they replace. Extraordinary ideas are only so because they are more than the ordinary ones that they succeed. You have to move from every day to this day, to the new special condition that your proposal can make into the new every day. You are sending your hero on a journey into the imagination, into the unknown space of your proposal. Every hero faces the unknown with a guide or mentor by their side. You, as the designer, are their mentor.

In the Hero's Journey, the hero is moved to begin this journey of transformation upon hearing a call to adventure. Something happens, something is observed that leads to a sense of questioning. Something about the everyday is no longer known in the same way as it was before. The hero is invited to see the next stage; and the hero's everyday world is about to change in some significant way.

For your project, the first threshold is in the definition of the problem to be solved. How have you framed it in your own mind? Somehow, a condition in the ordinary everyday of your site stands out as different, as a provocation to an adventure ahead for your hero.

The everyday normal of your site is now transformed into a mythical space of possibility, risk, and opportunity. Your hero may not see all of the trials ahead yet, but it is clear that they have moved into a new space – a space of the mind, a space of questioning and critique. Your client is with them on this journey and will need reassurance that the risks they are taking will be worth it. There are real money, time, and even physical risk ahead for your team. They will meet obstacles and opportunities. They will fight metaphorical dragons (or not, perhaps, depending on the nature of your local planning department) as, together with them, you learn the true nature of the project you are building together. It is difficult to follow the path you are on. To admit anything less will only set default expectations that you will have to unwind later.

In a presentation, you should prove that the risks are known and that you can face and overcome them as a team. A designer who doesn't acknowledge the dangers ahead is one who lacks credibility. You don't want to be seen as having your "head in the clouds" without also a sense that your feet are firmly planted on the ground below. If you make a grand claim about some abstract poetic moment in the space of the project, always show the physicality of that poetic. A soaring spiritual moment in a gothic cathedral is only so because the flying buttresses are physically capable of holding up the roof. Architecture is unique as an art in that it is nothing without the rational science of building to hold everything together. You have to learn to present the poetic using only the prosaic language of steel, wood, and concrete.

At some point, your hero will face the ultimate challenge. In your story, the transcended ordinary cup of coffee that people perhaps come to your project to drink now places your hero in a spiritual crisis that was unknown before your team confronted it. Coffee and a two-top table is now a place for a job interview, a first date, or a place to comfort a friend grieving a loss. The space you're designing is an embrace, a sanctuary, not for coffee, but for the life that is lived by your hero, coffee in hand.

If this sounds like too much, feel free to tone it down. Don't be heavy-handed in your storytelling, or you will lose your credibility for a different reason. If you see a subtle eye-roll or a furtive glance below the table to a hidden mobile device, you've lost your audience, and you need to pull attention back to the story. Remember that your construction team only wants to think about construction things. They build foundations, walls, and roofs. They want to know where the grease trap is going to have to go because it will take months to get that permit. But underneath all those everyday worries, they understand the same story arc that you do. Help them to join your vision by making it easy for them to do so.

You should prove to your audience that you understand everything there is to know about the tactical necessities of the project, but that above all of that, there is a higher purpose for the work. Every new Starbucks coffee shop that is built needs a grease trap. But the team never wavers from their understanding that the prosaic necessity of a grease trap is built *"to inspire and nurture the human spirit – one person, one cup and one neighborhood at a time."*[1]

Your team may be uncomfortable with the poetic of your work. Help them into that mindset by reminding them that the users of the project, the real people for whom you are doing all this work in the first place, are living their future lives within the work you are planning. In the darkest moment of value engineering, your team's ability to empathize with the users their work will serve may be the difference between a decision that just ticks a box and moves on and one which is considered with care and compassion.

With the hero of your story, you are helping your team know not just what they are doing, but why and for what reasons. If you can't help them to empathize with the real people they will serve through this project, you will struggle to make decisions with lasting value for the rest of your time working together as a team. In the abyss of despair that accompanies every creative project of any merit, you have to defend your user by telling their story with conviction and compassion. Even the simplest of cups of coffee is a part of the story.

Your presentation only progresses when the supreme ordeal has been met, acknowledged, and the stage is set for the solution you will propose. Hit the audience hard with the bright future you

---

[1] This is the mission statement of Starbucks, the coffee shop that sets the bar for coffee shops around the world. https://www.starbucks.com/about-us/company-information/mission-statement

propose. Now is the time for your best drawing, for the unveiling of that 30-second fly-through animation that you took a week to render out. Your hero has a crisis – don't leave them there by themselves. You have a proposal that will carry them through the nadir; will transform them and your project's site into its transcendent new state. Using only the simple tools of wood and steel, with the heroic efforts of a team of skilled craftspeople laying brick and running plumbing, you can, together, change this corner of the world.

Every presentation you make, large or small, should follow this basic outline. You need to state the original problem and establish the context for your presentation. Then you need to identify the problem that needs to be solved. Finally, and only after setting the stage properly, you have an opportunity to offer a solution to the problem. If you have been good enough in your presentation, your audience will begin coming up with ways to solve the issues that they see ahead. If you haven't, they will lob some intellectual grenade in from the sidelines and watch you squirm in the face of all the things you didn't yet consider.

Time spent to get the team aligned behind the higher goals of the project, the *whys* that underpin the *hows* and *whens* of what should be done next; is going to be the difference between a team that solves its own problems together and one which looks forward to seeing you squirm. If you come out too hard in your storytelling, you will make people uncomfortable, and uncomfortable people aren't receptive to thinking creatively together.

The best presentations are those that empower the audience to engage personally with the ideas you are presenting and to make them their own. You should hope that you find people in your audience telling the story of the project that you presented to one another later in other contexts. If you can get a mason on your job telling her apprentice that the reason they are building this wall is so that people can have their spirits nurtured, you have won a thousand battles that you didn't even know were coming your way. People all across the project want their work to have meaning. It is your responsibility to give them a framework for that meaning. If you don't do that, they will find their own meaning somewhere else, and possibly not in ways that will help your goals for the project.

I always like to give a gift of some kind in a really critical presentation. The gift might be physical – a paper copy of a vital diagram or access to a digital proxy for the same thing. I want people to have all the same tools I have to when they do their own retelling of the story of our project. Sometimes, the gift I leave behind might just be a great story, a story that is easy to retell over and over again – a story that has within it the seeds of other stories and which is open enough to allow others to reinterpret it in their own way, for their own purposes, and to their own conclusions.

Sketching for BIM

If your project has a story for its hero, it also has a story to be told about its own making. With contemporary digital technology, particularly building information modeling (BIM), it is easier than ever before to create fully detailed constructible simulations of the project well before the first shovel touches the construction site. Used variously to describe everything from automated 2D construction documentation through 3D modeling and whole building cost analysis, BIM is best understood as a methodology for architectural design that blends 3D modeling, 2D drawing, and non-graphical simulation of performance. If you know something about BIM, chances are you think it will be the saving force behind our troubled industry, but perhaps you're not totally sure how or why. Maybe, you bought one of the BigBIM suites and have it sitting on your shelf, costing you money while you figure out what to do with it next. Most likely, you bought it because someone, probably one of your consultants, told you that you were missing out if you didn't buy it. There is a better way, and it won't require you to change everything about what you already know. I think of BIM as the story of the building's construction. How will you tell that story?

My friend and colleague, Ragnar Wessman, laments the fact that all presentations about BIM (even now, two decades or more since the first significant tools for BIM workflow automation have arrived on the market) still begin with a slide that defines what BIM is. I think it is time to give up on the definition of BIM, and just move on with the problems of building design. Designers have had to essentially solve these same problems for the last couple of thousands of years. Ragnar, when asked about the future of BIM, is fond of saying that the future of BIM is to disappear – to dissolve into simply the way we make buildings.

At some level, like all attempts at rationalizing the design processes that preceded it, BIM is an attempt to leverage new technologies (in this case, database technology) to help automate, simplify, and generally improve the plight of the architectural designer. Building construction, as we have observed many times now, is a wickedly complex problem. Even the simplest of design problems in architecture is filled with mutually contradicting requirements and self-intersecting solution spaces. Changes to the design, particularly while sketching, are dizzyingly frequent. This is maddening to those who just want to get on with the business of building the building, and they will reach for any tool that promises to add structure to that madness.

Computers are, in general, quite good at automating and simplifying processes that can be described in rational terms. BIM tries to do that for buildings. And at first blush, it seems likely that building construction will respond well to this treatment. Buildings are made of parts, individually identifiable and describable. Those parts are assembled in systems through methods of connection and understandable as networks. Diverse teams of people carry the responsibility for making decisions about the arrangement of these systems. They must come together to make sure they can all be built together in rational ways. On-time, and for a fixed and predictable sum of money.

These are all virtuous goals, and the outcome promised is one that members of design teams are universally interested in achieving. Where it all begins to go wrong, though, is when those teams lose track of the still-present necessity for a conscious design process and use their powerful tools to jump straight to the first design conclusion that presents itself.

Because that is what is promised by the developers of the BigBIM tools. With their tools, it is guaranteed, a single designer can jump straight to a wholly realized set of construction documents that are fully detailed, perfectly coordinated, and comprehensively ready for a construction team. Instead, what more often than not happens is that a single designer makes a half-hearted attempt at conceptual design, then lets the machine detail everything for him without much attention paid to real constructibility or appropriateness.

And, quite rightly, in my opinion, construction teams often receive drawing sets created by these folks in this way with befuddled puzzlement. It looks, to the contractor, as though the building they are to build was designed by an amateur. And in many respects, they aren't wrong. And it is then up to the construction team to "fix" the bad decisions made by the designer so that they can accomplish their goal, to actually build the building.

Unfortunately, construction teams may not be made up of people who have had a formal design training of any kind. And so the decisions they make are not always the optimal ones from a design perspective. They can make a building constructible and can make it more likely that construction will be completed on budget and on schedule. But they may not end up making a building that delights and enriches the lives of its future residents. And that is a shame.

Good designers don't hide behind automation promised to them by tools that don't understand them or the way they work. A potent digital designer is one who understands the ways of their tools and can leverage them to do things the world has not seen before. A great designer is one who can pull those tools into their flow state, into their reflection-in-action design practice without being pulled down to the way someone else thought the project should be done.

Make no mistake – the designers of BIM software tools have in mind nothing less than changing the way buildings are built. They have at their disposal reams of research suggesting that the way we make buildings today is busted and wrong. And maybe, just maybe, the reason why is that designers don't know how they do what they do well enough to be able to rationalize it sufficient that software can automate and enhance its deliverables.

There are undoubtedly advantages to be found in a BIM workflow, but you don't need to buy some expensive, monolithic application suite that will bend you to its will when you're using it. You're the designer, and your tools should serve you. Not the other way around. At its heart, BIM is just an attitude about pre-construction building simulation, based on the notion that it is cheaper to design a building in simulation, where all the problems can be found and resolved before the first shovel hits the site. Unquestionably, this represents a revolutionary improvement to come in the way buildings are built. But none of the promised affordances of BIM obviate the need for good design to be done. A fully constructible level of detail in a rich BIM simulation of a poorly designed building will lead to an efficiently constructed . . . bad building.

Done well, BIM is about getting away from drawings and spending more time figuring out what the design for the building should be. This is what everyone on the project wants you to do for them, and so you'd better figure out how to make it happen if you're going to prove your real value. You are the professional responsible for making the building a magnificent building, not just one which is built on time and under budget.

BIM, like all styles of building design, begins with conceptual design. Tempting though the modeling tools make it to jump straight to a conclusion, BIM-centric processes have to progress through all the same phases of design that any project does. To skip any of them dooms you to return to them later when things are much more expensive and difficult to change.

BIM Levels of Development

Building information models that are richly detailed and ready for construction do not start in that way. While it is tempting to use BIM-capable modeling tools to jump straight to the highest levels of constructible detail in a project, experienced designers know that this is a mistake. Without spending time at the lowest conceptual levels of detail, architectural mistakes can be made, which are almost impossible to unwind later. You really have to be careful about this.

Projects naturally progress from low levels of certainty to higher levels of certainty. From high levels of abstraction to low ones. From low levels of detail to high ones. The key to a successful design process on a team practicing BIM-like workflows is in the handoff of data from one level to the next. Projects that skip consensus-building at the most abstract, sketchy levels of detail are doomed to return to them over and over again until consensus has been reached. If you jump to a highly detailed model before agreeing on what the higher goals of the project are, you run the risk of losing all the detailed design effort to rework required to unwind your team's poorly-thought-out preliminary work. Either that, or you will have to explain to your client why their project isn't as good as it could have been.

The American Institute of Architects, recognizing the typical dilemma that all architects were facing as more and more project data was being collected in BIM databases, set a standard definition for levels of detail on a project in 2009. A few years later, that standard was adopted by the Association of General Contractors (AGC 2019) and extended for the needs of GCs on the same projects. This work was a formalization of ideas that had existed in individual practices for many years before, but formalizing it, a common framework for discussion was established.

This framework is the answer to how conceptual design for BIM should be done. You have to read between the lines in its formal documentation, as it has (as a standard) grown pedantically detailed. The latest version of the specification is over 250 pages long, and they are really tailored for large commercial and institutional projects. Nevertheless, formal LoDs are broadly defined as follows:

- LoD 100: The Model Element may be graphically represented in the model with a symbol or other generic representation, but does not satisfy the requirements for LOD 200. Information related to the Model Element (i.e. cost per square foot, tonnage of HVAC, etc.) can be derived from other Model Elements.

- LoD 200: The Model Element is graphically represented within the model as a generic system, object, or assembly with approximate quantities, size, shape, location, and orientation. Non-graphic information may also be attached to the Model Element.

- LoD 300: The Model Element is graphically represented within the model as a specific system, object or assembly in terms of quantity, size, shape, location, and orientation. Non-graphic information may also be attached to the Model Element.

- LoD 350: The Model Element is graphically represented within the model as a specific system, object, or assembly in terms of quantity, size, shape, location, orientation, and interfaces with other building systems. Non-graphic information may also be attached to the Model Element.

- LoD 400: The Model Element is graphically represented within the model as a specific system, object or assembly in terms of size, shape, location, quantity, and orientation with detailing, fabrication, assembly, and installation information. Non-graphic information may also be attached to the Model Element.

- LoD 500: The Model Element is a field verified representation in terms of size, shape, location, quantity, and orientation. Non-graphic information may also be attached to the Model Elements.

Design work can happen anywhere up to LoD 400 (500 is reserved for field verification that the building was built as designed), but the conceptual phase of work is completely represented by LoD 100. So that's what we'll look at in the most detail here.

Sketching in LoD 100 BIM

The biggest challenge to overcome with conceptual design anywhere in a BIM workflow is in the classification of objects. To a BIM data model, buildings are made of objects that, when assembled, create the spatial experience of the building. In BIM, buildings are made of walls, floors, windows, stairs, and so on. There are special tools to help you model each of those classes of object and deep, rich databases of attributes that help to manage decision making all the way through the project. This is great, so long as the building you are imagining is made exclusively of the kinds of objects that the author of your BIM tool was able to imagine on your behalf. In other words, you are going to be successful using these tools so long as the architecture you imagine is made of the things that they imagined you might make.

There's a philosophical problem here, and it cascades all the way through every BIM workflow. It comes from the notion that there must be, a priori, a universal taxonomy of building components that is capable of naming and describing everything you, as a designer, might be able to imagine. To be fair, this is at least naively true, and you should be able to find a place in the taxonomy for everything you need to describe. Except when, for whatever reason, you can't.

And this is the essential problem with any rigidly defined logical ontology. While most things have a place in the taxonomy, some things don't. Consider, if you will, the platypus (Figure 7.1). To the scientific classifiers of the nineteenth century, the platypus was a real poke in the eye. Within the taxonomic ranks of biological classification, it had attributes that placed it in at least two separate branches. It nursed its young with milk like a mammal, but it gave birth to them through eggs like a bird. It also had poison-bearing stingers, which are used to defend itself. It really isn't like any other creature on Earth. Exasperated taxonomists had no practical way to account for it in their a priori system of classification. A new category (*monotreme*, which now includes as well both the short and long-beaked echidna) had to be defined to accommodate it.

The a priori taxonomy of the day was so rigidly defined that British taxonomist George Shaw (who is responsible for the initial classification)

Figure 7.1: A platypus.

speculated that the specimen might simply be a hoax. He simply could not imagine that his rationalization of the tree of life couldn't find a home for the creature.

It is impossible not to entertain some doubts as to the genuine nature of the animal.
George Shaw

Rigid taxonomies are among the most valuable gifts in rationality that we learned from the Enlightenment. Without Carl Linnaeus and his formalize binomial nomenclature for naming biological organisms, it would be difficult, if not impossible, to study biology with any real precision. I wouldn't want you to assume I think there's any fundamental problem with applying a similar system to building design. Quite the contrary – standardized taxonomies have been instrumental in the ongoing rationalization of construction.

However, if the intrepid naturalists who classified the platypus had been thinking differently about their work, perhaps more creatively, a different system might have been applied to help them answer the question, "What the heck is this thing?" When I'm in the heat of battle with any design project, the last thing I want to worry about is organizing my work. Many systems exist to help with this, some more effective than others. Taxonomies are valuable when they fit, but they must be flexible enough to allow expansion or redefinition when something new is invented that just doesn't fit.

Rather than a top-down, a priori organization, which will almost certainly not contain every category you need and will probably instead leave you with dozens of unused branches in the taxonomy that you'll never fill, the most useful classification systems let the organization emerge organically over time.

The tireless effort of BuildingSmart over many years of refinement in committee has led to the definition of the Industry Foundation Classes (IFC) – unquestionably the best, richest and most universal ontology ever assembled to describe the component parts of a building and the relationships that logically bind them all together. IFC4 (the latest version of those standard classes as I'm writing this book) is capable of defining hundreds of distinct building entities. And the BigBIM tools have specialized tools to make them all. Walls, slabs, stairs, and curtain walls. And so on and so forth. These objects can be drawn into a model in a highly efficient way, and once defined, they have all the correct attribute data attached to them ready for refinement to follow.

But I want to tip this a priori methodology over – it isn't compatible with every conceptual designer's flow, and it isn't a requirement for a tactical BIM mindset. You shouldn't be constrained by your design tools during the concept phase to conceptualizing only those elements your tool is capable of describing. Maybe, more importantly, you should only have to apply classification to your design when you're ready to do so – certainly not right away at the beginning.

I propose an alternative system of classification for BIM; a posteriori instead of a priori. Simply put, you should be free to draw anything you can imagine in your modeling environment. When (and only when) you are ready, you might choose to give it a classification so that you can carry its design forward to the next level of detail. And likely most of what you design will fit well in the known restrictions of a system like IFC. And, occasionally, some parts will not. Let's imagine a scenario where you are an architectural designer working in this way.

SKETCHING IN LOD 100 BIM

In your first client meeting (hopefully over a nice cup of coffee), you capture a couple of sketches in your sketchbook that defines a general scheme for the project you will be working on together. When you get back to the office, you warm up your computer, load your favorite 3D modeling environment, and spread out the sketches you made on the desk next to you.

In a fresh new model, you capture a quick representation of the sketch you made by hand- maybe it is only a couple lines in space and some abstract boxes representing the major functional areas of the project to come. At this stage, your model is made entirely of unclassified raw geometry. It is just lines, surfaces, and volumes. Arranged in space, with only an arbitrary sense of scale and proportion.

Since you already know that the boxes represent programmatic spaces of some kind, encapsulate them and give them a name. "Conference Room," maybe. Or "Reception." Whatever – these are just names and will be easy to change later. At this early (massing) stage, you don't need to decide much more about them. Maybe you are worried about how big things will get, and you might want to drop some kind of floor-area sensor into your model. Touch the bottom surface of each block and turn on an area calculation. Now, as you reshape your massing blocks, you can track changes in the overall floor area.

Or maybe what matters to you are the connections between blocks. You know the "Break Room" should be next to the "Bathroom" (whoops. . . better add a block for that next) and so you capture that. Now you have the beginnings of an adjacency graph that you'll be able to add detail to later on. Maybe you take this opportunity to snap a couple more connections into the model. Quickly capture whatever you know right now, and it is OK if what you know is incomplete. Just start anywhere.

You know the abstract boxes you're adjusting in the model are going to have an inside (protected from the weather) and an outside (that can get rained on.) Because, well, you're designing a building, and that's something all free-standing structures have to solve. This means the surfaces that make up your abstract model are going to have some special significance. Some of them will be on the outside, some of them will be on the inside.

Maybe, for now, you can start thinking about them as walls, but there's no need to jump to that conclusion right away. Some of them might need to be transparent (you remember that the client wants to preserve that great view of the mountains), and some might need to be made out of stone. For now, touch all of the vertical surfaces in your model, and classify them as "wall" – you'll come back and work out the details later. And if you have some faces classified as "wall," you might as well also tag the bottom surfaces as "floor" and the top surfaces as "roof." It is OK if you don't know anything else about any of these objects yet at all.

Another thing you know for sure is that gravity will point in some constant direction, probably down. This means you can start thinking about how the structural system might work. Something will have to stand on the thing you just classified as a "floor" and hold up the thing you just classified as "roof." Maybe it will be made out of stone, perhaps steel, maybe concrete, perhaps wood. Too early to tell, and ultimately not relevant to your conceptual design work. But you know it will be coming at some point, and your experience tells you that at least some of the things you tagged as "wall" are going to have to be structural in some way. Take note of that in your model as well.

While you're working away, you get an email from your surveyor – she has completed her analysis of your client's site and has a 3D terrain model ready for you to reference in. So you do – pulling it into the same model you've been working on all morning. With the surveyor's site model as a reference, you immediately notice a couple things. First, your massing model is all out of whack from a scale perspective. You quickly resize some of the massing boxes you had been playing with to match the new realities of your site. Your floor area sensors confirm that in doing so, you haven't lost the basic floor area ratios you promised to keep to with your client.

Also, you now have to choose the design's north orientation. You rotate it around until that seems right for now. And with that reorientation, you also recognize you have less space for your building's footprint than you thought you would. There's a utility easement you didn't know about before, and it looks like you're going to have to try some schemes where the project is on two floors, lifting some of your massing boxes up above others. You fiddle the boxes around until that seems to work better.

At this point, you may have more than one scheme in mind that could work. Log both the single-story and two-story versions into your version control – you'll probably come back to both of them later. But the two-story version seems like the best fit for you now. And no, maybe you can't say precisely why . . . you just know that's true and so you go with it for now.

With the addition of a second floor, you're going to have to make some more changes. Somewhere, you'll need to add a way to get from the first to second floors. Without committing to what that might be (perhaps you imagine a free-flight zone where your client can use anti-gravity belts to ascend gracefully . . . anything can be

envisioned at this stage), stick in another block to represent that vertical circulation zone. You know it will take some space, and you should plan for it accordingly. Classify it as "stair" if you're ready for that now.

You also know now from the referenced site model that there is a significant slope to the site. You'll need to deal with a height difference of three feet from the front yard to the back. And the site drops off quite a bit more than that to the back edge of the lot. You might need to bring in a civil engineer to help you figure out the drainage. Still, for now, you know that the front door (aha! You're going to need a front door. . . better add a placeholder for that now. . .) is going to be about three feet higher than the back might be (add a placeholder for that too. . .) and so you'll need to come up with a resolution for that.

Additionally, because you now know which way is north, you can begin to make some guesses about how the rising and setting sun will impact the light quality in the spaces you're designing. Maybe it would be nice to be able to watch the sunrise from a table in the kitchen (add a placeholder for that table now, because it would be a shame to lose that association) and there will have to be some transparency (not too much) in that wall so the light can come in.

Because you know your site is located in a desert climate zone, the afternoon sun in the summertime is brutally hot. You're going to want to manage that somehow. Probably some kind of awning, maybe a wall. Or a brise-soleil. Or perhaps you'll plant some sort of shade tree. Except . . . desert. Anyhow, you'll want to solve this eventually, so drop a surface of some kind into your model, and tag is as a "shade" object. Turn on your modeling tool's solar model and check to make sure your new shadow surface is blocking the mid-afternoon sun on the fourth of July.

SKETCHING IN LOD 100 BIM

Coincidently, as you're looking for that optimal shading solution, you notice that, because of the way your site is oriented and how you happen to have arranged the program so far, that there is an opportunity for a nicely shaded and protected outdoor space just off the dining room. You might not have planned for this per se, but now that you see it, you want to explore the idea further. Draw another surface and tag it "patio" for now. Drop another sensor on the surface so you can track annual temperature changes on it and other environmental properties as your model develops.

You also know, from your last client meeting, that there are several programmatic elements that they are really excited to keep in their new office (there – I just classified the project as "office" for the first time.) You want to be sure you get them logged in the model early on as well, so everything else can be formed up around them.

First, your client has a lot of books. Big, glossy art books. So many of them, in fact, that you may well have to reinforce the foundation under the bookshelves, you'll have to design to hold them. Clearly, books are an essential part of your client's identity, and they will need a way to keep them integrated into their daily lives.

Books live most naturally on shelves, and shelves typically go up against walls. Maybe what you need is some special kind of wall that is also a bookshelf. A "book wall" object. You know from looking at your client's collection that none of their current shelves are deep enough for every book, and you thoughtfully measured what looks like the largest book they currently have in their collection. You also have a rough approximation on the number of books they have, and a sense for how often something new gets added. Also, because you know your client, you know that nothing, once it has been added, will ever be removed.

Add another box into the model that represents the total area of bookshelf you're going to need – the number of books multiplied by the average thickness of a book will give you a linear approximation; you know you're going to have to wrap that line somehow, but maybe you don't know how to commit to that yet. Pick anything for now and give yourself a block that represents your client's "book wall."

You need to take some time to figure out how this is all going to stand up on the site. You recognize you have some walls that will hold up a second floor and a roof. You have a difference in elevation from the front of the office to the back, and you will need some kind of vertical circulation inside. You have an idea where the kitchen will be, and where the rest of the plumbing will probably need to go. And you know where there are some unusually heavy things (a wall of books) to support. Time to give the foundation a little thought. Probably some kind of perimeter beam, since that's what you did on the last office in this neighborhood. Quickly sketch that in underneath your massing model and tag it "foundation" for now.

You now have enough information in the model to do a quick cost analysis. Nothing you'd want to commit to, but surely enough to judge whether the project is feasible within the restrictions of your client's budget. You know how much floor area, and you know how much of it will have unique (more costly) programs like kitchen or bathroom. You know how much exterior surface you're going to have, and how much interior. You know something about the roof and the foundation. That's enough to take your first wild-assed-guess (WAG) at a cost.

And that's not all you have. You also have a serviceable 3D model, a preliminary performance analysis, and a refined understanding of the affordances that will come from the site. And you've discovered a new piece of program (an outdoor room), which your client can get without significant additional cost that will make the office a really lovely place to work. You're ready for your next meeting, prepared to present your first formal ideas about the project.

This is a formal BIM methodology, but it is operating outside the restrictive ontology of most BigBIM tools. You have identified some things which are clearly identifiable in the most common building taxonomies ("walls," "floors," "stairs," and others). Still, you have also identified some new types which weren't obviously part of the IFC taxonomy. A "book wall," for example, or a "shade," or even a "patio." These might be object types that only make sense to you for now, though, of course, later you'll convert them into the kinds of objects that a contractor can build. Your "book wall" might, in the future, be split into a structural wall, a built-in bookshelf, and a reinforced foundation. But for now, it is more convenient to think of it as a single unit. And it makes it easier for you to tell the story of its necessity to your client in that next meeting.

In formal IFC BIM Levels-of-Detail, all this work is encapsulated in "LoD 100," which is defined as *predesign* by most construction teams. But think carefully about all the decisions that were made in it – major decisions like siting, the number of floors in the project, and the general layout of the plan. These are the things that your client most wants to hear about, and the things that will most affect their use of the building, which is eventually built. Imagine what would happen if this phase of work were glossed over or in any way neglected. Disaster, right? This isn't "predesign," it is "design." Most of the significant, defining decisions for the project are being made right here.

In the work that follows, additional refinement will happen to every decision made at the beginning. New challenges, new dependencies, and even new opportunities will emerge as the design iterates. You'll have to accommodate some new code interpretation, some change in the cost of materials

and labor, or just some new ideas from your client about how they want things arranged. All of this is healthy, all of this is good. Embracing it will surely make the project better. But never neglect the early conceptual phase, because that is where the most significant decisions get made.

Teams that think they are skipping this phase are really just tacitly accepting conventions and rules-of-thumb that represent their preconception about how buildings of this type should be built. BigBIM tools encode many of these preconceptions in the form of convenient to use tools and ready-to-wear documentation and reporting tools. So long as you are designing things like the tool designer has imagined them, you are going to find these tools very convenient to use. But the minute you step outside the lines of their definition, you'll find you have a lot of trouble ahead.

Sketching for Construction

An architect who keeps a constructible 3D model up-to-snuff right through construction is one who can be trusted by the construction team to have answers to every question. You are probably already pretty good at keeping a model of the project in your head – most architects do this intuitively and with scientifically shocking degrees of precision and clarity. But a model that is only in your head requires you, personally, to be responsible for communicating the details. A digital model shared openly with the construction team is an enormously valuable asset for construction. In this section, we will talk briefly about designing for constructibility.

It is easy for a construction team to undervalue the hard work you've done during the early design phases of your project. To them, this work is mysterious, vague, and too abstract to follow. Really, they just want a constructible design (in its final form) that is good enough to meet the owner's needs, not too hard or expensive to build, and complete enough that they don't have to make many costly changes in the field. To most construction teams, designers are a risk during construction, not a benefit. Improvements to the design are still changes, and changes cause trouble to their fragile schedules.

To understand why this is true for them, you have to know how their business model works. While contractors work habitually with enormous budgets, the actual profit margin they get to keep to run their business is minimal. For the many contractors, one major screw-up on the job can pull their profitability down for the whole year, maybe even put them out of business. It is a tough business model in which to operate.

As a designer, you're much more likely to work on an hourly rate. This is a good match if you think of design as a value that you deliver throughout the project rather than a fixed deliverable that you work on until it is done, then hand over to the next team in line. But there are hundreds of different ways that designers, clients, and builders can all work together contractually, which is a strong indication that the industry still hasn't really figured out a single best way to make their business models work out. If there were a single best way to share the risks and rewards of a large construction project, probably all of us would be working in that model. But it hasn't happened that way yet.

The most efficient contractual models for construction projects are those in which the designers, builders, and operators are all bound together in a single financial entity. Commonly called integrated project delivery (IPD), this idealized state allows for a minimum of self-preserving behavior around the conference room table and usually leads to better, more collaborative outcomes in the final analysis.

Unfortunately, this business model is still relatively uncommon in real professional practice.

More common, but also practical, is a formal business integration between the designers and builders – working together to deliver a finished building for their client as a single company. Design-build is prevalent in many segments of the construction industry, especially residential construction, where it can speed time-to-construction and reduce overall costs to the client in remarkable ways. But it is not without its downsides as well. Contractors and architects can find they have trouble working together, even when they work against the same balance sheet. The designer's need to continue optimizing a design to be the best it can be still runs counter to the contractor's desire to just get the project done and quit changing everything all the time.

By far, the most common working relationship is one in which the architect is hired by the owner to design the building, then the design is handed off (as a "finalized" asset) to a contractor to build. If you're a working architect today, chances are, this is how you're working. Design-bid-build keeps risk carefully subdivided, but it requires the designers to be 100 percent complete on their design work well before contractors are brought onto the project to start really thinking through its construction. This is always a source of strife on a project, as the construction team loses a chance to contribute ideas about how the project should be built and, as a consequence, feels a loss of agency over those decisions. It is all too easy for them to come to resent the architect's early choices because they had no agency over them but are now stuck with implementing them in the field.

Most US architects work either alone or in offices of fewer than 10 people. In Europe, close to 90 percent of architects work as sole practitioners. There are good reasons why this has become the norm. Despite frequent calls for a more collaborative process, design remains a singularly solitary act. To be sure, once you have settled on some ideas you think are worth sharing, working with others will improve and extend them effectively. In fact, you are missing enormous opportunities to improve the design if you can't find any way to include construction and engineering collaborators in even the earliest conceptual phases. But the critical "reflection-in-action" flow state that you crave is not something easily done together with others. Design is a solitary act; analysis, critique, and coordination happen with others.

While you are probably working alone while you're designing, you are unquestionably coming together with others to review, coordinate, and analyze your design proposals. You're likely going to bring in design guidance from a structural engineer, for example. And many others, as the design, resolves over time. How will you share your work with these people?

Traditionally, and in many ways still the best practice, you would put together sets of drawings at a couple key milestones in the development of the project. Maybe, even, you'll have your payment schedule keyed to these deliverables. Regardless, traditional design processes really sweat the creation of these design deliverables. Obviously the delivery of a bid set sets the expectations from the owner and contractor about what will actually get built. Formal construction documents are legally binding definitions of that as well. But even your earliest design intent documents set expectations and you want to be careful not to write a check that the team won't be able to cash.

Traditional conceptual design sets might include a couple of your most photogenic sketches to help you tell the story of your design proposal. Still, generally speaking, the kinds of sketches I've been describing in this book have not been intended for public consumption. You sketch to help yourself understand the design, not to help you communicate it to others. For that, you need a different kind of work.

An effective design presentation, even the very earliest of those when almost nothing of detail has yet been worked out, can win even the most grizzled contractor to your owner's cause. People want to be associated with fantastic projects – they want to be able to drive by a building with their kids and point to something great they added to the world. But there's a lot of mistrust and suspicion on even the most outwardly friendly construction teams. A great construction project begins with a great design presentation.

When I worked for Communication Arts in the late 1990s, we were almost exclusively hired as design consultants and were never responsible for any stages past design development. Our sole vehicle for communicating design intent was a package of renderings, diagrams, and basic drawings that were pulled together into an 11″ × 17″ spiral-bound book that was a delight to hold and flip through. It was designed to tell the story of the project, not to document its every nuts and bolt detail.

To transition your design effectively to construction, you need to learn how to pull the team into the future world that will exist once the project is complete. You want to build confidence in yourself and the vision you're bringing to the project. And you want all eyes in the room on you, all ears and minds open to what you have to say. Obviously, this is not easy. But the success of the project depends on it. If you have the best design the world has ever seen, into which you've poured endless hours and gallons of blood, sweat, tears, and espresso... but you can't sell it to the rest of the team, you're sunk. The best construction team in the world is one that has internalized your big design idea and though it has become able (and willing) to answer their own questions when unexpected situations come up in the field.

If you haven't been able to win the team to your vision, you're going to face constant struggles every day going forward until the job is complete. And, unfortunately, this is the more common reality. It will lead you and everyone on the team to engage in some degree (even an extreme degree) of "ass-covering" behavior as passive-aggressively formal "Requests for information" are passed around the team.

Let's be completely honest here, not one single construction project has ever proceeded from conceptual design to final construction with perfect information available about every single condition that will be discovered in the field. Buildings, even the smallest and simplest of them, are incredibly complex to build. Every structure is unique, made for the first time by a team that has never worked together before. Every site comes with some unexpected discoveries. And the sequence of construction is so complex that it just isn't possible to predict every single thing that is going to come up.

But there is hope on the horizon, and a reason to be optimistic for the future. While I have spent many pages talking now about how useless computers are at solving the early, highly abstract, highly subjective problems of design, they are totally brilliant at managing high levels of detail in large simulations.

When I met Frank Gehry, I expected him to want to talk about the poetics of space, about the soaring artistic achievement of buildings like Guggenheim Bilbao and the Walt Disney Concert Hall. Instead, he wanted to talk about his ability to deliver projects like those on time, on budget, and with only a minimum of change orders in the field. To me, that is a significant and inspiring thing. Gehry Partners can accomplish this because of the detailed building simulation they do in the office, well before the first shovel hits the site.

Modeling to a constructible level of detail makes it easy to know exactly, down to the individual anchor bolt, what your future building will be. You can pre-visualize every single condition in the building, coordinate every trade, and rehearse every complicated installation. Your construction bids will be bulletproof – you know exactly how much the steel package is going to cost because you haven't "ballparked" anything. It is all right there, exactly how it should be built, in one giant model that can be shared across the whole team.

To be sure, this is the future of our construction industry. Buildings will eventually be built from digital models, not from 2D construction documents. Together we can think through this ultimate future in Chapter 8, but make no mistake, this is an inevitable outcome of the digitization of the architecture, engineering, and construction industry. And as a designer, you have an exciting role in that, maybe even the most exciting. But for now, you still have the same challenge that you've always had in communicating your design intent to the construction team so that they can build it the way it should be built. So what are some near-term strategies you can use for now?

Remember to question the promises of all-in-one BIM software packages. For many reasons explored previously in this book, you know how dangerous it can be to jump straight to the conclusion in your design process. You may be promised the ability to jump right to fully constructible levels of detail in your design. Once detailed, your models can then be automatically dropped into coordinated, detailed, and constructible CD sets ready for the field. This is alluring, but not yet particularly realistic. Frank Gehry manages to make models of incredible complexity, but he's using some pretty exotic software (Catia) from the aerospace industry.

It takes a lot of extra time, and it isn't cheap. In fact, it is "if you have to ask, it isn't for you" expensive.

At some point or another, you will be confronted with a demand that you adopt one of the BigBIMs (ArchiCAD or Revit, probably, depending on where you work) so that the team can work seamlessly together without worrying that interoperability can't be maintained. It is incredible to me the lengths a design firm will go to ensure that its most precious design data – the encoding of its design intent, is stored in a file format on disk that someone else can read.

In the 1990s, Microsoft had a similar hold on the digital office workers of the world. If you wanted to use a computer to write documents, to edit spreadsheets, or to make presentations on a big screen, you used some kind of office productivity suite like Microsoft Office. Unfortunately, only Microsoft Office could read documents created by other Microsoft Office users, and seemingly in an instant, every office worker in the world was using Microsoft's product for their work. Anybody remember Word Perfect? I used to use Word Perfect back in the day.

In time, these proprietary document formats were eroded, and today you have more agency over the tools you might choose to (for example) write a book like this one. Personally, I'm using a text editor and a highly compact markup language capable of giving me bold italic and underlined fonts. And not much else.

To me, what matters when writing a book isn't the formatting but the content of the writing. Just because, thanks to the miracle of desktop publishing and (even better) direct digital publishing, I can flow thoughts more or less directly from my keyboard to the physical book you may be holding in your hand right now. But the simplest tool is really the one that is the most powerful, even in that context.

I'm writing this book in an application called "Ulysses," which I bought for only a couple of bucks online. I can run it on every computer I have, my main writing Mac mini, my iPad, even my phone. I can use it on a Windows PC, too. I'm managing my data in a GitHub repository in the cloud, so I'll never have to worry about keeping backups or older versions. My editor needs only point a browser at a URL to see the "latest" version of what I'm working on. This is great.

Don't fall prey to the idea that your conceptual design work, your digital sketches, need to seamlessly flow into your project's eventual detailed design models. Sketches are meant to be ephemeral, to guide you and help you reach the best design ideas you can in the time that you have available. There will still be work to do in converting the ideas you've sketched into compelling design presentations (through rendering and visual storytelling) as well as into construction documents that you'll share with contractors in the field. Adjust your expectations if you think it can happen automatically without any further work from you.

Every time you switch media while you're working, you may have to redraw something. This is good and virtuous – every time you draw something again, you learn something, and the design gets better. You are doing incredibly valuable work as a designer every time you iterate your design. Embrace that, don't fear it. In fact, if you aren't reworking your design regularly, you aren't really doing your job. Clients hire architects to give them excellent design services, and the best designs, as we know, aren't ever hatched fully formed from the brow of the designer. Moving a conceptual design to construction prematurely will only lead to questions, disappointment, and accusations of incompetence later on in the project. Get it right at the beginning, and improve it for as long as the team's process will let you. Remember, design isn't an asset, it is an activity.

My best and most practical advice, the most effective way to move a design to construction, is to keep it simple. If you made a sketch on a piece of paper, scan that into the project and (maybe) use it later to help tell your story. If you made a model, feed it into a rendering engine if you need to really sell your project, but know there's still a lot of work to do to turn that model into a beautiful, evocative storyteller. You might be able to feed that model as well into a set of construction documents, but don't be afraid to throw it away, either, and start from scratch in your new CAD system. The ideas embodied in the sketches you do are going to live on, no matter what. They helped you make decisions. They helped you figure out what to do next. Thank them for their service, stuff them in an archive, and move on to the next phase.

But don't ever let yourself come to the conclusion that because you can't seamlessly convert your sketch work into more detailed future-state design documents, the effort you spent making them is wasted. Don't think, maybe next time you should just skip ahead to the conclusion and save everyone some time and money. This is a critical mistake, and one you're come to regret over and over again. Great projects have great bones, a conceptual framework that everything else depends on. How do you design for constructibility? The same way you design for everything else. Start simple, refine by sketching, test your decisions with others, and iterate, iterate, iterate until you're out of time.

chapter 8 Epilogue

I've made an effort to describe so far in this book ways of working that are accessible to you as a designer already today, with minimal investment of time, technology, and training. In fact, much of what I have described here can be accomplished with a budget of less than a couple thousand dollars, including a computer that you likely already own. This should be well within reach of even the smallest of professional architectural practices. To be sure, you can spend more. But you likely don't have to.

I have tried to explain a different way of working, which is orthogonal to the tools of your practice. In fact, I hope I've inspired you to be somewhat suspicious of new tools, except where their affordances are just so remarkably different from what you've previously had access to that it is impossible to ignore them. I'm also keenly aware that technology has a habit of keeping in motion; in many respects, it moves faster than most professionals can adjust, and so again, it seems appropriate to show some restraint.

But I am at heart a technologist, and I see a future that seems both inevitable and exciting – a future that motivates me and keeps me at the job every day, working in my small way to make it come to life. I hope you'll forgive me for a bit of self-indulgent science fiction for a peek over the horizon to what might come to be in the practice of architecture and design. Because what I see coming is a world where design and construction have become inextricably bound together in a cohesive, seamless process of creation. You might say I'm a dreamer, but (I hope) I'm not the only one.

Technology Is Incremental – Even Revolutionary Technology

The problems most AEC industry technologists seem to focus on today are, like always, somewhat more incremental than they might be. There is no shortage of software developers working on tools to help resolve the complex team dynamics on construction projects. Many companies are building rich collaboration platforms that promise to bring AEC-specific sense to the vibrant barrage of digital communications that govern the rest of our lives. This is essential work, to be sure. Work that will have a significant impact on the fundamental challenges of constructibility.

Additionally, many folks are working on ways to leverage algorithms to automate design tasks – we discussed some of these opportunities in Chapter 6. I think these tools are interesting to consider, but I see them as part of a broader tradition in architectural design rather than as something profoundly new from a technological perspective. There have always been designers who liked to work by building up a design system, winding it up, and then seeing where it goes on its own. Implementing such systems in code allows the designer to consider iterations on the theme more quickly, and to leverage machine learning and other AI optimization techniques to solve for an objective "best" solution to a rationalized design problem.

What really lights me up, though, are robots. I for one, will welcome our new robotic overlords when they arrive on the construction site. I welcome them because I think they will improve construction quality, reduce rework, and enhance job site safety. Also, they will make it easier for a skilled design team that is capable of driving design from concept to constructibility to take the design one step further – into construction.

I'm not talking about robots in the sense of abstract automation of some idealization of professional work, though it is undoubtedly true that any professional job that can be rationalized down to a set of repeatable activities will inevitably be automated by information technology. To be sure, we're already seeing the mid-stage development of such systems. It seems to be a foregone conclusion that any professional job that can be automated will, in time, be automated. It just makes economic sense.

This likely includes some work from all of the so-called "major" professions. Law, medicine, finance, and business management all have characteristics of their work, which can and in some ways already are automated. Legal contracts, including even highly personal documents like your last will and testament, can already be reasonably written by artificial intelligence. Most people don't really need an accountant to file their taxes for them either. Thanks, TurboTax.

But that's not all. Automated trading algorithms regularly meet or even exceed the performance of their human counterparts on Wall Street. Medical diagnostics for routine procedures like choosing a new prescription for your glasses have been automated. Even some surgeries, notably Lasik correction, are performed by robotic surgeons. And the list goes on.

However, there are apparent hard limits to what digital technologies can accomplish on their own, and in that is hope for the future of architectural design. In fact, it is likely the very characteristics of our profession which colleagues find most vexing that will eventually be the things that make us the most valuable. Until our scientific understanding of human "design thinking" advances to a point where it can be simulated in code, we will still need humans who are trained and capable of reflection-in-action thinking that is found at the heart of all design activity.

Computation Is Powerful, But It Isn't Limitless

It is in the nature of information technology that everything that can be made digital will be done so. It sometimes seems as though computers are consuming everything. But at the end of the day, they are designed by humans in our own mental image. We make them capable of everything that we know how to describe in our own capabilities. And they can do very little else. As a simulation of human cognition, computers are still very limited. Machine intelligence is still "artificial."

I spent several weeks studying with Stephen Wolfram and his core "New Kind of Science" team back at their first "NKS Summer School" in 2003, and it gave my computational worldview a hard reset. Seldom have I been surrounded by so many profoundly bright people working on such existentially weighty problems. I learned from this gang what the real, deep capabilities are for computational systems. There are some beautiful, seemingly universal truths at the heart of computation, and by expansion, at the heart of mathematics.

Computational systems are capable of the richest of complexity, behaviors that point deeply at the way the universe really works. There are computational models of physics, of chemistry. Even some systems that begin to poke at existential ontological problems in philosophy. But the fact that humans still struggle to describe how human thought really works means that our computers are still only the palest shadow of ourselves. They are capable of solving every rational problem we know how to solve – and they can do so faster, more reliably, and with far greater efficiency. But that's all.

An Imaginary Scenario about Construction

The robots I welcome in our future are those that will extend my reach as a builder. Robots that will lift heavier loads than I can, that will weld steel beams with greater precision and pour concrete to finer tolerances than I can. I welcome the robots that will shape the earth for my projects, lift and assemble the components and build the building. I think these robots will be faster, more efficient, and will touch the Earth more lightly than the humans they will replace. Imagine this with me.

As an architect, I have worked closely with my client through a detailed design process for their new home. Together, we explored many options and picked the one we think is the best fit for their needs. The design started as just an idea in conversation over a cup of coffee a little under a month ago, and in the time that followed, we gradually cranked up the level of detail, making decisions where needed, and freely exploring opportunities as they presented themselves.

At every step of the design process, we knew with increasing certainty not just what the project might cost or how long it could take to complete, but we also knew how it would feel, what life it would have and how it might delight its future inhabitants. We knew how much the building would cost to operate and knew that we had done everything that could be done to make the building efficient to work with only a minimal impact on the environment.

We were able to make decisions on time for items with long lead times (the cabinet order, a few exclusive structural members, and the high-efficiency energy systems that would control the temperature and air quality). We were able to resist changes after those decisions that would lead to trouble later when they arrived on the construction site. Knowing where the hard constraints would be during construction while we were sketching on even the original design concepts meant that we weren't trapped later by the construction process.

The build team's work construction site – in fact, their entire construction process from groundbreaking to handover – was rehearsed multiple times in digital simulation before we even pulled the construction permit. By the time we were ready to go to construction, the design had been so thoroughly tested, coordinated, and validated that we knew there would be no changes in the field at all. Simulated to full constructibility, but only after considering the design carefully at all levels of detail.

And when the time is right, and the design is constructibility complete, we hand it off to the robots to put it together. Construction barricades would be raised to protect against interfering human incursions, but also to preserve the predictability of an informationally pristine construction site. There could be no surprises on the job site because nothing exists there that wasn't logged into the project by design.

Concrete extruding robots on tractor treads wander the site like roaming 3D printers, leaving behind completed structural foundations. Robots with magnetic grippers capable of climbing to the top of unfinished frame bolt-together structural members handed to them at just the right time by an automated lift; picked just in time for erection from an automatically managed layout yard. Once topped out, waves of new robots armed with spools of sheet metal roll just the right studs for each interior wall. Drywalling robots follow just behind as if in formation, lifting in full sheets of sheetrock, mudding, and sanding to perfection. Glazing robots specialized in setting windows in frames complete the exterior facade, never worrying that any component might have been manufactured to the wrong size because the automated factory that fabricated them was only capable of building the rights ones.

Mistakes during construction in this scenario would happen only due to a failure to simulate constructibility to sufficient detail. A robot without clear instructions might stop work frequently if it didn't know what to make of an unforeseen situation. But the wrong decision wouldn't be possible – an automated system has no judgment of its own it is, in fact, harder for it to make the wrong decision than it is for it to follow directions.

How Close Is This Future in 2020?

These robots I'm imagining won't arrive on job sites for some time. The challenges of a real construction site are unlike anything that industrial automation has tackled before. Unlike the pristine certainty of an enclosed factory floor, construction sites are open to the weather, with rain, snow, baking sun, and high winds. Unlike industrial automation, where the same object is assembled hundreds, thousands even millions of times, a construction site is assembled exactly once, with the intention of creating exactly one finished product. No physical rehearsal, and no redoes. And no repeats ever again. Construction sites are hostile places for robots. Digital rehearsal, from a fully constructible whole-building simulation, is the only way to be sure the project will work the first (and only) time it will be built.

All great science fiction has a basis in contemporary reality, and this is no exception. While you may not recognize or fully appreciate it, there are already construction robots at work on construction projects all around you. You don't have to look far to see what I mean, nor to gain some appreciation for just how close this vision may turn out to be. As William Gibson taught us, *"The future is already here – it's just not very evenly distributed."*

I have a robotic vacuum cleaner in my house that has taught me many things. Every morning, as I'm tapping away at my keyboard, my Roomba sings a little wake-up song and sets off on its merry way to vacuum my floor. It has some of the most advanced robotic systems available to consumers. It built itself an onboard map of my house, which it created by bumping around blindly until there wasn't anything left to bump into. It carries a full array of contact sensors around its perimeter for close-contact sensing as well as an array of infra-red distance sensors to keep it from falling down the stairs. It even carries an onboard camera that it uses to orient itself with machine learning and stops it from running over my dog. Who, for his part, is unthreatened and even oblivious to the robotic invasion.

Despite all its intelligence, I have to rescue my robot almost every single day. And for a different reason practically every time, it seems. Just the simple elements of my daily life befuddle it with discouraging regularity. I am accustomed to hearing its merry little "uh oh," song about an hour after it heads out from its base station. Usually, the problem is that it has gotten stuck behind a door, or maybe just tangled in a rug. Once, I rescued it from under a pile of shoes. And once it was sitting right in the middle of a room it had vacuumed dozens of times before. . . just lost in plain sight.

But I'm not disappointed by this, nor discouraged in any way. That robot that shares my home is so many orders of magnitude more capable than its ancestors from even five years ago. And the problem it is trying to solve is so profoundly hard. We keep a pretty tidy home – it is unbelievably pristine when compared with even the most organized of construction sites. Imagine how hard just the navigation problem will be for robots if the very waypoints they use to navigate, walls, floors, and ceilings, are changing every day in nonsubtle ways. We will need many more orders of magnitude improvement in our robots before they are really able to take control of construction sites.

There is a growing labor shortage among the construction trades, which will necessitate an acceleration of the development of automation technology. More than half of the skilled tradespeople in every trade are over the age of 55, and they are not being replaced in the industry at a sufficient rate to keep up with market demand. Skilled trades are just not an exciting job opportunity for a young person entering the workforce. We're actually going to need robot assistants before very much longer if we want to keep up the current pace of construction.

And they are coming. Robotic assistance is already available in construction tasks like field layout of location for holes to be drilled for hanger bolts. It doesn't take much imagination to see a follow-on robotic assistant that is capable of actually drilling the holes. On significant civil construction sites around you every day, there are already robot-like capabilities built into heavy earth-moving equipment that is used to cut and shape the land for new roads, to layout railway lines and pave airport runways. These tools are not sci-fi. In fact, they already represent the best way to get the job done in today's competitive civil construction market. Already today, a civil designer can create a constructible digital simulation of a road and download it directly to a bulldozer as machine control instructions that automate the blade position so the dozer can only cut the right path across the landscape.

As a designer, you have a responsibility to recognize these coming changes and learn how to take maximum advantage of them in the work you're doing for your clients. Don't lose sight of what it is that you do and the value it has for the projects you work on. Design, especially at the early, fuzzy, messy, and abstract stages, sets the bones of every project you work on. Your conceptual design is the foundation on which the rest of the project, all the way to ultimate constructibility, rests. Weak foundations make for poor buildings. Never forget that your ability to think beyond obvious solutions to something better is unique, all the more so because you probably are incredibly bad at explaining how you do it.

The work of architectural design is still among the most essential work of our time, the most impactful at every scale. From the city to the front porch, from the largest of airports to the humblest of strip shopping centers, architects are shaping the world to better and more appropriately fit the needs of real people. They are taking responsibility to enhance their client's work, to house them from the rain and the cold, and embrace them and their families in warmth and hearth, helping them grow into our better future selves.

As a technologist, I could probably make more money working on the next great transport layer for funny cat videos. Ours is an unusual industry – high technology, but with muddy boots. Those muddy boots are the driving force behind more than a 10-trillion-dollar market that touches the lives of every single person on Earth. I'm honored to instead have a chance to help architects shape the world into something more sustainable, more accessible, and . . . just better. I hope you're with me. Because there's a lot of work to be done.

Because, no matter how this all works out in the future, someone will always have to design the building. Predictable construction outcomes are achieved through standards and practices that favor sameness and consistency. But that doesn't always help us to build a delightful and surprising world that people will want to live in. With access to digital design tools, construction automation, and knowledge of the best ways to employ them, the designers of the future will be able to build almost anything they can imagine. Why don't we free them to do what they do best – to build the fabric of a better world?

APPENDIX A

What Computer Should I Buy?

By now, it seems pretty unlikely that you don't have some kind of computer in your life. Probably, you have several. Maybe even more than you care to admit. The personal computing revolution of the 1980s has matured many times over now, and the benefits that have come to the lives of ordinary people around the world are stunning and . . . largely outside the scope of this book.

Some Basic Assumptions

I'll assume you have a smartphone. A device that is capable of taking photos, sending and receiving messages, telling you the time, capturing notes, and setting reminders for you. And so much more. I think, also, you can still place plain old voice phone calls with it – but who really does that anymore?

The computing capacity of your smartphone, even if yours is a couple years old, so stunningly leaves all other computers you have bought over the years in the dust that it is more or less beyond ordinary human capacity to understand it. It's OK, I know you're using it to send funny cat videos to your friends. No shame in that. Can you remember how you did that before 2007?

I point this out, mainly because the answer to your question, "What computer should I buy?" is no longer straightforward. Probably, you already have a small fleet of computers, and you are using them in every aspect of your life in deep and profound ways that you barely even recognize anymore. Probably, in addition to your smartphone, you also have at least one more traditional computer. Maybe a laptop, perhaps a workstation, perhaps both. When people ask for a new computer buy recommendation, this is usually what they mean.

Usually, it is because they tried to do something with their "old" computer that it just wasn't up to doing efficiently. Or maybe, they just feel like a new piece of gear might leverage them into being a better designer. This is what I want to focus on, and I think there are some simple answers to consider that might surprise you.

But if you're reading this book, you're probably thinking about your relationship with computing again, maybe questioning some decisions you made the last time you thought this question through. So, let's consider the question mainly in terms of your design practice. What computer should you buy to support your work as a designer?

I want to convince you that the computer you buy should be able to empower you as a designer, and as such, it is a reasonably personal decision. There are no other designers in the world exactly like you, and nobody can tell you precisely what you should buy for your digital work any more than they can recommend the right sketchbook or fountain pen.

"Which computer should I buy" is one of the most common questions on user forums where architects gather to talk about technology. And it makes sense you would want to be careful to make the right choice – computers, while cheaper and more powerful than ever, are still a pretty big expense for most architects. And it can be confusing if it isn't something you think about every day.

Since the computer hardware landscape is still changing much faster than the publication schedules of books like this one, I'll resist making specific recommendations in favor of talking you through the things that will likely make a difference to you no matter when you buy your next upgrade.

And there are some factors to consider that have nothing really to do with the most obvious computer choice questions you'll want to answer for yourself.

Focus on the places where you're going to be interacting with your computer – your monitor, keyboard, and mouse. With a few exceptions, these three pieces of technology are the parts that will last you the longest and will provide you with the most significant opportunities to get comfortable with the inevitable upgrade cycles that are going to happen everywhere else in your system. Virtuosi digital sketchers are those for whom the computer is an extension of their body, people who are so kinesthetically at home with their tools that they no longer think about them at all. Focus first on the parts of the computer that you touch, feel, and listen to.

Get a Great Monitor

Before you buy anything else, buy yourself a great monitor. Prepare to spend an uncomfortable amount on money on it, but never look back after you have. Get a large monitor and get the highest resolution you can find. Get one that can render more color than your eyes can see. Get one that refreshes so fast you don't even know it is happening. Spend money here because you are going to be staring at it a lot and because nothing your computer does makes any sense to you at all if you can't see it.

As an architect, you work visually more than anything else. Your work depends on a clear depiction of your designs. If your design work doesn't look good on your screen, you can't be sure it will look good anywhere else.

While we are surely on a path toward a paperless future in every profession, it seems likely to me that construction sites are going to be operated with paper drawings for a very long time to come. You are going to end up printing large drawings in black and white, probably only see them plotted and reproduced onto all kinds of grotty, coffee-stained, and torn pieces of paper. But they should look great to you when you're working on them. Because if you can't make them read well, nobody will ever be able to build your vision.

At Google, we bought two enormous monitors for every programmer, recognizing that being able to track two full pages of code at one time led to immediate and measurable productivity improvements. Something on the order of a 50 percent increase in productivity came from this one purchase alone.

For architects, the problem is a bit different. We habitually communicate with the world through vast sheets of paper. Depending on the kind of projects you do, or maybe just some local standards or norms, you are probably working on drawings of at least 24″ × 36″ or larger. At a minimum, you probably put together drawing sets for design presentations that are tabloid (11″ × 17″). Rarely, I bet, do you

APPENDIX A

work any smaller than that. Plan for a monitor that can show you a full 1:1 representation of at least a tabloid sheet, with a little room left over for menus, toolbars, panels, and whatnot.

In 2020, the best choice for this is going to be a 27" monitor. You can go down to 24", but I wouldn't go smaller than that. Surely there will be larger monitors available in the future, maybe all the way up to something on which you can see a full 24" × 36" sheet without zooming or panning. We aren't really quite there yet, though. At least, not at the resolution you should try to reach.

The resolution of your monitor can have a much more significant impact on your work than you might realize. When Apple began shipping "Retina" displays in their iPhones and iPads a few years ago, something transformative happened between me and my computer. Simply, I stopped seeing pixels on my screens. The resolution of the screen surpassed my eyes' angular resolving power, and I could no longer see the pixels. This was an amazing development that wholly reset my expectations for how a computer could represent my work.

Early computing experiments with cathode-ray tubes (CRTs) repurposed from oscilloscopes were rapidly replaced by much more versatile scan-line techniques derived from TV broadcasting. In the 1990s, these were replaced by flat panels using individually addressable pixels of liquid crystal. The last big CRT I used was purchased (at great expense) before the turn of the millennium. It has been flat screens exclusively since then.

All modern computer screens are raster displays. This means everything that your computer draws on them for you to look at must be broken down to individual pixels to be displayed. This is usually fine for images (like photographs) that have continuous tonal variation from pixel to pixel. Still, for things with sharp edges, like text and the lifework in a set of architectural drawings, those individual pixels lead to "aliasing" where curved or diagonal lines look like they are made out of little stair steps. I find this incredibly distracting when I see it.

Human eyes (with some variation) have a resolving power of about one arc minute, which means they are generally capable of telling the difference between two things that are about 1/60th of a degree apart. For a quick and dirty reference, a standard #2 pencil, held at arm's length from your eyes, will represent about half a degree of arc. Your eyes (assuming 20/20 vision) are mechanically capable of seeing features on your screen that are 30 times smaller than that.

To understand how this makes a difference to your particular usage, you really want to map the pixel density of the display into degrees of arc. This depends on your distance from the display surface. For a typical computer monitor, your eyes are probably somewhere between 18 and 24 inches away from the display surface. For a handheld device (like your phone or tablet), it may be as close as 10 inch. At these distances, your eye's resolving power is somewhere around 300 pixels per inch.

The original Macintosh display (in 1984) had a resolution of 72 pixels per inch. This was chosen to match the size of a printer's "point," the traditional unit of measure for movable type. It was close enough that graphic designers immediately felt more at home on a Macintosh display than anything else. They intuitively "got" the scale of the display. However, 1/72 of an inch is well above the resolving power of human eyes. Everyone could easily see the pixels. It was OK if you thought of the screen as a preview for the ultimate product, which was to be printed on a piece of paper.

When the artwork they made on those 72 dpi displays went to print, it was usually done on printers that could print at 300 dpi or higher. If you held up a piece of paper printed at 300 dpi, even if you held it right up to your face, you couldn't really see the aliasing. Even though printers, like monitors, are all raster devices as well.

Computer monitors gained the ability to draw in color well before their basic resolution increased. With continuous variation in color across a monitor, you can play tricks on the human eye to fool it into thinking lines are rasterized more smoothly than they really are. By rendering pixels around the stair steps in varying shades of gray, eyes can be fooled into thinking there is more data on the screen than there really is. Called *anti-aliasing*, it was used to sharpen the visual appearance of lines and text drawn on-screen considerably. Not everyone liked this approach, however, thinking that it made the text look blurry rather than smooth. I happened to like the effect, but your mileage may vary.

That has all changed now. Practically speaking, you can now buy a monitor capable of displaying graphics at resolutions previously only possible in print. That's pretty neat. That means your eyes will be able to tell the difference between line weights in a drawing from 6/0 to 7 (using the classic Rapidograph line weights), twice as many choices as you had with a 96 dpi display. You can't see aliasing anywhere, no matter what angle the lines are drawn. There is no need to resort to tricks like anti-aliasing anymore. A really high-resolution display is a more straightforward solution that makes a big difference.

A large display also makes a big difference. And this is one area where paper might still beat even the most expensive monitor for some time to come. A 24″ × 36″ architectural drawing sheet is about equivalent in size to a 43″ monitor but monitors of that size are not really designed to be viewed as closely as you are accustomed to viewing a computer monitor. You'll probably need at least an 8k (maybe even better) resolution to render a 6/0 line without seeing the pixels. In 2020, this is still a very exotic technology. Imaginable, but still unusual.

Your eyes have about a 140° horizontal field of view. Vertically, it is around 90°. Your eyes don't really work like traditional camera lenses, though. So, your ability to comprehend visual events at the extreme edges of this maximum isn't the same as it is in the center. Your retinas, in physical terms, resolve to lower angular resolution at the center than they do around the edges. This will become more important when we talk later about head-mounted displays (for virtual reality), but for now, just plan on getting the biggest monitor you can stand having on your desk.

I have an ultra-wide monitor as well, which I love for different reasons. I can see the pixels on it more easily; the raw pixel resolution on it is almost half that of my high-resolution display, but it wraps around my field of view much more completely. I feel like I'm much more immersed in it than my high-resolution screen. For immersive modeling where I want to feel like I'm completely inhabiting the space of my design, this thing is excellent. The critical point of differentiation for me is probably the kind of content I'm viewing. Things that are meant to communicate like a drawing or page of text benefit from having a sharp, pixel-free presentation. Things that are intended to communicate more like a photograph with continuous tonal variation and few sharp high-contrast edges are OK on lower-resolution displays.

When you're working on things like photographs (including photorealistic renderings), you should care about getting the color rendition right. Human eyes vary in their ability to resolve subtle differences in color, but most of them can differentiate between colors that are only a few nanometers apart. This means, given top and bottom thresholds

APPENDIX A

at the above the infrared and below the ultraviolet, most humans can see about 1 million distinct colors. Most computers address color values with 24–32 bits of precision today, accommodating a color space of more than 16 million distinct colors. But not all monitors can display the same range of colors. And none of them can display them all.

All monitors have a color bias of some kind. If you are really sensitive to color in your work, you'll want to spend time getting your monitor correctly calibrated for your particular workspace and needs. And finally, it matters how fast your monitor can refresh itself. If the refresh rate is too slow, you might feel like it is strobing. I haven't had a problem with this effect in years, though – it seems almost all monitors have resolved this. Even the cheap ones.

By now, you're probably beginning to glaze over a little, and I haven't even started to answer the question you wanted me to answer – which computer should you buy? But bear with me, there's more to consider before we get to that. I promise it will be worth it.

Get a Great Keyboard

When you bought your first desktop computer, it probably came with a keyboard, which you have been using for years without even thinking about it. Keyboards are among the least considered components in a computer, and yet they are the part you touch more than any other component. Your keyboard is worth getting right.

Partly, I think, the problem with keyboards is that people aren't really clear to themselves about what the purpose of the thing is. If you think about your keyboard as the thing you type out occasional emails and search strings, then any old keyboard will do. If you are a programmer or writer who bangs out pages of words for a living, you probably already have a better keyboard because you know how much of a difference it really makes. But if you are someone who sketches for a living, your keyboard might well be in the way most of the time.

And that's because you're thinking about it all wrong. Really, your keyboard is just a convenient collection of special programmable buttons that can make things happen on your computer. I'm talking about keyboard shortcuts, and they are the centerpiece of any great reflection-in-action flow state with computer applications.

Some keyboards are big, clacky, and kinesthetically rewarding to hammer on. Some are wafer-thin and barely register a keypress when you touch them. I've tried many of them. Right now, I'm typing on a small mechanical keyboard with distinctly clack-y keys. I love it. Feels like I'm accomplishing something with every keystroke. I think my family is less amused – it is pretty noisy.

But this is how I touch my computer, something that is otherwise a relatively cold and impersonal process. The touch, the feel of a thing, is somehow quite importantly human. A good keyboard is one of the most humane of interfaces, full of rich history and tradition as well.

What's neat today, though, is that it is straightforward to build up precisely the keyboard that you like, just the one you need. It can be a very personal experience today, thanks in no small part to the growth of the open-source hardware hacking community. The keyboard I'm typing on today has individually programmable keys and is capable of unimaginable customization. If I want to switch from QUERTY to DVORAK layouts, I can do that. Tired of unexpectedly hitting caps lock? I turned mine off and replaced it with an extra function key.

Among programmers, there is a text-editing environment called EMACS, which has justifiably earned an almost cult-like following. Actually, there are two of them (there's a competing environment called Vi that, if you like it, is clearly the better one) – and you don't want to be on the wrong side of the argument about which of them is the best. What makes both of them unique, though, is their leverage of the keyboard. When you're using EMACS, your hands never leave the keyboard. You never have to mode shift to grab a mouse.

I'm sure this sounds amazingly nerdy, and for sure, it is. But I don't think it is trivial. I'm already tooling up a new keyboard just for SketchUp keyboard shortcuts, arranged just the way I like them. Because clicking virtual buttons on a toolbar in SketchUp's GUI is so slow, so tedious. Now, if only I could figure out how to turn these things into foot-pedals like a pipe organist. . . I could get some real productivity.

Get a Great Pointing Device

This is amazingly hard. I don't think I've ever found a mouse that I really thought was great. And really, mice are an odd sort of user interface device. They aren't particularly ergonomic and most of them have far too many buttons on them. I like to leave the buttons where they belong – on my super clack-y keyboard. But as long as "WIMP"[1] graphical interfaces are the norm, you are going to need a good mouse.

For most of what I do, a simple mouse with two buttons (left click, right click) and a scroll wheel works the best. I have never had a favorite one, and typically the cheapest one seems to be the best for me. To be honest, I often lose mice, especially when I try to travel with them, and so I have more or less adapted my habits away from them. At home, where I am writing this book, I have been using one of Apple's "Magic Mouse" devices, and have grown to like the few extra gestures that it offers for managing basic operating system functions. If you have an Apple workstation, the Magic Mouse is pretty nice.

If you are running Windows or some other operating system, and you head down to your neighborhood computer store to buy a new mouse, you will likely be confronted with a wall of very sci-fi gaming mice. The needs of a gamer are really not wildly different from those of a daily digital sketch modeler, but the visual design may make you pause. Colorful pulsating LED lighting and cyberpunk graphics abound, as do walls of impressive sounding statistics about pointing precision and response time. In general, I would ignore all that and pick up the simplest device you can find. I like mice without wires, which means a Bluetooth device is your best option. And you should look for a mouse that uses optical sensors (rather than a mechanical rolling ball) to sense position.

Don't get a mouse with more than three buttons. Gaming mice peppered with shortcut buttons make sense for games where your fast-twitch response time can make the difference between winning or losing a battle are not that useful for daily work in design computation. Every one of those buttons has to be manually programmed for every application in which you'll them, and the configuration will be more complex and brittle than you expect. Stick with something durable and simple.

In general, I prefer trackpads to mice, and the one on my MacBook Pro is really great. Trackpads remain attached to my laptop at all times, so I am

---

[1] "WIMP" is an acronym for "windows, icons, mouse, pointer" and it refers to the dominant graphical user interface paradigm for our time, wherein users are offered the ability to manipulate commands in the computer by pointing at and clicking on icons that cause things to change in data presented to them in individual windows. You need a mouse to do all the pointing and clicking.

unable to lose them. A sideboard trackpad (again, Apple makes a great one in the "Magic Trackpad") for a desktop computer is handy to have, and if you are typing a lot, you can save some extra movement of your pointing hand by using one. If you are considering a trackpad as your main pointing device, look for the largest one you can find. Tiny outboard trackpads that barely fit the tip of your finger have long ago been replaced by devices the size of an index card with plenty of space to spread out your hand.

What about a tablet and stylus? To anyone who has spent the time to grow comfortable drawing with traditional media – and this should include by now everyone who has picked this book up to read, a stylus is a clean, natural interaction device for drawing on a computer. They are, however, less useful in general desktop computing for the basic operations of picking, clicking, and manipulating icons and windows in your operating system. And for a simple reason. When you are drawing, you want every position on the tablet to have an absolute coordinate that you can return to predictably every time you tap in that spot with your stylus. If you want to tap something in the top left corner of your screen, you have to tap in the top left corner of your tablet. On a big screen, this can rapidly become onerous. You really need something more gestural for basic pointing efficiency. For drawing, on the other hand, by all means use a tablet.

And, if you are serious about digital sketching, you are almost certain to want a tablet computer like and iPad or a Microsoft Surface in the future. Whether you will make this your daily driver or use it as a sideboard device for a larger desktop computer, a computing device that lets you draw right on the display is invaluable. It isn't really a pointing device (though categories are increasingly blurry) in a traditional sense, and for the near term, you're still going to want a basic mouse as a pointing device. Maybe in addition to a tablet.

Get Fast, Reliable Connection to the Internet

I have been a citizen of the internet for as long as I have had a personal computer. My family's first Apple computer had a modem and I quickly found ways to bridge from our local dial-up bulletin-board system (BBS) out to other connected systems. As an undergraduate student at Cooper Union in the early 1990s, I hung out in the computer lab primarily because that is where I could get on the internet. I built my first worldwide-web browser from source on an SGI workstation. I have had leased lines into my house since before that was a good idea. And I simply cannot imagine a world today where computers aren't connected together in meaningful ways to share information with one another.

You should connect every computer you care about doing work with to the internet with the highest bandwidth you can afford. I've made the case for this amply throughout this book, I hope, but in this appendix I want to offer some practical advice for how to actually manage that connection. In most parts of the world, access to the internet is beginning to be treated like a public utility. Just like you must set up electricity and water service to a new home when you move in, so must you set up internet access.

But access does vary from region to region, and universal high-speed internet access, especially for folks who work from remote locations, can still be somewhat challenging to maintain. For most of you, though, fast and reliable internet access is available today through either your local telephone company or a provider of cable or satellite television service. Today, I have a contract with my telephone company that provides internet access through a technology called "digital subscriber line (DSL)" that provides TCP-IP networking over the same lines that serve traditional telephone service to my house. Thanks to a recent neighborhood-wide infrastructure upgrade, I now have fiber optics replacing the old copper twisted-pair cable to my house, and so I am about to get a 10× increase in available bandwidth. Wahoo!

Regardless how you choose an internet provider, and no matter what technology they use to deploy it into your home or office, the endpoint (formally a router – though your provider may still try to call it a "modem," thinking that will make you more comfortable) it presents to you will be a wired ethernet connection. Hardwired connections to the internet are still the best for all your computers. If you are setting up a home network for the first time, you may be tempted to use wireless networking (wifi) for everything. This is certainly the most convenient way to work, and you will find that today's wifi protocols are very capable and easy to set up. I recommend against using the router your service provider gives you for this. If you want to use wifi for local networking, buy yourself a dedicated wifi base station that you control.

When you connect an office of computers to the internet, you will be bridging between your (private) local-area network (LAN) and the public internet. Doing this the wrong way can leave computers that should be private exposed to anyone who wants to snoop them from the public internet. You are right to be concerned about this, but in practical terms, all internet-connected routers are preconfigured to make it hard for attackers to break in from the outside. A *firewall* (your internet service provider has undoubtably included a basic one inside your router) is a device that implements security protocols that make it easy for you to look out to the internet from your LAN but make it hard for folks to snoop in the opposite direction. You really don't need to worry about this much any more.

Get a Fast, Stable, and Reliable Computer

You'll notice I have talked quite a bit about the parts of your computer hardware that you probably haven't been thinking much about. What about all the megabits and the gigahertz? Isn't it important to get the fastest possible computer?

In my experience, I stopped worrying about whether my computer was fast enough about five years ago. Every computer on the market is now more or less fast enough for what I want to do with it. With some notable exceptions, of course, but processor speed and memory are plenty fast enough for almost anything these days. Local disk capacity is plentiful and fast as well. Spend your money on other things, I would suggest. This is a controversial position, I am sure.

My stance on raw computer performance is probably surprising to many, especially those who think that having the right computer can make or break them as a designer. I think of the computer as a tool for my creativity, and as such I really want to think about it as little as possible. The parts of it that I touch should be delightful and rewarding to use.

Acknowledging all of that, you will need to pick a central processing unit (CPU), some memory, a graphics processing unit (GPU), and some persistent local storage. And you'll need to pick an operating system to run it all.

The old argument of Mac vs. Windows (no wait! What about Linux!) seems to matter less and less every year. Practically speaking, most of the software you're going to be running as a designer is equally at home on Windows or macOS machines. Linux is still kind of a tough choice for professional work, but totally credible if you know what you're doing. And there's Google's ChromeOS to consider as well, which is growing in capability all the time.

In addition, the basic OS-supplied capabilities of modern operating systems are all kind of the same as well. They all connect reliably to the internet, load applications, and make them run.

I have, personally, been a heavy user of Apple's Macintosh computers since the day they were launched back in 1984. I've wandered widely across the available operating system landscape over the years, including Windows and Linux, but also Silicon Graphics, Sun Solaris, and NeXTSTEP machines as well. I've installed countless builds of operating systems and even once recompiled a kernel from source, just so I could say I'd done it. Through it all, even during their darkest days as a company, I've had a Mac somewhere in my fleet.

Apple's macOS offers me just the right combination of simplicity of setup (low barrier to entry) and deep UNIX-nerd hackability. When all of this is coupled with an integration of the rest of Apple's iOS ecosystem (because, I also carry an iPhone), basic personal computing things just work without me having to fuss much at all. I like to preserve my technical futzing for things that serve my design goals, first. Rather than futzing just to make everything work on a basic level. As a general-purpose operating system, macOS is (in my experience) tough to beat.

On top of my operating system, and this is the part that is really going to change the most in the next 10 years, most of my tools are delivered to me over the internet through my web browser. This trend is unlikely to reverse any time soon. All major new application development that I know about (and I am watching the industry as closely as anyone) is building for the web before building a dedicated desktop client application. This changes the balance of performance from computation done on your desktop computer to computation done in the cloud. Your local computer will increasingly become a simple terminal to large computation being performed elsewhere.

Design computation will likely lag the farthest in this eventual transition to the cloud, however, and for that reason, spending a bit more than the average consumer on desktop performance still makes a lot of sense. But there's no real reason to go overboard with it. Pick a computer with the fastest single-core clock speed you can afford (multiple cores matter if you are rendering, but won't really help you model faster in 3D) and, if you plan on working on really big models, get as much memory as you can practically afford. If you care about the performance of real-time rendering applications, you may want to invest in an aftermarket GPU to speed that up. Gaming-class GPUs are usually fine.

What I Use, FWIW, in 2020

For what it's worth, and for the interest of future aficionados of computing history, here's what I bought myself. I have a Mac mini (the 2018 model) with a small Apple keyboard, a Magic Mouse, and a big trackpad. I bought a gorgeous 5k monitor that I love more than any monitor I have ever used. In my entire life. I have 64 GB of memory. Don't judge me, it was cheap. And I have about 1.25 TG of SSD for persistent storage. By the time this book goes to publication, I'll have a gigabit connection from my house to the internet.

I keep my computer at the latest OS release, and reflexively install every system update that is offered to me without hesitation. I use Google's Chrome web browser for 80 percent of my daily work, including my work in SketchUp. I have off-site (in the cloud) backups running redundantly to three cloud storage services, including one dedicated backup service called "Backblaze." All of my important projects are also managed via version control on either GitHub or Trimble Connect. I run a Google Apps domain for personal email and basic document management, anchored on a domain I have owned since the mid-1990s.

I carry a MacBook Pro with me for work, which echoes most of the basic logic I described for my personal computer above, though it is centrally managed by my employer's IT services folks. I have always been careful to keep a clear division between work for my employer and work on personal projects, so I seldom mix personal and professional computing together. I like to know which of my creative projects are mine alone and which ones belong to my employer.

I carry an iPad (v.6, several years old and soon to be replaced) with an Apple Pencil for sketching. I also carry an iPhone for on-the-go communication and to provide mobile internet access when I'm away from home of office networks. Generally speaking, I upgrade my technology about every three years – though practically I could go longer now than I did in years past.

Beware of Gadgets

There are seemingly endless numbers of new tech gadgets bouncing up and down in the periphery of my computing news every day. I have tried them all, or at least I've tried to try them all. But they are almost always false friends.

Doug Brent, Trimble's former VP of Innovation, gave the best advice for me about gadgets. I now try to live by this principle. He said he never bought a new tech gadget unless in doing so he could replace two or more gadgets he already had. For example, that new smartphone is a great thing to buy because it replaced an old phone, camera, music player, and GPS navigator. By adopting one new gadget, he could unload four others.

Because the real expense of gadgets isn't in the actual cost of buying them. It is really in the time you spend faffing about learning how to use them and integrating them into your daily practice. Nobody has time to menu dive and manual surf just to figure out how to move a piece of data from one device to another. Your attention is so fought over by technology today that it is almost impossible to find focus to get real work done. Don't feed the machine willingly.

Don't Throw GAS on the Fire (Gear Acquisition Syndrome)

Coined originally by musicians, gear acquisition syndrome (GAS) is the intense feeling of desire that you feel when you see some new piece of tech that is just out of reach for you right now. For guitarists, it is for that new effect pedal, or maybe a new amplifier. Even though you already have a half dozen guitars in the closet and so many pedals that the electricity in your neighborhood browns out when you plug them all in. GAS is like falling in love – you feel an emotional flush every time you see a new review posted online. Remember that you are buying technology to help you be more creative. You know that new piece of gear will take time to learn, time to integrate, and (ultimately) will not make you a new virtuoso right out of the box. There is a time and place for exciting new gear, but try to temper your initial enthusiasm.

Buy Things Without Cables Every Chance You Get

While the front side of your computer may be a carefully designed minimalist perfection of glass, aluminum, and blinky LEDs, the back side is probably a rat's nest of tangled cables full of dust bunnies and old coffee cups. Cables are the bane of my computing existence, the kryptonite to my digital sketching superpowers. I hate cables and have never ever found an effective way to deal with them.

You aren't going to be able to avoid dealing with some cables. You will have to plug into some source of electricity with a cable. You'll have to connect together every peripheral device you add to your system with more cables. You might even want to connect to an office network (or just to the internet) with a cable as well. Actually, that is probably the best way to do it.

But I try to take every chance I can get to lose cables from my configuration. I rarely cable to my network today. Wifi is plenty fast enough for everything except my most powerful desktop computers. I use a wireless keyboard and mouse. I use offsite (cloud) storage instead of my old stacks of hard drives and other storage media. Right now, my computer has only two visible cables coming out of it, one for the monitor and one for power. My monitor has its own power supply, so it gets its own power cable as well. That's only three cables, which I have tied up as neatly as I can and tucked out of the way.

And one more thing. As a computer user, it is inevitable that you will spend at least some portion of your day every day troubleshooting things about your computer that don't seem to be working quite right. More often than not, if you have a rat's nest of cables, you'll find yourself under your desk faffing about with them when something isn't working right. Fewer cables means less time under the desk, which it a great improvement in quality of life.

APPENDIX B

Integrated Development Environments (IDEs) for Architects

Computers think differently than the humans who designed them. Actually, they don't really think for themselves at all. What they do really well is take directions and act on them. Over and over again, at lightning speed. But they have to be given directions in very specific and detailed ways. Humans take directions in general ways, and they are capable of making judgment calls in the absence of clear direction. Computers are terrible at this. You really have to tell them exactly what you want them to do. And this, it turns out, is a pretty hard thing to do.

Learning to Talk to a Computer

To tell a computer what to do, you have to learn how to speak to it in a language that it will understand. The best programmers are those who can think natively in the computer's language without needing to translate it into their own human language much at all. Like native speakers of a human language (like English) they don't need to translate in their minds while they are communicating – they simply "think" in the language of the computer. Most of us, however, aren't so facile with computer languages, and so we need extra help working with them.

The first computer programs were written directly in a mathematical language that needed no translation by the computer. Instructions ran directly on the computer's processing unit, making every program specifically (and only) tailored to work on a single type of computer. But as computing machines grew and matured, it became desirable to write computer programs in a more general way. A program that ran on one computer should be reusable on another computer without necessitating a complete rewrite from first principles. From this desire came the first general-purpose programming languages and the first "compilers."[1]

Most of the large software applications you run on your computer today are written in languages that sit somewhere between a conversational human language and a language that a computer can understand. The process of converting the "programming" language (the language that humans read and write) into binary machine code the computer can execute is done by a special program called a "compiler." Programmers write pages of code in some language (the "C" programming language remains a popular choice, though there are countless others.) and then process that into executable machine code with a compiler. Every time the programmer wants to make a change, the "source" code must be re-compiled again.

Compiling is a long and complex process, and as a consequence, programmers have developed all kinds of strategies to help them manage the

_____

[1] Grace Hopper, John Backus.

amount of time they have to wait for the compiler to chew through their latest iteration so they can see if it worked. By breaking large programs into smaller logical chunks of code, they can modularize their efforts and reduce the length of compilation. Modularization also makes it easier for programmers to share their work easily with others, each of them working on one smaller part of the project. But with modularity comes complexity, and some kind of system is needed to manage the complexity. Enter the integrated programming environment, or IDE.

An IDE is a (usually quite large) software application that brings together a compiler, a source-code management system, a code editor and other related programming tools into a single, integrated workflow of some kind. Think of an IDE as being like that big building information modeling (BIM) application you bought to manage all aspects of the design and documentation work you're doing for your architectural practice. IDEs are rich, deep, and phenomenally complex to learn completely, but for a beginning programmer, they really do put everything you need to be successful in one place. Including, for the really complete ones, a full range of sample code and documentation to show you as quickly as possible how to make the whole system work for you.

If you really want to learn how to program, eventually you are going to be faced with learning one of these. But for a beginner, they are pretty daunting. Especially if you are learning to code primarily because you want to sketch design ideas with it. You may not be planning to actually ship production-quality applications right away. So rather than tackle Microsoft Visual Studio, Apple XCode, or any of the myriad other IDEs on the market today, you might want to try working with something simpler.

Processing, born in John Maeda's "Aesthetics and Computation Group" at the MIT Media Lab, is one of the simplest traditional IDEs built specifically for visual design. Rooted in the Java programming language, Processing adds some high-level object types specifically targeting common 2D, 3D, and interactive visual design.

As I have written in Chapter 6. "Sketching in Code," I like working in the Wolfram Language, using the Mathematica Notebook interface. Available today for folks tinkering in code (like you, presumably), a Mathematica subscription costs as little as $172 per year. Running in just your web browser, it requires minimal setup and is instantly at your fingertips from any computer on the internet. For a little more money, you can get a version that runs right on your local computer, giving you all the performance your system can muster. The Wolfram Language doesn't require compiling; you can write a line of code in the Notebook interface, then interpret it right away to see results. Very convenient. And the language is very large, containing something like 5000 specialized functions, all well documented with clear examples. And many of those functions are tied dynamically to data sources on the internet, like weather data, demographic data, maps, and other such things. If you are just getting started, you will be able to explore your most complex ideas very quickly. You can even build basic user interfaces around your code and share them with others online.

But I am a bit of an outlier in my affection (as a designer) for Mathematica. It is more commonly used in scientific computing circles, where it competes against large IDEs like Matlab or Maple. And for those who bristle at the thought of tying their futures to a commercial entity like Wolfram Research, the Jupyter Notebook interface to Python offers a nice open-source alternative, provided you are willing to spend the extra time setting the environment up and managing the many interdependencies between shared libraries in the

APPENDIX B

Python ecosystem. But most designers, I think, when they consider learning how to sketch with code, gravitate toward graphical programming languages like Grasshopper.

Visual Programming Languages

Unlike a traditional IDE, or the notebook-centric environment of Mathematica, graphical programming languages, wired directly into a modeling application like Rhino, provide a quick experimental form synthesis environment that feels more natural to visual thinkers than tapping away pages of code in a text editor. Visual programming languages replace pages of code with an infinite canvas filled with boxes connected together with wires like a flowchart, rather than a wall of text. They are easier for beginners to grok, but also quite powerful and capable of profound computation if care is taken to manage the visual clutter of a wall of boxes and wires.

Visual programming languages have be most strongly adopted in problem spaces where data flows through a network of operations sequentially. My first experience with visual programming was with the IRIS Explorer programming environment on the old Silicon Graphics workstation we had in Cooper Union's computer lab when I was a student (Yohanan n.d.). With it, I could read a data set in, manipulate it with various filters and surfacing functions, then render it out to the screen for dynamic visualization. Pretty powerful, but, as it ran only on SGI computers, it never really took hold in the wider computing community. My next interaction with visual computing came in the music and sound manipulation world, first with the composition environment Max/MSP and (later) with the open-source Pd (Pure Data) environment. Flow-based thinking, captured in a flowchart-like visual programming language, works well for time-based things like sound.

But for visual design work, time-based flow analogies are a bit harder to grasp, though for certain the forms associated with architecture do have a time-like quality to them. Architecture is of course always in a state of becoming, either through construction and reconstruction, or through the phenomenological flow of an occupant through the space of the building. Beginning, perhaps, with Houdini (hugely influential in the visual effects community in Hollywood), visual programming languages have found significant traction in the 3D modeling community. There are several choice to consider, all of them good in their own ways.

Grasshopper is likely, in 2020, to be the one you have heard of before. Tied directly into the Rhino modeling kernel, Grasshopper allows designers to sketch in code to create all kinds of complex parametric objects. If you already know Rhino and are comfortable navigating its modeling environment, Grasshopper affords you a well-integrated and complete visual IDE for form synthesis. And, like all of the most popular programming languages, Grasshopper has an active developer community that shares code, examples, and documentation to help newbies climb the learning curve as efficiently as possible. For those who like visual form synthesis, Grasshopper is a solid choice for beginners and experts alike.

The most significant problem with visual programming languages seems to be that programs created with them quickly become too cluttered to be readable (and therefore reusable) by anyone other than the original programmer who created them. As Peter Deutsch famously observed, *"Well, this is all fine and well, but the problem with visual programming languages is that you can't have more than 50 visual primitives on the screen at the same time. How are you going to write an operating system?".* A text editor, perhaps because we are more accustomed to managing complex data visually in the text of books, is capable of holding and communicating much more information per square inch than a visual graph can support.

Which takes me back to Mathematica. It has everything I need, lots of headroom for exploration, and a very convenient notebook interface that I can use to keep sketches neatly organized and accessible in the future. If you really want to sketch in code, I think Mathematica remains an ideal choice.

APPENDIX C
Sketching in Virtual Reality

Among the most exciting new technologies to arrive on our already overstimulated, oversaturated technological landscape, head-mounted displays offering a mixture of immersive spatial modalities may really change the way in which architectural ideas are experienced at all stages of the design process.

For conceptual phases of design, for the processes of design thinking by sketching, there are some unique possibilities to consider. I wrote about this briefly, in abstract terms, in Chapter 3. But there is more to say, particularly when considering the rate at which new technology is being released to the market. This technology is surely still in its infancy, with few commercial products outside of gaming and entertainment yet available, but it is fun to imagine where it may go in the future.

What Makes a VR Display "Immersive?"

The design of buildings and other kinds of physical space differs from other design subjects in that you need to represent both the insides and outsides of the things you're designing. If you are designing an electric toothbrush, you really only need to be able to design from the outside of the object; there is no real value in "inhabiting" the toothbrush from the inside unless you are trying to work out some complex mechanical placement. Most architects can readily switch (in their minds) between a point of view above and outside the building, thinking of it in its exterior, sculptural sense, and an interior, first-person point of view that is more closely matched to the human experience of the building's future inhabitants.

Modern video gaming has done a lot to encode these different points of view into the general population's basic human experience. First-person vs. third-person points of view are readily understood by most people today due to their use in popular games. But these games have struggled to map their screen-based projection paradigms into the immersive simulations afforded by VR headsets. Apart from their demanding technical requirements, the basic interaction paradigms are still in development. The way in which the user selects an object, picks it up to examine it, then does it through space to some new location are still undeveloped. And the way that the user moves their virtual body through space are still making people motion sick.

The immersive quality of a VR headset comes from the close placement of a high-resolution video display (actually, of at least *two* displays; one for each eye) directly in front of the wearer's eyes. To be truly immersive, those video displays should completely fill the wearer's field of view, so that no black space is apparent anywhere the wearer might turn their gaze. And the resolution of the display should be high enough that the rendered scene they are experiencing should feel convincingly natural.

As I am writing this appendix, the state of the art for a head-mounted display resolution is about 1440 × 1600, with a horizontal field of view of about 120 degrees. There are higher-resolution displays available, and wider fields of view, but they typically come with technical requirements that make them harder to manage for most people. Refresh rates, which are much more important for a head-mounted display, are generally in the range of 60–120Hz, though this depends quite a bit on the capability of the graphics card in the computer that drives the display.

A fully immersive head-mounted display, one that convincingly replaces the natural reality your eyes are immersed in every day, is still somewhat out of reach. The human visual sensing system, defined narrowly as the camera-like optical capability of your eyes, will only be filled by a display that fills more than a 200 degree field of view with greater than 80 million pixels per eye. We aren't even close to that yet.

And of course, the experience of seeing involves much more than just the optical characteristics of your eyes. Seeing is really a whole-body experience, bound to your body's general sense of spatial awareness and memory of prior experience as much as it is to the physics of a photon landing on your retina. But even though a truly immersive experience may still elude technology for a decade to come, you can have quite a bit of fun with the gear available to you today on the consumer market.

Materials for Sketching in Virtual Reality

If you are ready to experiment with VR in your design practice, you are going to need to gear up a bit. The computer you already have, even if you have invested in an aftermarket graphics card with more power than usual, may not be sufficient to drive a high-end VR display. You will find that VR puts heavy demands on your system, and you may need an upgrade.

Get a Great Graphics Card

Chances are, the graphics hardware that came with your computer isn't up to the task of rendering to a VR headset. You are going to need to buy a powerful aftermarket video card to drive almost any head-mounted display on the market today. This will mean spending at least $400 and maybe even as much as $1000 for a really capable one.

Hardware graphics acceleration moves about as fast as any part of the technology market today, driven by an ever-increasing thirst for realism in consumer gaming. I won't try to recommend specific cards here or even to suggest which vendors are the best. By the time this book gets to press, any recommendation I make will be wrong. I would suggest paying attention to cards that are capable of performing "real-time raytracing" like NVIDIA's RTX graphics. This is new, but speaks directly to the deepest requirements for convincingly immersive VR.

Pick a Comfortable Head-Mounted Display

To me, until there are head-mounted displays capable of a completely convincing visual simulation, comfort is the most important feature of a head-mounted display. If I am going to use it for design, I will need to be able to wear it for hours at a time. Every device I have ever tried feels heavy on my face, and it gets sweaty. That heaviness and sweatiness converts to real discomfort over time, and you aren't going to be able to stand it after a while.

This means you will want to actually try some headsets on before you buy one. For me, the HTC

Vive Cosmos is comfortable. Valve's Index is pretty nice, too. But the most comfortable for me at the time I'm writing this is probably the Oculus Quest. By the time you're looking, there will be many other choices available, and only you can know which ones fit your face the best.

Consider Your Controllers Carefully

In VR, you can't use a traditional mouse and keyboard in the same way you would with a traditional display. You are going to need to pick a specialized VR controller to interact with the virtual space in which you will be sketching. And there are some great choices to choose among today.

There are few standards, and pointing is still evolving rapidly. I have tried at least three different controllers, and each has really been paradigmatically better than its predecessor. The critical thing to consider is that the controller in your hand is really the *only* way you can interact with the virtual world around you.

For me, the Valve Index controllers were the most valuable improvement to my VR experience. They have a unique capability (I don't know how it works; I think it is magic) to sense the position of each of my fingers when I am holding them. That means I can begin to imagine acts of much more precision than I can with a device that just simulates a single point of interaction like I am pointing a laser beam into space. That said, I haven't yet found a piece of sketching software that lets me use this dexterity for anything useful.

Find a Big, Empty Space

You are going to be moving your body around a lot when working in virtual reality. And if you have a display completely covering your eyes, you are going to end up bumping into hard things in the real world around you all the time. The more space you can clear around you, the better. It is easy to forget this requirement when you are shopping for a new headset online, but you are going to have to get well back from your computer when you are working in virtual reality.

The most precise VR headsets today still rely on some externally placed hardware beacons to provide positional awareness to the headset. Called *lighthouses* by manufacturers like HTC and Valve, you will have to place these somewhere at the perimeter of the space in which you'll be moving around. And they aren't capable of tracking your head position in infinitely large spaces. VR space is basically Cartesian in all implementations I have tried, which means you are going to be interacting within a cubic space about the size of a typical bedroom. Plan accordingly.

Manage Your Wires, If You Have Them

Most high-performance VR headsets available today require a wired connection back to your computer. If you think of them as essentially a fancy, super-high-resolution computer monitor (they are a lot more complex than that, of course) you'll appreciate why most of them still require a wired connection. There is just so much data being transmitted that a wireless connection can't reliably keep up.

There are, of course, wireless VR headsets as well, and they are getting better and better every year. A wireless headset is always going to be the most comfortable to wear, the most convenient to use, and the easiest to set up in a physical space. Unfortunately, you will have to make some significant concessions in raw performance if you choose to use one.

A big thick wire trailing out the back of your VR headset can be a real distraction, even a hazard, when you're immersed in simulation. As you move around in physical space, you will find the wires trailing behind you are always getting tangled around something, caught underfoot or even turning into a tripping hazard. Nothing wrecks the sense of immersion like tripping over a wire you can't see. Or having it pull the headset off your face because it has gotten wrapped around something.

There are some simple and cheap ways to manage the cables, and it is worth your time and money to invest in them. I have experimented with a variety of them, but in the end I settled on a small array of cheap cable retractors, similar to the kind you might use on your car keys. I attached a half dozen of them to my VR headset's cable, attached to the ceiling, and spread along the shortest line from the center of my VR space to my computer. Now, as I stand, turn, and wander (cautiously) around my virtual space, the wires are tucked safely up in the ceiling, out of my reach.

Try All the Sketching Apps

There are not many commercial applications that encourage free sketching in virtual reality. And those that do exist are still pretty immature. Like all aspects of virtual reality, the technology is developing quickly, but you aren't going to find tools that are anywhere close to the maturity of the 3D modeling tools you can find for a traditional desktop computing environment. Mostly, VR headsets today are not used by professional designers. They are used by gamers or for some other entertainment purpose.

But you can have a lot of fun with the tools that are available today, and they may even find a way into your daily sketching practice if you're persistent in learning how to use them. You shouldn't expect to find a fully featured CAD system, with snaps, inference, or parametric modeling features like you might be using on your desktop computer – the overhead is just too much for VR and the endless detailed heuristics of how tool interaction works haven't had time to grow up. Yet.

Instead, you should expect to find a collection of delightfully painterly 3D environments that are unlike anything you have used anywhere else before. Imagine standing in the middle of a room with a fully loaded paintbrush. But rather than laying down a stroke of paint on a canvas in front of you in that space, you can paint in the air all around you. It is a delightfully creative experience, and quite intuitive once you get the hang of a few basic controller interactions.

There aren't many sketching applications for VR yet, though I expect there will be more to come in the future. At the time of this writing, there are only a half dozen or so available options, and none of them really feel fully featured yet. In the bunch, I have found three that are worth playing around in, though of course your interests may vary from mine.

Google's Tilt Brush was the first one that I tried, and I found it quite intuitive and easy to learn right out of the box. Just like painting with light in the air around me, which was a startlingly novel experience. Quill, a painting program similar to Tilt Brush, turned out to be my favorite tool. Similar in interaction to Tilt Brush, but with an additional set of basic animation tools and an overall more polished and complete feeling about it. And for something that is about as close to a desktop 3D modeling application as I have tried so far, Gravity Sketch is hard to beat.

Using VR for Design

If you have successfully navigated all the technical complexities that surround a functional VR headset

APPENDIX C

and the space in which to use it effectively, you will probably use it to play some games. There's no shame in that; VR gaming is pretty fun. Also, games offer great environments in which to experiment in a general way with the new affordances of VR as a medium. You'll need to log quite a few hours in VR before it loses its novelty and becomes, for you anyway, just another computing modality that has its own benefits and detriments. Valve's 2020 game "Half Life: Alyx" is just about the richest and most complex VR game I have ever played, and if you can get past the zombie head-crabs jumping out from the shadows, you'll learn a lot about how VR really can work as a medium.

To use VR for design, you have to learn how it works well enough that you forget the details and become able to act instinctively within its constraints – really, just exactly like any other design media that you learn to use. VR is, in a sense, just another pencil and paper. It has more constraints in some ways (you are going to have to deal with cables attaching your head to your computer) and fewer in others (you can draw on air, in 3D all around you.) In time, and with practice, sketching in VR can become a valuable new tool in your sketching toolbox.

Using VR for Clients Who Can't Read Drawings

It is an unfortunate reality that many architects and designers work for clients who have trouble imagining a building from only a set of orthographic drawings. Traditionally, you might have worked with clients like this by sketching perspective views, making renderings of models, or even physical sketch models that would help them use their imagination to inhabit the space you designed for them. If they are willing, and if you have some patience for them as they stumble around your office, VR headsets can be a mind-expanding wonder for your clients.

Most 3D modeling environments today include convenient export paths from the modeling application you use to make them into a VR viewer of some kind that you can use to show your client around in the model. The experience is a spectacular one for most folks who haven't tried it before, and the experience of actually walking around inside their as-yet-unbuilt project is full of advantages. Your client can really see how the view from their kitchen table, understanding immediately why you made the design moves you made to open it up.

Beware, however, the lure of photorealistic rendering and premature physical simulation. Just like more traditional modeling, models experienced in VR can feel like they are much more complete than your design may actually justify. Be careful that the renderings you present aren't showing things you can't really deliver in the real construction project to come.

The use of a VR headset to help your clients understand the space you're proposing to build for them is probably sufficient to justify the expense of tooling yourself up. Even on a single project, if it saves you an extra round of design revision or just the emotional rollercoaster of a client who is disappointed with a project in construction that they secretly never really understood, then it will be worth it.

Using Your VR Sketches Outside Your VR Headset

There aren't many ways to extract a usefully portable 3D model from any of these tools, and I think you'll find it is best just to think of them as pure sketching environments – just a place to experiment with rough ideas that you will carry on to deeper levels of detail in another tool. Think of them as being useful, like sketching on paper is useful, and don't spend a lot of time trying to figure out how to make them into something more than that.

I have found the experience of sketching in 3D all by itself to be a worthwhile design activity. When I am my own audience, and when I'm not worried about sharing my sketches with anyone else, it's enough to view them in the headset, by myself, and not share them with anyone.

Sketching in VR Will Be Pretty Great. Eventually

VR is not really mature enough to rely on for everyday work yet, and it may well be years before it is. But it isn't hard to imagine a future where VR becomes an increasingly essential part of the designer's toolbox. It is just too compelling to ignore. Being able to walk through a simulation of a future space, experiencing the proposed space in full binocular fidelity, will give you an appreciation for the qualities of that space on a kinesthetic level. You can know what it feels like to stand on tip-toe to peer over an obstruction, or to crawl on all fours under a table where the kids may want to play.

You should expect to see a continuing stream of improvements to basic VR headsets in the years ahead, and to the dedicated graphics hardware it takes to render to them at interactive frame rates. Headsets will get lighter and more comfortable, and will lose the wires that tether them to a desktop computer. Resolutions are going to increase, at least incrementally, for years to come. And the optical field of view is surely going to continue to expand. This is not a technology that is going to fade away, even though it still has many years of core development research and design ahead.

If you are ready to experiment with VR, you won't be disappointed with the capability of today's technology. Sketching with light in the air around you is a beautiful, creative experience. You owe it to yourself to try that out. But if you just need a way to justify it as a business expense, using VR to help your clients understand your design proposals works well enough today that every architect should probably already be doing it for every project.

BIBLIOGRAPHY

Adami, V. (1973) *Adami*. Derriere le Miroir, n° 206.

Alberti, L.B. (1435) *De pictura*. Translated by Spenser, J. R., (1976) Greenwood Press, Westport, Conn.

Angelico, B. (1420) *Thebaid*. The Uffizi.

Baumgart, B.G. (1972) *Winged Edge Polyhedron Representation*. Memo AIM-179, STAN-CS-320, ARPA Order No. 457, Computer Science Department, School of Humanities and Sciences, Stanford University.

Beck, K., Beedle, M. & Bennekum, A. (2001) *Manifesto for Agile Software Development*. Available at: https://agilemanifesto.org/ (Accessed: 10 June 2020)

BIMForum. (May 2020) *Level of Development (LOD) Specification*. Available at: https://bimforum.org/lod/ (Accessed: 10 June 2020).

Boole, G. (1847) *The Mathematical Analysis of Logic*. Cambridge: Macmillan, Barclay, & Macmillan.

buildingSmart (2020) *Industry Foundation Classes (IFC) for data sharing in the construction and facility management industries* ISO 16739-1:2018.

Byrne, O. (1847) *The first six books of the elements of Euclid, in which coloured diagrams and symbols are used instead of letters for the greater ease of learners*. Taschen.

Cage, J. (2015) *John Cage: Diary*. Catskill, NY: Siglio.

Catmull, E. and Clark, J. (1978) Recursively generated B-spline surfaces on arbitrary topological meshes. *Computer-Aided Design*, 10, pp. 350–355.

Chaikin, G.M. (1974) An algorithm for high-speed curve generation. *Computer Graphics and Image Processing*, 3, pp. 346–349.

Chatwin, B. (1987) *The Songlines*. Elisabeth Sifton Books, Viking, New York.

Ching, F.D.K. (1979) *Architecture, Form, Space and Order*. New York: Van Nostrand Reinhold.

De Ruijter, G. (2019) *Gerco De Ruijter*. Rotterdam: Nai010 Publishers. rier Corporation.

Descartes, R. (1637) *Des matures de la geometry*. Translated Smith, D. And Latham, M. (1954) *The Geometry of René Descartes*. Courier Corporation, New York.

Doyle, M.E. (2006) *Color Drawing*. John Wiley & Sons, Hoboken, New Jersey.

Dürer, A. (1525) *The Painter's Manual*. Translated by Strauss, W.S. & Strauss, W.L. (1977) Abaris Books.

Eco, U. (1988) *The Open Work*. Harvard University Press.

Frampton, Kenneth; Peter Eisenman et al. (1975) *Five architects*. Oxford University Press, USA, New York.

Gladwell, M. (2008) *Outliers*. New York: Little, Brown.

Hanabusa, I. (1888) Blind monks examining an elephant. Library of Congress, https://www.loc.gov/pictures/item/2004666374/.

Hockney, D.; Philip Haas et al. (1988) *A day on the Grand Canal with the emperor of China, or, Surface is illusion, but so is depth*. Milestone, Harrington Park, NJ.

Houdini (2020) Toronto: SideFX.

Huerta, S. (2006) Structural design in the work of Gaudi. *Architectural Science Review*, 49, pp. 324–339.

Hull, C. (1989) (STL) *STereoLithography Interface Specification*. 3D Systems.

Jansen, L. (2013) *Vincent Van Gogh, Ever Yours: The Essential Letters*. Yale University Press, New Haven.

Kant, I. (1892) *The Critique of Judgement*. New York: Hafner Publishing.

Kronos Group. (2020) *COLLADA Overview*. Available at https://www.khronos.org/collada (Accessed 10 June 2020).

Killen, G. (2017) *Ancient Egyptian Furniture*. Oxford & Havertown, PA: Oxbow Books.

Leggitt, J. (2009) *Drawing Shortcuts*. Hoboken, NJ: John Wiley & Sons.

Lévi-Strauss, C. (1974) *Structural Anthropology*. New York: Basic Books.

Lightwave (2020) San Antonio, Tx: NewTek.

Lobachevsky, N.I. (2019) (Translated by Petkova, S.) *The Foundations of Geometry: Works on Non-Euclidean Geometry*. Minkowski Institute Press, Montréal.

Lockard, W.K. (1965) *Drawing as a Means to Architecture*. New York: Van Nostrand Reinhold Company.

Lynn, G. (1998) *Folds, Bodies and Blobs: Collected Essays*. Bruxelles: La lettre Volée.

Lynn, G. (1999) *Animate Form*. Princeton Architectural Press, New York.

Maeda, J. (2001) *Design by Numbers*. Cambridge: MIT Press.

Mäntylä, M. (1988) *An Introduction to Solid Modeling*. New York: Computer Science Press, Incorporated.

Martin, D. (1991) *Book Design: A Practical Introduction*. New York: Van Nostrand Reinhold Company.

Maslow, A.H. (1987) *Motivation and Personality*. New York: Pearson College Division.

Moleskine Srl. (2020) *About Us*. Available at: https://us.moleskine.com/en/about-us (Accessed: 4 March 2020).

Mueller, P.A. and Oppenheimer, D.M. (2014) The pen is mightier than the keyboard. *Psychological Science* 25, 1159–1168.

NVIDIA (2020) *Corporate History*. Available at: https://www.nvidia.com/en-us/about-nvidia/corporate-timeline/ (Accessed, 20 June 2020).

Oles, P.S. (1979) *Architectural Illustration: The Value Delineation Process*. New York: Van Nostrand Reinhold Co.

Panofsky, E. and Wood, C.S. (1997) *Perspective as Symbolic Form*. New York: Zone Books.

Papert, S.A. (1993) *Mindstorms*. New York: Basic Books.

Pare, E.G. (1997) *Descriptive Geometry*. New York: Peachpit Press.

Peddie, J. (2013) *The History of Visual Magic in Computers*. London: Springer.

Plato. (375 B.C.) *The Republic*. Translated by Jowett, B. (1892) Project Gutenberg, http://www.gutenberg.org/ebooks/55201 (Accessed: 4 March 2020).

Procreate (2020) North Hobart, Aus: Savage.

Puckette, M. (2020) *Pd Documentation*. Available at: http://msp.ucsd.edu/Pd_documentation/ (Accessed: 4 March 2020).

Raymond, E.S. & Steele, G.L. (1996) *The New Hacker's Dictionary*. Cambridge, MA: MIT Press.

Rosenberg, S. (1992) Virtual reality check digital daydreams, Cyberspace nightmares. *San Francisco Examiner*, Style, C1.

Rittel, H.W.J. and Webber, M.M. (1973) Dilemmas in a general theory of planning. *Policy Sci*, 4, pp. 155–169.

Sacks, O. (2009) *What hallucination reveals about our minds*. TED: Ideas worth spreading, https://www.ted.com/talks/oliver_sacks_what_hallucination_reveals_about_our_minds (Accessed: 4 March 2020).

BIBLIOGRAPHY

Saint-Exupéry, A.D. (2010) *Wind, Sand and Stars*. New York: Houghton Mifflin Harcourt.

Sanchez, A. and Chavez, N. (1979) *A drawing manifesto from Barcelona*. Translated by Moia, M. Available at http://rixjennings.com/a-drawing-manifesto-from-barcelona (Accessed: 4 March 2020).

Schön, D.A. (2017) *The Reflective Practitioner*. New York: Routledge.

Schumacher, P. (2008) *Parametricism as Style - Parametricist Manifesto*. Available at https://www.patrikschumacher.com/Texts/Parametricism%20as%20Style.htm (Accessed: 4 March 2020).

Sketchers, U. (2019) *Urban Sketchers: Our Manifesto*. Available at: http://www.urbansketchers.org/p/our-manifesto.html (Accessed: 4 March 2020).

Solidworks (2020) *Waltham*, MA: Dassault Systèmes.

Sutherland, I.E. (1968) A head-mounted three dimensional display. Fall Joint Computer Conference, 757–764.

Tarkovsky, A. (1987) *Sculpting in Time*. Austin: University of Texas Press.

Trimble (2020) *3D Warehouse*. Available at: http://3dwarehouse.sketchup.com (Accessed: 4 March 2020).

Various (1998) Comp.Lang.Visual - Frequently-Asked Questions List. Available at: http://www.faqs.org/faqs/visual-lang/faq/ (Accessed: 4 March 2020).

Vasari, G. (1550) *The Lives of the Artists*. Translated by Bondanella, J.C. (2008) Oxford University Press, Oxford.

Vesalius, A. (1543) De humani corporis fabrica libri septem. Ex officina I. Oporini, Basileæ.

Walker, J. (1989) *The Autodesk File*. Que Pub. Thousand Oaks, CA: Sage.

Wolfram, S. (2018) *A New Kind of Science*. Champaign, IL: Wolfram Media, Inc.

Wright, F.L. (2010) *The Essential Frank Lloyd Wright*. Princeton, NJ: Princeton University Press.

Yohanan, S. (n.d.) *IRIS Explorer: Collaborative Scientific Visualization through Visual Programming*. Available at: http://yohanan.org/steve/projects/iris-explorer (Accessed: 4 March 2020).

Zicarelli, D. (2020) *Max/MSP*. Walnut, CA: Cycling '74.

INDEX

NOTE: Page references in *italics* refer to figures.

A

Adami, Valerio, *48*
Adobe (company), 18, 29, 56
    Creative Cloud (application), 29
    Illustrator (application), 56
    Photoshop (application), 17, 29
Agility, 19–31
    agile development, defined, 20
    attention to technical excellence/good design for, 23–24
    avoiding over-organization, 30–31
    backing up work and, 27–30, *30*
    changing requirements and, 20–21
    for client satisfaction, 20
    cooperating with stakeholders/other designers, 21–22
    for credible design proposals, 22–23
    critiquing at regular intervals, 25–26
    delivering complete design proposals and, 21
    for design emergence, *19*, 19–20
    design team and design ownership, 26
    integration and co-location of design teams, 22
    for motivation and trust, 22
    practicing, 26
    self-organizing teams and, 24
    simplicity and, 24
    for sustainable practice, 23
Allegory of the cave, *72*
American Institute of Architects (AIA, organization), 192, 199
Angelico, Fra, 86, *87*
"Animate Form" (Lynn), 157–158
Apple Computer (company)
    iCloud, 29
    iPad, 106, 117, *117*, 119, 213, 220
    iPhone, 213, 220
    MacBook Pro, 216–217, 220
    Macintosh, 17, 213, 218–219
    Magic Mouse, 216, 219
    Pencil, 106, *117*, 119, 220
ArchiCAD (Graphisoft), *150*, 202
Architectural scale, 76–79, *77*, *78*
Archiving (tablet sketching feature), 119
ArchViz. *See* Advanced architectural visualization
Arcs, 49–51, *49–51*
Association of General Contractors (AGC, organization), 192
AutoCAD (Autodesk), 15, 17, 159, 181
Autodesk (company), 18, 29
    AutoCAD (application), 15, 17, 159, 181
    DWG (file format, DraWinG), 29
    Fusion (application), 150
    Maya (application), 148, 159
    Mel (programming language), 159
    Revit (application, Autodesk), 202
    3DSmax (application, Autodesk), 148

B

Backblaze (application), 220
Backup of data, 26–30, *30*, 119, 220
Bacus, John, *106*
Barcelona Manifesto (1979), 110
Basílica de la Sagrada Família (Gaudí), 52, *53*
Beck, K., 20–25
Beedle, M., 20–25
Bézier, Pierre, 148
Bézier curve, *54–56*, 54–57, 146, 178
Bézier surface, 57–58, *57–58*
BIM. *See* Building information modeling
Bit rot, 16
Blender (application), 148
Boccioni, Umberto, 40, 42, 98
"Boneyard" (physical backup), 28
Boolean logic, 60
Boundary representation (BREP) forms, *63*, 63–64, *64*
Braque, Georges, 98
Brent, Doug, 220
Bricolage, 128
Brunelleschi, Filippo, 88
Building information modeling (BIM)
    for building objects and higher-order primitives, 64–70, *66*, *67*, *69*, *70*
    expectations for, 201–202
    levels of development (LoD), 191–192
    sketching for, 189–191
    sketching in LoD 100 BIM, *193*, 193–198
BuildingSmart (organization), 194
Industry Foundation Classes (file format, IFC, BuildingSmart), 29, 194

C

CAD systems, inception of, 17
Cage, John, 6
Campbell, Joseph, 186–189
Candela, Felix, *52*

INDEX

Candido, Anthony, 112, 113
Cardboard, sketch-modeling with, 62–64, *63, 64*
Cartesian coordinate system, 38–39
Cartesian space, 35–40, 44
Casteljau, Paul de, 54–55, *54–56*, 148
Catia (Dassault), 149, 202
Catmull, Ed, 58
Chaikin, George, *53–54*, 53–55, 58, 181–182
Chatwin, Bruce, 105
Chrome (application, Google), 218–220
Cinematic perspective, 92–93, *93*
Citröen (company), 52
Cloud backup, 27–30, 119
Code, sketching in. *See* Programming
Collaborative sketching, 108, 119
COLLADA (file format, COLLAbortive Design Activity, Khronos Group), 29
Color
   Color Drawing (Doyle), 120, 121
   as element of design, 42
   sketching with watercolor, *114*, 115
Communication Arts, Inc. (company), 120, 123, 186, 200
Computers, 211–221
   author's preferences for, 219–220
   choosing materials/tools for digital sketching, 159–161, *161*, 211–212
   computer "intelligence," 11
   "gear acquisition syndrome," 220
   internet connection for, 217–218
   keyboard selection, 215–216
   monitor selection, 212–215
   pointing device selection, 216–217
   precaution about gadgets, 220
   reliability of, 218–219
   wireless peripherals for, 221
   See also Software
Conceptualization of design, *33*, 33–35, 198–203. *See also* Presentation; Sketching for conceptual design
Conics, 51–52 *52, 53*
Construction documentation, 185–189. *See also* Presentation
Controllers, 229
Cooper Union School of Architecture, 4, 109, 111, 127, 181–182, 225
Cost, as "fifth" dimension of space, 39–40
Creative Cloud (Adobe), 29
Creativity
   process of, 6–10, *7, 8, 10*
   of programming, 181–184
   rigor in design and, 1–3
Creo (application, PTC), 149
Critique of Aesthetic Judgement (Kant), 34
Cutting planes, *79*, 79–80

D

Dassault (Company)
   Catia (application), 149, 202
   Solidworks (application), 149, 150
Data backup, 26–30, *30*, 119, 220
A Day on the Grand Canal with the Emperor of China (Hockney), 84
"De Architectura" (Vitruvius), 157
De humani corporis fabrica (Vesalius), *100*
Denver Art Museum, 18
Descartes, Rene, 35
Descriptive geometry, 73–76, *75, 76*
Design, professional practice of, 3–5
Design By Numbers (Maeda), 182
Design elements. *See* Elements of design
Design thinking, 5–6
Deutsch, Peter, 225–226
Digital modeling. *See* Sketching in 3D
Digital tools
   avoiding over-reliance on, 14–18
   future of, 205–210
   learning to use tools, 10–12
   See also Computers; *individual names of hardware; individual names of software*
Dixon Ticonderoga No. 2 pencils, 105, 114, 117, 119
Doyle, Mike, 91, 120–121
Drawing. *See* Sketching in 2D
DWG (Autodesk), 29
Drawing Shortcuts (Leggitt), 121
Drive (Google), 29
Dropbox (application), 29
Dürer, Albrecht, 50, *50*, 73, *74, 88*, 100

E

Eco, Umberto: *The Open Work*, 132
Elements (Euclid), 34
Elements of design, 33–70
   building objects and higher-order primitives, 64–70, *66, 67, 69, 70*
   conceptualization of design, *33*, 33–35
   freeform curves, 52–56, *53–56*
   geometry and form, 44–52, *44–53*
   measurement of space for, 35–40, *36–39*
   qualities of space and, 40–44, *41–44*
   surface, 56–64, *56–66*
Elevations, 76, *76*, 82–83, *83*
EMACS (application, text editor), 216
Engineering, mindset of, 19
Euclid, 34, 39, 44–47, 49, 51, 55–56, 58–60

F

"Fifth" dimension of space (cost), 39–40
Flow state of design, 9, 108
Form
   joined surface forms, 62–64, *63, 64*

INDEX

measurement of, 59
primitive forms, 59–60, *60*
sketching in 3D, *147–149*, 147–150
swept forms, 60–62, *61, 62*
See also Geometry and form
FormZ (application, Auto-des-sys), 18
Fourth dimension of space (time), 39
Freeform curves, 52–56, *53–56*
Freeform surfaces, *145*, 145–146
Free perspective, 93–95
Froebel blocks, 130, *130*
Fusion (Autodesk), 150

G

Gadgets, precaution about, 220
Gaudi, Antoni, 52, *53*
"Gear acquisition syndrome," 220
Gehry, Frank, 109, 126, 149, 201–202
Geometry, descriptive, 73–76, *75, 76*
Geometry and form, 44–52
   arcs, 49–51, *49–51*
   conics, 51–52, *52, 53*
   Euclid's postulates on, 44–45, *45, 46*
   freeform curves, 52–56, *53–56*
   lines, *47,* 47–49, *48*
   points, 46, *46*
Gibson, William, 208
Giotto (painter), 49
Git (application), 31–32
GitHub (application), 202, 220
Gladwell, Malcolm, 4
Google (company)
   Chrome web browser, 218–220
   Drive, 29
   Tilt Brush, 230
   work process at, 212
Graphics cards, 228
Graphics processing units (GPUs), 133–134
Graphisoft (company), 150
   ArchiCAD (application, Graphisoft), *150*, 202
Grasshopper (McNeel), 159, 161, 183, 225

Gravity Sketch (application), 230
Guggenheim Museum Bilbao, 201
Gussow, Sue, 112

H

"Half Life" (application, Valve), 231
Hallaka, David, 112
Head-mounted displays, 228–229
Headsets, wireless, 229–230
Hejduk, John, 4, 186
"The Hero with a Thousand Faces" (Campbell), 186–189
Heuristics, 141
Hockney, David, 84, *84*
HoloLens (Microsoft), 97
HTC Vive Cosmos headsets, 228–229

I

ICloud (Apple), 29
Industry Foundation Classes (IFC, BuildingSmart), 29, 194
Illustrator (Adobe), 56
Integrated development environments (IDEs)
   Grasshopper IDE, 159, 161, 183, 225
   Mathematica IDE (see Mathematica)
   See also Jupiter Notebook
   overview, 223–226
Internet connection, 217–218
IPad (Apple), 106, 117, *117,* 119, 213, 220
IPhone (Apple), 213, 220
Iris Explorer/Iris GL (Silicon Graphics), 91, 181, 225

J

Java (programming language), 224
Jennings, Rix, 112
Johnson, Phillip, 147
Joined surface forms, 62–64, *63, 64*
Jupyter Notebook (IDE, Python), 161, 224–225

K

Kant, Immanuel, 34
"The Kanxi Emperor's Tour of the South" (Wang Hui), *84,* 84–86, *86*
Keyboard selection, 215–216

L

Lanier, Jaron, 97
Leggitt, Jim, 91, 121
Lego Mindstorms, 128, *128,* 160
Levi Strauss, Claude, 154
Liebeskind, Daniel, 18
Light as design element, 44
Lightwave (application, Newtek), 120
Lines
   as element of design, *47,* 47–49, *48*
   linear measurement (one dimension), 35–36, *36*
   parallel postulate, 44–46, *45, 46*
   programming for, 163–170, *164–170*
   sketching in 3D, 141–144, *142–144*
Linnaeus, Carl, 194
Linux kernel, 31, 32
Lisp (programming language), 159
"Lives of the Artists" (Vasari), 49
Lobachevsky, Nikolai., 45
Lockard, W. Kirby, 121
Level of Development (LoD, Level of Detail). *See* Association of General Contractors
Lovelace, Ada, 159
Lynn, Greg, 157–159

M

MacBook Pro (Apple), 216–217, 220
Macintosh (Apple), 17, 213, 218–219
Maeda, John, 182–183, 224
McNeel (company)
   Grasshopper (application), 159, 161, 183, 225
   Rhinoceros (Rhino, application), 150, 159, 161, 183, 225
"Le Magasin Pittoresque" (hand shadows), *73*

MagicLeap One, 97
Magic Mouse (Apple), 216, 219
Manifesto for Agile Software Development (Beck, Beedle, & van Bennekum), 20–25
Mathematica
  AnglePath in, 166–167, *168–169*
  BSplineCurve, 178, *179*
  BSplineFunction, *147*
  BSplineSurface, 178, *179*
  graph visualization functions, 104
  Line, 163–165, *164*
  notebook interface in, *161*, 224
  overview, 224–226
  Random, *166–167*, 180
  for sketching 3D surfaces, *170–177*
  surfaces imported to SketchUp, 176–178, *177*
  Table, *165*
  Wolfram language and, 160–163, 224
  See also Programming
Mathematics, emergence and, 19
Maya (Autodesk), 148, 159
Measurement of space, 35–40
  Cartesian spaces, overview, 35
  "fifth" dimension of space (cost), 39–40
  fourth dimension of space (time), 39
  linear measurement (one dimension), 35–36, *36*
  planar space (two dimensions), 36–37, *37*
  three-dimensional space, 37–39, *38*, *39*
Mel (Autodesk), 159
Mesopotamian architectural drawings, 80, *81*
Microsoft (company)
  HoloLens, 97
  Office, 202
  Surface tablet, 106
  Visual Studio, 224
Mignot, Jean, 2
MIT
  Lincoln Lab, *97*, 132
  Papert and, 128

Modeling. *See* Sketching in 3D
Monitor selection, 212–215

N

A New Kind of Science (Wolfram), 19
NeXTCUBE computer, 29, *30*
Normal surface, 64
NURBS (Nonuniform Rational B-Spline Surfaces), 52, *52*, 57–58, *146*, 146–147, 159, 178
NVIDIA (company), 17, 133, 228
NX (application, Siemens), 149

O

Oblique projection, 83–88, *84–87*
L'Oceanogràfic (Candela), 52
Oculus Quest headsets, 229
Office (application, Microsoft), 202
Oles, Steve, 91
One dimensional space (linear measurement), 35–36, *36*
OpenGL, 91–92, *92*
The Open Work (Eco), 132
Organization
  avoiding over-organization, 30–31
  backup of files for, 26–30, *30*, 119, 220
Orthographic projection, 36, 80, *81*
O2 workstation (Silicon Graphics), 120

P

The Painter's Manual (Dürer), 50, *50*, 73, 100
Palladio, Andreas, *76*, 157, *158*
Papert, Seymour, 128, *128*
Parallel postulate, 44–46, *45*, *46*
Parts, for sketching in 3D, *150–152*, 150–155, *154*
Paulos, Basilios, 112
Pencil (Apple), 106, *117*, 119, 220
Pencils for sketching, 105, 114, 117, 119
Perspective
  cinematic perspective, 92–93, *93*
  free perspective, 93–95
  projection, 88–92, *89–92*

  stereoscopic perspective, 95–97, *95–97*
Photoshop (Adobe), 17, 29
Physical presentation models. *See* Sketching in 3D
Picasso, Pablo, 98
Planar space (two dimensions), 36–37, *37*
Plans. *See* Representations of space
Plato, 59–60, 71–73, *72*
Platonic solids, 59–60, *60*
Pointing device selection, 216–217
Points, 46, *46*, 136–137
Presentation, 185–203
  construction documentation (telling a story) for, 185–189
  levels of development (LoD) and, 191–192
  sketching for BIM, 189–191
  sketching for construction, 198–203
  sketching in LoD 100 BIM, *193*, 193–198
Primitive forms, 59–60, *60*
Procreate (application), *89*, 117
"Program" for projects, 186–189
Programming, 157–184
  choosing materials/tools for digital sketching, 159–161, *161*
  creativity of, 181–184
  getting started with, *162*, 162–165, *163*
  for lines, 163–170, *164–170*
  rules-based design and, 157–158, *158*
  for 3D surfaces, 170–181, *171–182*
Programming Languages
  Fortran, xxx
  Java, 224
  Lisp, 159
  Mel, 159
  Python, 161, 224–225
  Scratch, 160
  visual programming languages, 225–226
Wolfram Language, 160–163, 224. *See also* Mathematica

INDEX

Projection systems
   orthographic projection, 80, *81*
   perspective, 88–92, *89–92*
   unfolded projections, *98–100*, 98–101
Python (programming language), 161, 224–225

Q
QoR watercolors, 115
Qualities of space, 40–44, *41–44*
Quill (application), 230

R
Reflection-in-action, 107
The Reflective Practitioner (Schön), 4
Renault (company), 52
Representations of space, 71–104
   architectural scale, 76–79, *77, 78*
   cinematic perspective, 92–93, *93*
   cutting planes, *79*, 79–80
   descriptive geometry, 73–76, *75, 76*
   elevations and vertical sections, 82–83, *83*
   free perspective, 93–95
   oblique projection, 83–88, *84–87*
   orthographic projection, 80, *81*
   perspective projection, 88–92, *89–92*
   plans, 80–82, *82*
   projection systems, overview, 71–73, *72–74*
   sketching in diagrams, 101–104, *102–104*
   stereoscopic perspective, 95–97, *95–97*
   unfolded projections, *98–100*, 98–101
Republic (Plato), 71–72
Revit (Autodesk), 202
Rhinoceros (McNeel), 150, 159, 161, 183, 225
Right-hand rule, *66*

Rittel, H. W., 19
Robots, 206–210
Ruled surfaces, 146, *146*

S
Sacks, Oliver, 6
Saint-Exupéry, Antoine de, 24
Schön, Donald, 4, 5, 108
Scratch (programming language), 160
Sculpting in Time (Tarkovsky), 93
Search engines, 31
Sections, 76, *76*, 82–83, *83*
Sharing (tablet sketching feature), 119
Shaw, George, 193–194
Silicon Graphics (company)
   Iris Explorer/Iris GL, 91, 181, 225
   O2 workstation, 120
Sketching for conceptual design, 1–32
   agility and, 19, 19–31, *30* (See also Agility)
   avoiding over-reliance on tools, 14–18
   conceptualizing process, *33*, 33–35, 198–203
   creative process for, 6–10, *7, 8, 10*
   creative rigor and, 1–3
   design as professional practice, 3–5
   design expertise and, 12–14
   design thinking and, 5–6
   digital tools for, 10–12
   good design and, 1
   sketching apps, 230
   technical pens vs. charcoal, 40–44, *41–43*
   version control systems and, 31–32
   virtual reality and, 227–232
Sketching in code. *See* Programming
Sketching in diagrams, 101–104, *102–104*
Sketching in 2D, 105–121
   conceptual design and, 110

   digital sketching tools, 116–120, *117, 118*
   physical and digital sketching together, 120–121
   physical sketching with pencil and paper, 110–115, *113*
   sketchbooks for, 105–106, *106*, 115–116
   sketching with purpose, 106–110, *107*
   with watercolor, *114*, 115
Sketching in 3D, 123–155
   digital modeling, *132*, 132–138, *133, 135–137*
   forms, *147–149*, 147–150
   lines, 141–144, *142–144*
   from parts, *150–152*, 150–155, *154*
   physical modeling, 127–132, *128, 130, 131*
   process of 3D modeling, 138–141, *139, 140*
   sketch modeling, overview, 123
   sketch modeling benefits, 123–127
   "sketch-rendering," 125
   surfaces, 144–146, *145–147*
Sketch-modeling with cardboard, 62–64, *63, 64*
Sketchpad (Sutherland's device), 116, *132*
Sketch to production. *See* Presentation
SketchUp (Trimble)
   ant model example, *149*
   backing up, 220
   components in, *151*, 151–152
   cross-eyed perspective projection, *96*
   Dürer's shadow casting algorithm in, *74*
   3D Warehouse and, *66, 67, 69, 70*, 178
   flexibility of, 29
   inception of, 17
   inference system in, 142–143, *143*
   keyboard shortcuts for, 216
   kit-bashing in, 153, *154*
   lines for 3D modeling, *144*

SketchUp (Trimble) (*Continued*)
   line tool in, *136*, 142, 147
   Mathematica surfaces imported to, 176–178, *177*
   modeling in 3D with, 133, *133*
   outliner in, 140
   popularity of, 18
   PushPull tool, 61, *62*, 147, *148*
   right-hand rule in, *66*
   Ruby, 159
   scale figures in, 36, 79
   surfaces in, 144–146, *145*
   Wall tool, 153
   winged-edge data model in, 64, *65*
   See also Sketching in 3D
Software
   choosing materials/tools for digital sketching, 159–161, *161*
   rationalized processes automated by, 6
   as "un-intuitive," 134
   wicked problems of, 19–20
   See also individual names of software programs
Solid forms, 62, *63*
Solidworks (Dassault), 142, 150
The Songlines (Chatwin), 105
Space. *See* Representations of space
Stereoscopic perspective, 95–97, *95–97*
STL (file format, STereo-Lithography), 29
Sub-divisional surfacing (Sub-D), 58
Surfaces
   as element of design, 56–64, *56–66*
   programming for 3D surfaces, 170–181, *171–182*
   sketching in 3D, 144–146, *145–147*
Surface tablet (Microsoft), 106

Sutherland, Ivan, 18, 96–97, *97*, *132*
Swept forms, 60–62, *61*, *62*

T
Tablet computers for digital sketching, 116–120, *117*, *118*
Tarkovsky, Andrey, 93
Taxonomies, 30–31, 64, 153, 193–198
Technical pens vs. charcoal, 40–44, *41–43*
"Thebaid" (Angelico), 86, *87*
Thomas Aquinas, Saint, 185
Three-dimensional space, 37–39, *38*, *39*
3DSmax (Autodesk), 148
3D Warehouse (application, Trimble), *66*, *67*, 69, *70*
Tilt Brush (application, Google), 230
Time as fourth dimension of space, 39
Torvalds, Linus, 31
Triangles, 56, *56*
Trimble (company)/SketchUp (application). *See* SketchUp
Trimble Connect (application), 29, 220
3D Warehouse (application), *66*, *67*, 69, *70*
Trimble Connect (Trimble), 29, 220
2D, sketching in. *See* Sketching in 2D
Two dimensional space (planar space), 36–37, *37*
3D, sketching in. *See* Sketching in 3D

U
Ulysses (application), 202
Undo (tablet sketching feature), 119
Unfolded projections, *98–100*, 98–101

Urban Sketching movement, 112

V
Valve (company)
   "Half Life," 231
   Index controllers, 229
Van Bennekum, A., 20–25
Van Gogh, Theo, 12
Van Gogh, Vincent, 12
Vasari, Georgio, 49
Version control systems, 31–32
Vertical sections, 82–83, *83*
Vesalius, Andreas, 100–101
"Villa Rotunda" (Palladio), *76*, 157
Virtual reality, 227–232
Visual Programming Lab (VPL, company), 97
Visual programming languages, 225–226
Visual Studio (Microsoft), 224
Vitruvius, 157

W
Walt Disney Concert Hall, 201
Wang Hui, *84*, 84–86, *86*
Wessman, Ragnar, 190
Wicked problems, 19–20
WIMP graphical interfaces, 216
Windows (operating system, Microsoft), 216
Wires/wireless peripherals, 221, 229–230
Wolfram, Stephen, 19, 160, 178, 206–207
Wolfram Language (programming language), 160–163, 224. *See also* Mathematica
Woodworking tools, 130–131, *131*
Wright, Frank Lloyd, 130
WriteNow (application), 29–30

Y
Yahoo! (company), 30